Dextrous Robot Hands

S.T. Venkataraman T. Iberall
Editors

Dextrous Robot Hands

With 95 Figures

Springer-Verlag
New York Berlin Heidelberg
London Paris Tokyo Hong Kong

Subramanian T. Venkataraman
Jet Propulsion Laboratory
4800 Oak Grove Drive
Pasadena, CA 91109
USA

Thea Iberall
Center for Neural Engineering
Department of Computer Science
University of Southern California
Los Angeles, CA 90089
USA

Library of Congress Cataloging-in-Publication Data
Dextrous robot hands / edited by Subramanian T. Venkataraman. Thea Iberall.
p. cm.
"This book grew out of the Workshop on Dextrous Robot hands that occurred at the 1988 IEEE Conference on Robotics and Automation in Philadelphia . . . co-sponsored by the IEEE Computer Society and the Office of Naval Research"—Pref.
Includes bibliographical references.
ISBN 0-387-97190-4
1. Robotics—Congresses. 2. Manipulators (Mechanism)—Congresses. I. Venkataraman, Subramanian T. II. Iberall, Thea. III. Workshop on Dextrous Robot Hands (1988 : Philadelphia, Pa.) IV. IEEE International Conference on Robotics and Automation (1988 : Philadelphia, Pa.) V. IEEE Computer Society. VI. United States. Office of Naval Research.
TJ211.D49 1990
629.8′92—dc20 89-21999

Cover illustration by Jacqueline Weir.

Printed on acid-free paper.

Camera-ready text prepared using LaTEX.
Printed and bound by Edwards Brothers, Inc., Ann Arbor, Michigan.
Printed in the United States of America.

9 8 7 6 5 4 3 2 1

ISBN 0-387-97190-4 Springer-Verlag New York Berlin Heidelberg
ISBN 3-540-97190-4 Springer-Verlag Berlin Heidelberg New York

Preface

Dextrous manipulation has been a topic of interest in industrial assembly, prosthetic hand design, and in the study of human movement. In recent years, dextrous robot hands, developed in the United States, Europe, and Japan, have become available as research tools. While significant progress is being made in their design, construction, and low level control, the true potential of dextrous mechanical hands has yet to be realized. Roboticists are up against fundamental problems in developing algorithms for solving real-time computations in multi-sensory processing and motor control.

One possible way to develop better dextrous robot hands is through the study of human hands. After decades of research into the control problem of human arm movement, motor behaviorists are discovering key features of sensorimotor integration in the central nervous system. In a sense, they are reverse engineering the controller of a highly versatile, multiple degree of freedom, sensor-based 'machine', of the type that roboticists would love to build.

The aim of this book is to explore parallels in sensorimotor integration in dextrous robot and human hands. While it is a view of the state of knowledge in 1989, we feel that one way the robotics community can design and learn to control more sophisticated robot hands is for them to work with motor behaviorists. As seen from some of the work presented in this book, such alliances have already proven fruitful.

The significance of this text is that it brings together researchers in dextrous manipulation, and is a generally accessible resource for both experimentalists and engineers. In a computer science department, it would be a good companion text for a graduate level course in robotics, as it deals with integrating sensory information with motor control for performing a task. It would be a useful resource for a mechanical engineering course in robot hand design, as students could use it to gain insights into functional requirements for a dextrous hand. However, a strong feature of the book is that it could be used as a text for an interdisciplinary graduate course on sensorimotor integration for both experimentalists and engineers.

This book grew out of the Workshop on Dextrous Robot Hands that occurred at the 1988 IEEE Conference on Robotics and Automation in Philadelphia, Pennsylvania. Highly successful, the workshop was attended by over 150 people, and was co-sponsored by the IEEE Computer Society and the Office of Naval Research.

The first part of the book focuses on the functionality of human hands, discussing results from studies in prehension and apprehension. The first four chapters describe models of human hand function and present evidence for why these models are valid. It is interesting to note that knowledge-based expert systems are described in Chapters 1 and 4 for codifying those models. The second part of the book focuses on control and computational architectures needed for dextrous hands. Chapters 5 through 7 describe architectural design decisions for three different robot hands: the Stanford/JPL hand, the Utah/MIT hand, and the Belgrade/USC hand. The third part of the book focuses on the two key features relevant to the actual usage of dextrous robot hands. Chapters 8 and 9 describe grasp and manipulation planning algorithms, while the subsequent three chapters describe the use of sensors in dextrous hand usage. The fourth part of the book is an edited transcript of the panel discussion that occurred at the end of the workshop. Emerging from the discussion was a clear sense that key problems in dextrous robot hand development and control include hand geometry, hand functionality, development and placement of sensors, sensory fusion, and the need for better actuators. And while the psychologists were able to offer ideas about how these are accomplished in human systems, they stressed the need for better criteria for evaluating effective dextrous robot hand design and sensorimotor control.

The ultimate goal of this book is to make the motor behaviorists' literature more accessible to roboticists, and vice versa. The human brain solves many motor problems in real-time and seemingly without much effort that engineers would like their robots to solve. As well, psychologists could use many of the tools that roboticists have developed for analyzing motor control. Solving these fundamental engineering problems, and perhaps even gaining insights into the human brain, will take a combined effort of researchers coming top-down from the task level and bottom-up from the servo-control level. We hope that this book will facilitate interdisciplinary communication and make a contribution toward advancing knowledge on dextrous manipulation.

For their insights, and encouragement, we would like to thank Michael Arbib, Ruzena Bajcsy, George Bekey, Alan Desrochers, Virginia Diggles-Buckles, Krithi Ramamritham, and Rukmini Vijaykumar. We would also like to thank Harry Hayman of the IEEE Computer Society for his help in assuring the success of the workshop, as well as Alan Meyrowitz of the Office of Naval Research for his support. We thank E. Colleen Scott for her assistance in manuscript preparation, and Jacqueline Weir for her own artistic dexterity. Our special thanks to the authors for their invaluable contributions and timeliness, and to our families for their constant encouragement and support.

August 1989

SUBRAMANIAN T. VENKATARAMAN
Pasadena, California
THEA IBERALL
Los Angeles, California

Contents

Part I

Lessons Learned from Human Hand Studies

Thea Iberall

For thousands of years, the human hand and its neural controller has been a constant source of fascination to philosophers. Aristotle pointed out that the hand can grasp a variety of weapons and tools, noting that:

> *"...Nature has admirably contrived the actual shape of the hand so as to fit in with this arrangement. It is not all of one piece, but it branches into several pieces; which gives the possibility of its coming together into one solid piece, whereas the reverse order of events would be impossible. Also, it is possible to use the pieces singly, or two at a time, or in various ways. Again, the joints of the fingers are well constructed for taking hold of things and for exerting pressure."* – Aristotle, Parts of Animals, IV, X.

From these philosophical underpinnings, it was not until the end of the 19th century that psychology distinguished itself as an exact science of behavior [Schultz 1975]. Using carefully controlled observation and experimentation to study human behavior, tools, techniques, and methods have been developed and further refined in order to achieve precision and objectivity in measurement. One of the earliest systematic studies of rapid arm and finger movements was done by Robert Sessions Woodworth [1899], who argued that not only must the response to a stimulus be considered, but also the structure and condition of the organism doing the responding. Combining introspection, observation, and experimentation, he described goal-directed movements as being two-phased: an initial, ungoverned adjustment followed by a final, controlled adjustment. In the first demonstration of accuracy-speed measurements, he showed that visually-guided slow movements were more accurate than fast ones or eyes-closed slow movements. These early experiments led to studies by other researchers that addressed questions such as: how much time is needed for processing visual feedback, what happens if visual feedback is distorted or suppressed, what happens if the target is moving, and what are the different contributions of central and peripheral vision.

At the same time that the field of experimental psychology was emerging, physiological experiments were being conducted that began to shed light on the neural control of movement [Schmidt 1982]. Dominating the early work in this field, the physiologist Sir Charles Sherrington coined the term proprioception for the collective position sense stemming from sensors in the muscles, joints, and tendons. Ultimately, this work has led to questions such as: in what frame of reference do movements occur, how does automatic regulation of movement occur, what happens when different parts of the brain are lesioned, and what are the physical characteristics of muscle.

Today, these fields blend as tools and techniques for measurement become more and more sophisticated, and the complexity of the central nervous system realized. An important influence on today's current trend toward understanding the processes that support movement came from the psychologist Ulric Neisser [1967]. As Schmidt [1982] points out, Neisser was able to legitimize the study of underlying processes, which ultimately can only be inferred from behavior and not directly observed.

Struggling to develop testable hypotheses for how the neural controller might work, studies in motor behavior address many of the same problems faced by roboticists: how does the brain coordinate multiple degree of freedom limbs, how are sensors used and modalities integrated, how is sensory information integrated with movement, how does stable grasping occur, how does movement occur in spite of the complex forces acting on the system, how are objects perceived, and how does planning occur. A current controversy facing the study of human movement, relevant to feedforward and feedback robot control, is the question of the existence of motor programs [Keele 1968]; that is, are skilled movements stored as a prestructured set of muscle commands that are just read out as a movement sequence occurs? If so, how many patterns can the nervous system store? If not, what are the underlying control mechanisms for movements that are too rapid for feedback processing?

In this section, two important skills of the human hand are studied. The first is prehension, or the hand's ability to grasp or take hold of objects. The second is apprehension, or the hand's ability to understand through active touch, or haptics. Both involve highly complex sensorimotor integration, dexterity, compliance with the environment, and tactile sensibility. As well, both involve two separate components: stabilization and exploration. In apprehension, if the object is not stabilized (by the environment, by the other hand, or even by fingers from the same hand), then there is a problem with exploration: the fingers will push the object away. In prehension, forces for stabilizing the object in the hand are applied in opposition to each other; sensibility facilitates dextrous manipulation. Engineers interested in understanding the dexterity of the human hand are encouraged to read these four chapters carefully.

As an introductory bridge between experimentalists and roboticists, the first chapter is by Cutkosky and Howe. Being mechanical engineers, they focus on the question of how to capture the skill of human prehension for the design of robot hands. By studying single-handed operations in small batch machining, a grasp taxonomy was constructed, extending earlier work done in the medical literature. Using this taxonomy, an expert system was then developed that attempts to duplicate the human brain's ability to choose a grasp posture given the necessary task and object parameters. An important result of their work is that they can tie analytic approaches to prehension with more abstract notions of human prehensile tasks. For readers interested in exploring the details of the analytic approaches, Chapters 8 and 9 are excellent sources.

Taking a less reductionist approach, in Chapter 2 Iberall and MacKenzie look carefully at human prehension in order to identify all its complexity. Pointing out that numerous constraints act upon this skill, they put together a comprehensive picture, drawing on current knowledge from hand anatomy, biomechanics, sensory physiology, perception of object properties, and task functionality. In order to relate this picture to dextrous robot hands, they argue that the asymmetries in the human hand, in terms of its structure, motor control, functionality, and distribution of sensors, are an important lesson to be learned from nature. In addition, they present an abstracted 'machine-independent' view of prehension that could map either into robot hands or human hands.

In Chapter 3, Wing follows the Woodworth tradition and looks at human prehensile movements when constrained to occur under conditions of sensory deficit. Reminiscent of non-sensor-based robot control, this situation arises in human motor behavior when visual information is lacking (due to rapidity of movement or eyes-closed conditions) or when proprioceptive information is affected (as in the use of prostheses without proprioceptive feedback). Ultimately, Wing shows that human prehension involves aspects of stabilization and exploration, and argues that engineers should take into account some of the underlying motor control processes in their designs of prosthetic devices. Related to this is an important point about variability: in robot control, engineers want to reduce variability in order to achieve absolute repeatability; however, in human movement, variability is an important feature, with the brain compensating as necessary to cope with a variety of uncertainties in order to ensure success in the task.

In the last chapter of this section, Klatzky and Lederman ask the question "What determines the next hand movement?" Unfortunately for engineers, they show that there is no one simple answer to this question. Coming from the tradition of psychophysics where researchers attempt to catalogue sensations and measure discrimination thresholds, Klatzky and Lederman systematically identify the components of object perception. In a concise overview of their research, they offer a theoretical model as well as their empirical evidence for how humans explore their environment. Objects are shown to be complex collections of properties, and the movements that humans make depend on the property or properties of interest. Numerous constraints influence the use of hands, particularly cost constraints on underconstrained choices. Roboticists interested in sensor fusion will note their discussion of "natural correlations" between object properties. Most importantly, by pushing their work into the field of knowledge-based expert systems for robot control, these researchers are establishing a bridge between their psychological results and sensor-based intelligent robots.

With over 25 degrees of freedom, the human hand is a highly complex structure that in many ways defies understanding. Questions abound as to whether its beauty comes from its own internal architecture or else from its controller, the human central nervous system. As the famed surgeon Frederick Wood-Jones [1920] said, "The difference between the hand of a man and the hand of a monkey lies not so much in the movements which the arrangement of muscles, bones and joints make possible...but in the purposive volitional movements which under ordinary circumstances the animal habitually exercises." Some day, it is hoped that the understanding of the human nervous system will let us capture the beauty of human movement in order to design and control highly versatile dextrous robot hands.

1

Human Grasp Choice and Robotic Grasp Analysis

Mark R. Cutkosky[1]
Robert D. Howe[1]

ABSTRACT In studying grasping and manipulation we find two very different approaches to the subject: knowledge-based approaches based primarily on empirical studies of human grasping and manipulation, and analytical approaches based primarily on physical models of the manipulation process. This chapter begins with a review of studies of human grasping, in particular our development of a grasp taxonomy and an expert system for predicting human grasp choice. These studies show how object geometry and task requirements (as well as hand capabilities and tactile sensing) combine to dictate grasp choice. We then consider analytic models of grasping and manipulation with robotic hands. To keep the mathematics tractable, these models require numerous simplifications which restrict their generality. Despite their differences, the two approaches can be correlated. This provides insight into why people grasp and manipulate objects as they do, and suggests different approaches for robotic grasp and manipulation planning. The results also bear upon such issues such as object representation and hand design.

1.1 Introduction

In the broadest terms, there are two approaches to the study of grasping: the empirical and the analytical. The empirical approach studies grasping by humans and animals, the only successful grasping systems in our experience. The motivation is the often expressed and occasionally justified hope to learn from natural systems how to make good artificial ones. Unfortunately, we are rarely able to deduce how humans perform so well, and even when we can, human methods may not be right for mechanical systems.

In contrast, the analytic approach models grasping from first principles. Interactions between the hand and grasped object are modeled in terms of motions and forces, using the laws of physics. Here the problem is that the

[1]Design Division, Mechanical Engineering Department, Stanford University, Stanford, CA 94305

grasping process is so intrinsically complicated that many simplifications are required to make the analysis tractable. But this means that important effects are left out, resulting in models that only apply to carefully structured laboratory experiments.

Fortunately, these two approaches complement each other. Study of natural systems helps to assure that the analytic models are not overlooking important effects. Analytic work helps to explain why people do what they do – which provides insight for design and control of dextrous robot hands.

In the following sections we first review the results of studies of human grasp selection. We also review analytic grasp models and examine the assumptions upon which these models rest. Next we compare these approaches, and use them to evaluate each other. Finally, we consider directions for further investigation in the design and control of robot hands.

1.2 Human grasp choice

We begin by examining previous studies of human grasping. Next we describe our work in this area, including the construction of a grasp taxonomy and an expert system. The taxonomy is a method of organizing the range of human grasp types, and the parameters used in this organization reveal some of the factors influencing grasp choice. The expert system uses knowledge coded in the form of rules to deduce the appropriate grasp from object and task descriptions. It is in constructing the rule base that we learn the important properties of objects and tasks. Finally, we discuss the lessons to be learned from these studies of human grasping.

1.2.1 Previous studies of human grasping

The study of human grasping has long been an area of interest for hand surgery, for designing prosthetic devices and for quantifying the extent of disability in individuals with congenital defects or injuries. As a result, there is a substantial, empirical, medical literature on the grasping capabilities of the human hand. Much of the literature refers to six grasps defined by Schlesinger [1919] and summarized by Taylor and Schwarz [1955]: cylindrical, fingertip, hook, palmar, spherical and lateral.

Such a categorization leads to associating grasps with part shapes. Thus a sphere suggests a spherical grip while a cylinder suggests a wrap grip. However, when people use objects in everyday tasks, the choice of grasp is dictated less by the size and shape of objects than by the tasks they want to accomplish.

Even during the course of a single task with a single object, the hand adopts different grips to adjust to changing force/torque conditions. When unscrewing a jar lid, the hand begins with a powerful grip in which the palm is pressed against the lid for extra torque. As the lid becomes loose, torque becomes less important than dexterity and the hand switches to a light grip in which only the fingertips touch the jar lid. This task dependence was noticed by Napier [1956], who suggested that grasps should first be categorized according to function instead of appearance.

In Napier's scheme, grasps are divided into power grasps and precision grasps. Where considerations of stability and security predominate (as in holding a hammer or getting a jar lid unstuck) a power grasp is chosen. Power grasps are distinguished by large areas of contact between the grasped object and the surfaces of the fingers and palm and by little or no ability to impart motions with the fingers. Where considerations of sensitivity and dexterity predominate a precision grasp is chosen. In precision grasps, the object is held with the tips of the fingers and thumb.

More recently, Arbib, Iberall and Lyons [1985] have developed the concept of "virtual fingers," an abstraction whereby any number of real fingers that work as one are modeled as a single entity. For example, when picking up a pencil from a flat surface, two or three fingers are typically used in unison to oppose the thumb; the thumb would be described as one virtual finger and the set of fingers would be described as another.

Iberall [1987b] describes human grasping in terms of "oppositions," which are the basic hand configurations for applying forces to opposing faces of a grasped object. All human grasps are formed from a set of only three oppositions:

- pad, for forces between the pads of the fingers and thumb
- palm, for forces between fingers and the palm
- side, for forces between the thumb and the side of the index finger.

These oppositions are independent and may be used separately or simultaneously in a task. Each opposition is formed from two virtual fingers, one of which is always either the thumb or the palm. Thus picking up a pencil uses a pad opposition, turning a key in a lock uses a side opposition, and pounding with a hammer uses a palm opposition. Iberall also correlates this opposition/virtual finger approach to the previous categorizations of Schlesinger [1919], Napier [1956], and Cutkosky and Wright [1986b]. In Section 1.4.3, we relate these concepts to some of the analytical work on robotic grasping.

Finally, it is important to note that the human hand has evolved over millions of years as an organ used as much for sensation and communication as for manipulation. Thus in an approach similar to these studies of prehension, Klatzky and Lederman [this volume] have studied how humans use hands for exploration and perception.

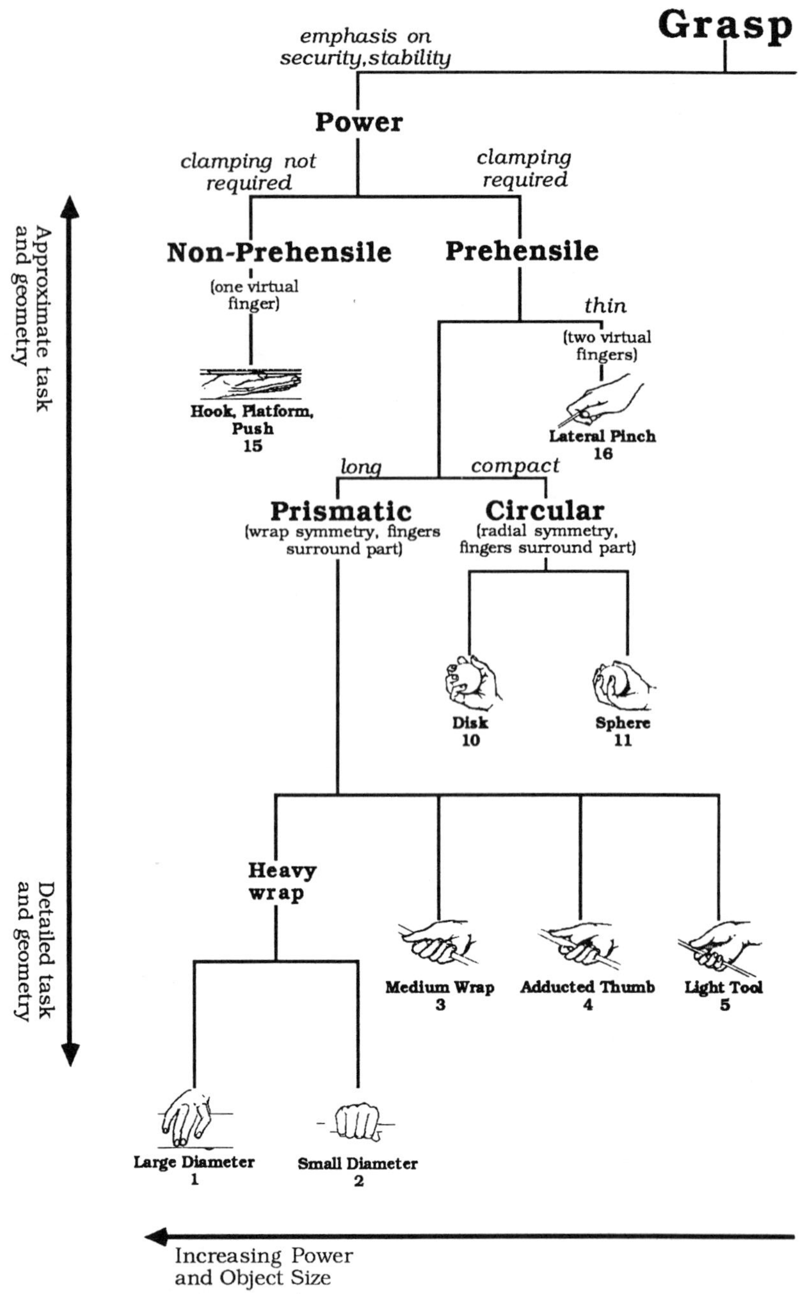

Figure 1.1: A partial taxonomy of manufacturing grasps, revised from [Cutkosky and Wright 1986b]. The drawings of hands were provided by M. J. Dowling.

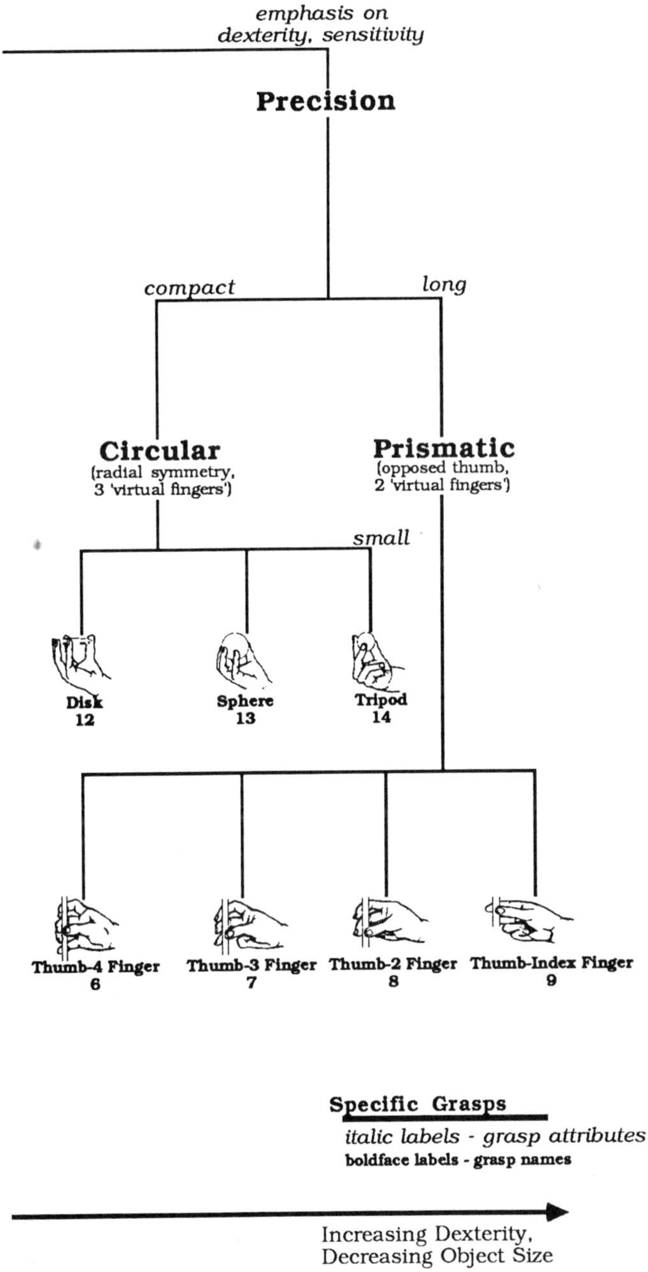

Figure 1.1 (continued): and are reprinted with the permission of the Robotics Institute, Carnegie-Mellon University.

1.2.2 A GRASP TAXONOMY

To study how humans choose grasps we have constructed a grasp taxonomy. The taxonomy is a systematic arrangement of the space of human grasps, and the organization of the taxonomy reveals some of the factors influencing grasp choice. Grasps can be placed on a continuum according to object size and power requirements. The taxonomy shows how task requirements (forces and motions) and object geometry combine to dictate grasp choice.

The taxonomy is based on observations of single-handed operations by machinists working with metal parts and hand tools. The machinists were observed and interviewed and their grasp choices were recorded as they worked. In addition, their perceptions of tactile sensitivity, grasp strength, and dexterity were recorded. Preliminary results of the study, and a resulting partial taxonomy of manufacturing grasps, were presented in [Cutkosky and Wright 1986b], and full details may be found in [Cutkosky 1989].

Starting with the two basic categories (power and precision grasps) suggested by Napier, we descend a hierarchical tree of grasps. As we move down the tree, details of the task and the object geometry become equally important so that in the final analysis, both task requirements and object shape play important roles in determining the grasp.

Once the basic choice between a power grasp and a precision grasp has been made, a combination of task-related and geometric considerations comes into play. Starting at the top of Figure 1.1, let us suppose that a power grasp has been chosen. The first question is: does the object need to be clamped to sustain forces from a variety of directions, or does it merely need to be supported? If it merely needs to be supported then a non-prehensile hook grasp (as used in carrying a suitcase) or a palmar support (as used by a waiter carrying a tray) may be adequate. If the object must be clamped, a prehensile grip is chosen in which the fingers and palm confine the object.

At this stage some basic geometric considerations become important: Is the object large? small? flat? thin? These subsidiary choices are illustrated in Figure 1.1. In the following, Grasp numbers refer to the numbers assigned in this figure. If, for example, a power grip is needed, and the object is small and flat (as in turning a key in a lock) then a Lateral Pinch (Grasp 16) will probably be used. If the object has a compact or approximately spherical shape then a Circular grasp (Grasps 10 and 11) is most likely. If the object is roughly cylindrical, then a wrap is chosen. Since many objects, including the handles of most tools, have cylindrical shapes, the power wrap represents a large family of grips.

While the different precision grasps appear to be motivated by part geometry, the decision to use one precision grasp instead of another may actually be task-related since many objects have several gripping surfaces with different shapes. For example, a light cylindrical object can be gripped either on the sides using the thumb and four fingers (e.g., Grasp 6) or it can be gripped by one end, using a circular finger placement (e.g., Grasp 12).

Trends in the taxonomy

Moving from left to right in Figure 1.1, the grasps become less powerful and the grasped objects become smaller. Thus the Heavy Wrap grips are the most powerful and least dextrous (all manipulation must be done with the wrist and even the wrist is restricted to a limited range of motions), while the Tripod (Grasp 14) and Thumb-Index Finger (Grasp 9) grips are the most precise. However, the trend is not strictly followed. A Spherical Power grasp may be either more or less dextrous than a Medium Wrap, depending on the size of the sphere.

Moving from top to bottom, or more precisely, from the trunk toward the leaf-nodes of the taxonomy, the trend is from general task considerations to details of geometry and sensing. Toward the top of the tree we are concerned with such basic considerations as whether the grasp must restrain the part, and with the overall shape of the object (long, compact, thin). At the lowest levels, we choose among similar grasps on the basis of geometric details (e.g., whether the object is closer to a disk or a sphere) and details of the task (e.g., whether sensitivity to vibrations is more important than the ability to apply rotations about the tool axis). However, the top-down trend is not strictly observed. For example, a small, flat object may provoke the choice of a Lateral Pinch near the top of the tree. In fact, the Lateral Pinch Grasp, though positioned above the Prismatic and Circular grasp types on the page in Figure 1.1, is essentially at the same level of detail.

The role of task forces and torques on grip choice is most apparent when the hand shifts between grips during a task. For example, in unscrewing a knob the hand shifts from Grasp 11 to Grasp 13 as the required torque decreases. Similarly, when holding a tool as in Grasp 3, the hand shifts to Grasp 5 as the task-related forces decrease and may adopt Grasp 6, a precision grasp, if the forces become still smaller. Task motions can similarly be used to distinguish among grasps with similar tools and force requirements. For example, when using a screwdriver to turn a screw, Grasp 4 is preferred over Grasp 2, because it orients the wrist better for applying rotations along the screwdriver axis.

The role of object size is most apparent when similar tasks are performed with different tools. For example, in light assembly work Grasps 12 and 13 approach Grasp 14, and finally Grasp 9, as the objects become very small. A related observation, brought out more clearly in developing the grasp expert system discussed in Section 1.2.3, is that sequences can be traced in the taxonomy, corresponding to adjustments that the machinists make in response to shifting constraints.

Limitations of the taxonomy

While the taxonomy in Figure 1.1 has proven to be a useful tool for classifying and comparing manufacturing grasps, it suffers from a number of limitations. To begin with, it is incomplete. For example, there are numerous everyday grasps, such as the grasp that people use in writing with a pencil, that are not included. It was also found that the machinists in our study adopted numerous variations on the grasps in Figure 1.1, partly in response to particular task or geometry constraints and partly due to personal preferences and differences in the size and strength of their hands. Such idiosyncratic grasps could usually be identified as "children" of the grasps in Figure 1.1.

1.2.3 An Expert System for Choosing Grasps

To clarify issues such as the roles of dexterity, sensitivity and stability in grasp choice, we constructed an expert system. This system, "Grasp-Exp," was designed to choose human grasps from initial information about the task requirements and object shape. Details of the system can be found in [Cutkosky 1989]. The system has gone through several iterations and has been tested informally with several machinists and researchers. In the tests, subjects were first given a tool and a task to perform and their grasp choice was recorded classified in terms of the taxonomy. Subjects were then asked questions about the tool and task by the expert system. Within the limited context of one-handed grasps in a manufacturing environment, we found that the system could usually predict how people would grasp parts or tools. Moreover, we found that where the expert system failed to identify the particular grasp that a person used, it picked a close relative that could also have been used to accomplish the task.

However, the purpose of the codification exercise was not to develop a program to predict what grasp a human would adopt under particular circumstances but to have a running, testable framework in which to try out hypotheses. In addition, the codification exercise forces one to be more careful about defining terms and organizing information. An expert system also makes it possible to ascertain what information about a task and object is required in order to deduce the appropriate grasp. Using this system also lead us to explore patterns or sequences among grasps, which provide insights for controlling robotic hands to manipulate parts.

Lessons from Grasp-Exp

Grasp-Exp shows that grasps can be uniquely and correctly determined by specification of object characteristics (e.g., size, shape) and *grasp attributes*, such as dexterity, precision, sensitivity, stability, and security. Rules in the system's knowledge base matched grasp attributes with the requirements of a task and characteristics of an object, as provided by the user. Several factors regarding geometric descriptions and grasp attributes became clear in the course of developing the system.

First, it is necessary to quantify grasp attributes so that different grasps may be ranked. We used scales of dexterity, precision, sensitivity, power, and stability, so that precision grasps tend to be at one end of the spectrum and power grasps at the other. Thus grasps in the center of the taxonomy show a mix of these properties. For example, the Light Tool grasp (Grasp 5) is classified on the taxonomy as a power grasp, as evidenced by its low dexterity and high stability and power. However, in discussions with machinists, and in our own experience with this grasp, we have found that it has good sensitivity to forces and vibrations – a characteristic of precision grasps.

Grasp attributes often occur in sets determined by the general type of task. Users found it difficult to assign quantities to terms like dexterity and sensitivity, so it was necessary to have the expert system ask additional questions about the force requirements, approximate object weight, and so forth. Often, it is easiest to ask such questions in terms of analogies: "Would you classify the task as most like a prying task? a tapping task? a pushing task? ..." This suggests that there are a relatively small number of prototypical tasks such as prying or tapping, each requiring a particular set of grasp attribute values.

We learned similar things about the description of objects. As with the grasp attributes, users found it difficult to quantify the size of an object (e.g., "Is the object large, medium, or small?"). A better reference is the size of the hand (e.g., "Is it bigger than a fist?"). We also found that the approximate geometric descriptions (e.g., compact, thin, prismatic) in the taxonomy were too vague. These descriptions have been extended to include the rough-object-shape and detailed-object-shape (long, thin, disk-shaped, rectangular, etc.). However, even this description is not sufficient, as the features of an object which are important for grasping may not be emphasized in this type of geometric information. In Section 1.5 we consider the question of appropriate representations for reasoning about grasping.

As with the taxonomy, Grasp-Exp permits us to examine the interaction between task requirements and object geometry in a sequence of tasks with a single object or in the same task with different objects. For example, consider precise manipulations of a light, compact object with a precision grip. Initially, Grasp-Exp may choose the precision disk grip. But as the size of the object grows smaller with respect to the hand, the selected grasp shifts to a tripod grip and finally to a thumb-index finger pinch.

Tactile sensing and grasp choice

Our work with Grasp-Exp shows that sensitivity to forces and vibrations must often be taken into account to correctly predict grasp choice. Iberall [1987b] also notes that humans balance the ability to sense with the

Figure 1.2: Although it is a power grasp, the Light Tool grasp (Grasp 5 in the taxonomy) shows good sensitivity (©IEEE, Cutkosky 1989).

ability to impart forces and motions during manipulation. Sensitivity is a characteristic of the precision grasps, which use small forces, limit contact areas to the ends of the fingers, and often separate finger positions widely.

In general, the hand becomes less sensitive to small forces and vibrations as more force is applied. One reason for this is the compliant structure of the fingers and palm [Howe and Cutkosky 1989]. When the hand is lightly loaded the skin effectively floats on soft subcutaneous pulp. The skin is thus decoupled from the more rigid bone and muscles of the inner structure of the hand. Small forces or motions from the object can cause substantial motion of the skin, where many sensors are located, resulting in good sensitivity. As the contact force increases, pressure beneath the skin increases and the skin and hand structure become coupled. The system becomes stiffer and task forces now produce smaller skin motion, so sensitivity decreases.

The placement of finger contacts also plays an important role in sensitivity. The mechanoreceptors responsible for our most acute tactile sensitivity are concentrated at the fingertips, so sensitivity improves if these areas are kept lightly in contact with the object. Also, separating the contacts from the center of compliance of the grasp means that when the object rotates, the motion at the contact location will be greater.

An example of these principles is the Light Tool Grasp (Grasp 5), shown in Figure 1.2. Unlike the other power wrap grasps, this one has good sensitivity, although the only topological difference is that the index finger is extended along the grasped object. This finger is not in opposition with the palm and experiences much smaller forces than the other fingers. The tip of the finger is kept in light contact with the object, and it is much closer to the origin of the task forces than the other fingers, which are located essentially at the center of compliance of the grasp. Thus small forces and vibrations produce comparatively large motions of the fingertip skin, resulting in good sensitivity.

1.3 Analytic approaches to grasp modeling and grasp choice

In this section we examine grasp analyses for robotics. These studies attempt to model the interaction between object and hand from first principles. For example, contact forces, kinematics, or compliance of the hand-object system are often included in a model. However, in order to make analyses tractable, many important factors are not included, and we consider some of these omissions and their implications.

Many analyses develop quality measures, which are used to evaluate the success of a grasp. We summarize the most useful measures in Table 1.1. Grasp choice is often then characterized as an optimization within constraints, with each author picking a different set of quality measures to optimize while treating others as constraints. These quantitative analyses, which have constituted the bulk of the work in grasping theory, do not treat large motions during manipulation, or prescribe a basis for selecting a particular type of grasp. Geometric and knowledge-based approaches have been suggested for this area.

1.3.1 Grasp modeling

As Figure 1.3 indicates, manipulation is complex, typically involving combinations of open and closed kinematic chains, non-holonomic constraints, redundant degrees of freedom and singularities. In addition, there are nonlinearities in the contact conditions between soft, viscoelastic fingers and grasped objects, and in the drive-train and actuator dynamics. To keep the analysis tractable, early analyses (e.g., [Hanafusa and Asada 1977b]) made the following assumptions, many of which are also found in current analyses of dextrous manipulation:

- rigid-body models with point contacts between the fingertips and the grasped object, often in 2-D
- linearized (instantaneous) kinematics
- quasistatic analysis (no inertial or viscous terms)

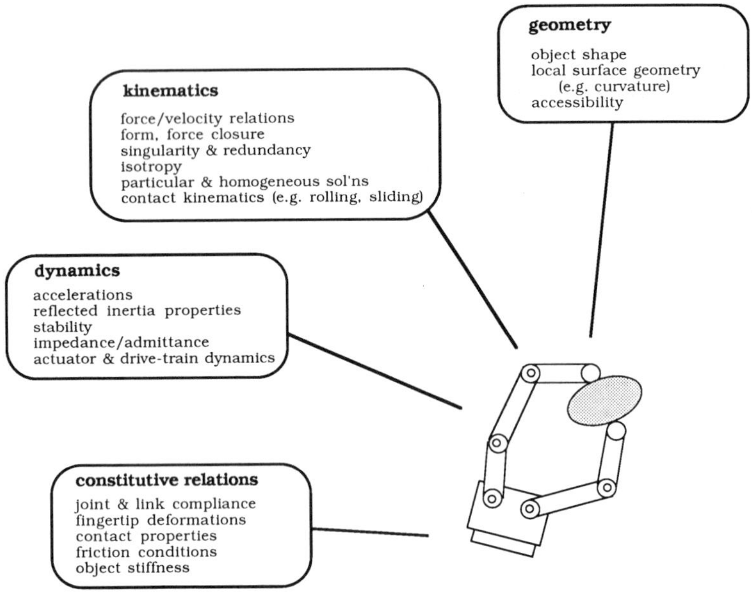

Figure 1.3: Issues in analytic modeling of grasping and manipulation (©IEEE, Cutkosky 1989).

- no sliding or rolling of the fingertips
- no cases with redundant degrees of freedom and no over-constrained grasps.
- assumed full knowledge of object and contact states, with no consideration for using sensory information during manipulation

Recent analyses, such as those by [Cutkosky and Wright 1986a; Nakamura, Nagai and Yoshikawa 1987; Ji 1987; Li and Sastry 1988], have relaxed some of these assumptions, although at the cost of greater complexity. Moreover, even the most sophisticated models involve the following simplifications:

- idealized models of the fingertips (e.g., point contact, or "soft finger" models with linear elastic deformation)
- idealized friction models (e.g., Coulomb friction) that ignore the effects of sliding velocity, material properties of the "skin," and the presence of dirt or moisture
- simplified actuator and drive-train dynamics, often ignoring elasticity, backlash and friction

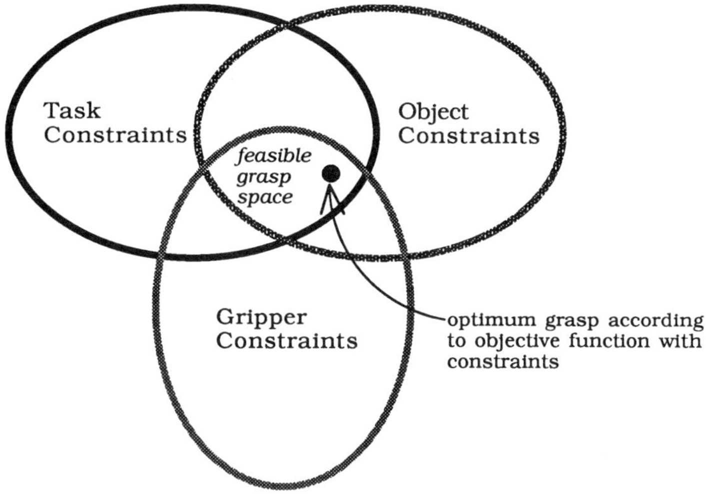

Figure 1.4: Analytic grasp choice: maximizing an objective function subject to task, object and gripper constraints (©IEEE, Cutkosky 1989).

- simplified representations of the grasped objects, treating them as smooth, rigid geometric primitives or polyhedra.

Based on the various analytic models of grasping and manipulation, a number of quality measures have been developed. Each measure provides a quantitative means of evaluating an important aspect of a grasp. For reference, these are summarized in Table 1.1. In Section 1.4 we evaluate the effects of the above simplifications and compare the quality measures with the empirically derived grasp attributes used in the expert system.

1.3.2 Analytic grasp choice as optimization

Many analyses treat grasp choice as constrained optimization. The problem of choosing a grasp based on analytic grasp models, quality measures, and constraints is illustrated in Figure 1.4. There are three overlapping sets of constraints arising from the task (e.g., forces and motions that must be imparted), from the grasped object (e.g., the shape, slipperiness and fragility of the object), and from the hand or gripper (e.g., the maximum grasp force and maximum opening of the fingers). Within these constraints is a space of "feasible grasps."

Choosing a grasp involves the definition of an objective function, which is optimized, subject to the constraints. The approach is conceptually straightforward, except that there is little agreement on which of the measures in Table 1.1 (along with additional geometric issues) should be in-

Compliance	What is the effective compliance (inverse of stiffness) of the grasped object with respect to the hand? The grasp compliance matrix is a function of grasp configuration, joint servoing, and structural compliances in the links, joints, and fingertips [Cutkosky and Kao 1989].
Connectivity	How many degrees of freedom are there between the grasped object and the hand? Formally, how many independent parameters are needed to completely specify the position and orientation of the object with respect to the palm [Mason and Salisbury 1985]?
Force closure	Assuming that external forces act to maintain contact between the fingers and the object, is the object unable to move without slipping when the fingers are locked? Formally, a grasp satisfies force closure if the union of the contact wrenches has rank 6 [Ohwovoriole and Roth 1981; Mason and Salisbury 1985].
Form closure	Can external forces and moments be applied from any direction without moving the object, when the fingers are locked? Formally, there is form closure, or complete kinematic restraint, if the intersection of all contact twists is a null set [Lakshminarayana 1987; Mason and Salisbury 1985]. For many tasks form closure is only required in certain directions.
Grasp isotropy	Does the grasp configuration permit the finger joints to accurately apply forces and moments to the object? For example, if one of the fingers is nearly in a singular configuration, it will be impossible to accurately control force and motion in a particular direction. Formally, the grasp isotropy is a function of the condition number of the grasp Jacobian matrix [Mason and Salisbury 1985; Kerr and Roth 1986]. Li and Sastry [1988] define similar grasp quality measures that are functions of the singular values of the grasp Jacobian.

Table 1.1 (Part 1): Definitions of analytic measures used to describe a grasp (©IEEE, Cutkosky 1989).

Internal forces	What kinds of internal grasp forces can the hand apply to the object? These forces are an important means of increasing friction at contacts. Formally, the internal grasp forces are the homogeneous solution to the equilibrium equations of the object. Thus, internal grasp forces can be varied without disturbing the equilibrium [Mason and Salisbury 1985; Kerr and Roth 1986].
Manipulability	While not consistently defined in the literature, a useful definition is: Can the fingers impart arbitrary motions to the object? Thus, a fully manipulable grasp must have a connectivity of 6, although some precision grasps have good manipulability in only a few directions, with the wrist and arm supplying further mobility. In addition, the rank space of velocities due to the joints must span the space of velocities transmitted through the contacts [Kerr and Roth 1986].
Resistance to slipping	How large can the forces and moments on the object be before the fingers will start to slip? The resistance to slipping depends on the configuration of the grasp, on the types of contacts and on the friction between the object and the fingertips [Kerr and Roth 1986; Cutkosky and Wright 1986a; Jameson and Leifer 1987; Ji 1987].
Stability	Will the grasp return to its initial configuration after being disturbed by an external force or moment? At low speeds, the grasp is stable if the overall stiffness matrix is positive definite [Nguyen 1988; Cutkosky and Kao 1989]. At higher speeds, dynamic stability must be considered [Nakamura, Nagai and Yoshikawa 1987].

Table 1.1 (Part 2): Definitions of analytic measures used to describe a grasp.

cluded in the objective function, and which should be used as constraints. Kerr and Roth [1986] establish a polyhedral region of "safe" grasps, bounded by friction limitations at the contacts. They define an optimal grasp as one that is furthest from the boundaries of the friction polyhedron, while also satisfying force closure and constraints on internal forces and actuator torques.

By contrast, Nakamura, Nagai and Yoshikawa [1987] search for a grasp that minimizes internal forces (and consequently, grasping effort) subject to constraints on force closure, friction and manipulability. If a safety factor is used in setting the friction constraints, this approach should give results similar to the approach that people seem to use, with forces a consistent percentage above the minimum required to prevent slipping [Ring and Welbourn 1968; Westling and Johansson 1984].

In a very different approach, Jameson and Leifer [1987] adopt a numerical hill-climbing technique in which a simplified three-fingered hand searches for positions that are most resistant to slipping, subject to constraints on joint torques and geometric accessibility. However, they cast the constraints as potential functions so that their effects are added to those of the objective function. In still other work, Li and Sastry [1988, this volume] define a "task ellipsoid," whose orientation and relative dimensions depend on the expected magnitudes of forces and moments during a task. Grasps are then compared according to the largest diameter of the task ellipsoid that they can encompass.

1.3.3 Other approaches to analytic grasp choice

The measures in Table 1.1 describe the kinematics and dynamics of a grasp, but there are clearly other factors involved in grasp choice. For example, if an object is to be picked up from a table, the grasp cannot place any fingers on the underside of the object. Other considerations include the size, shape and location of the center of mass of the object, the workspace of the hand, and the subsequent task requirements. Thus, a number of investigators have proposed geometric criteria for automated grasp selection [Lozano-Pérez *et al* 1987; Brost 1988; Tomovic, Bekey and Karplus 1987].

While there are numerous articles on grasp stability, force closure and quality measures for comparing different grasps, little has been proposed in the way of an overall strategy for grasp planning based on these measures. However, Ji [1987] outlines a sequence in which the first step is to find "grasp planes" defined by three fingertip contacts. He suggests several heuristics for locating possible grasp planes. The next step is to determine the arrangement of contacts about the periphery of the grasp plane. The grasps are checked for accessibility constraints (e.g., which parts of the object can the fingers actually reach?) and finally, task requirements are checked, possibly using a task-oriented quality measure such as that proposed by Li and Sastry [1988].

Several knowledge-based approaches, involving more detailed geometric descriptions than those in Grasp-Exp, are also being developed. Stansfield [1988] has built an expert system for robotic grasp choice using a two-stage model of grasping, with hand preshaping followed by object acquisition. This system generates candidate preshapes for grasps from symbolic, geometric object attributes, and considers the role of vision and tactile sensing in grasping.

Iberall *et al* [1988] are also developing an expert-system based approach to grasp choice. A high-level specification of a simple task is mapped into appropriate internal representations of the object, action, and hand. The system uses knowledge bases containing tool descriptions and rules for selecting oppositions and virtual fingers to generate a grasp.

1.4 Comparison of human and analytic grasp choice

By comparing studies of human grasping with analytical grasp models, we can evaluate the effectiveness of each approach. As a first instance, we consider the question "Given their numerous simplifications, how successful are the analytic models of grasps?" Since the simplifications used in most analyses represent approximations to real conditions, under particular circumstances any one of the analytic models may be a good approximation. The grasp taxonomy is useful here, making it easy to see which approximations apply to which sets of grasps. Study of the taxonomy reveals which grasps are likely to produce slipping or rolling of the fingertips, which have large areas of contact, or which have redundant degrees of freedom.

For example, the point-contact models are reasonably accurate for the precision Disk and Sphere grasps, where the contact areas are small compared to the diameter of the grasped object. On the other hand, a very-soft-finger model [Cutkosky and Wright 1986a] more accurately approximates the Tripod and Thumb-Index Finger precision grasps, where the finger pads conform to and even partially entrap the object.

Considering the range of grasps in the taxonomy, it is clear that the great majority of analyses have dealt only with fine-motion precision grasps. For the power grasps, most of the theoretical analyses are irrelevant since the fingers do not manipulate the part. Perhaps the best solution for power grasps is to assume complete kinematic coupling (with compliance) between the hand and the object, and to assign a set of friction and joint torque limits to the grasp.

Furthermore, with a few exceptions [e.g., Fearing 1986b], large motions of the object with respect to the hand have not been considered. It is also clear from the human studies that a combination of object, task, and hand properties must be considered. But few, if any, analyses have considered all three. Clearly, in order to determine which are the correct analyses and which are the fallacious approximations, we must test the predictions of these models in experiments with real robot hands.

1.4.1 HUMAN GRASPS IN TERMS OF ANALYTIC GRASP MEASURES

Although the analytic approaches to grasp choice involve many simplifications, the measures developed in the analytic work, and summarized in Table 1.1, can nonetheless be applied to human grasps. For example, all of the grasps (except the non-clamping hook or platform grasps) satisfy force closure, assuming extra contact wrenches due to friction. The non-clamping grasps are force closure, provided that external forces do not cause the fingers to detach from the object.

Many grasps do not satisfy form closure without friction. Particularly, if we know the direction from which the task forces are coming, then we do not need the added security of kinematic restraint. Figure 1.5 shows a person pulling on a wrench to tighten a large bolt. The principal task force is known and the hand adopts a "hook" grasp from the grasp taxonomy. There is no opposition between the fingers and the palm and therefore no ability to control internal forces. By contrast, if the same person were trying to free a sticky (perhaps rusted) bolt, a more conservative wrap grasp would be adopted to prevent the wrench from flying out of the hand if the bolt should suddenly come loose. When compared to the hook grasp, the wrap grasps have an opposition between the fingers and the palm, an ability to control internal forces, and an ability to apply both positive and negative forces with respect to the wrench handle. Only in extreme cases (e.g., washing dishes or handling a wet bar of soap) do we tend to use grasps that would satisfy form closure without friction.

In terms of the other analytic measures, human power grasps are less compliant, more stable and have a larger resistance to slipping than precision grasps. This is because power grasps are able to exert large internal forces, thus increasing the magnitude of frictional forces. Finally, power grasps have a connectivity of zero since the fingers do not manipulate the part. However, the precision grasps have high manipulability, and a connectivity between object and hand of at least 3 and often 6.

Many of the detailed grasp attributes in Grasp-Exp can also be correlated with the analytic measures. However, since the terms that people use for describing grasps are subjective, and depend on many subtle factors, the correspondence is rarely exact. For example, security decreases as the applied force increases. When people say they need a secure grasp, they implicitly take into account the magnitude of the forces in the task. Thus, manipulating a small object between the pulps in a two-finger precision pinch grasp may be perfectly secure (i.e. the object won't slip out of the grasp), but this is not considered a "secure" grasp.

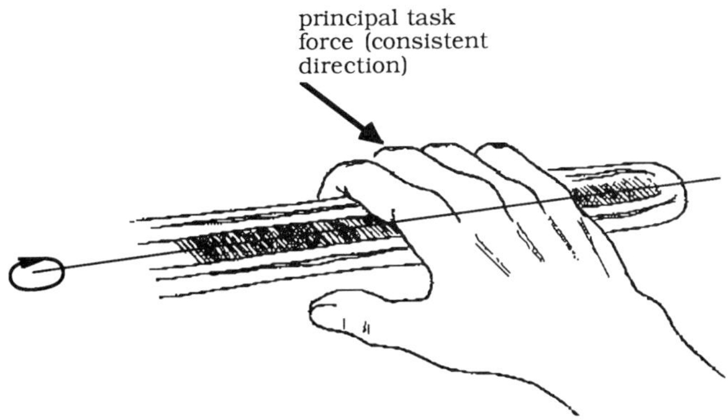

Figure 1.5: A Hook grasp is used to pull on a large wrench. The lack of closure is acceptable if the task force remains predictable.

As shown in Figure 1.6, grasp attributes used in the expert system correspond to analytic measures from Table 1.1 as follows:

- *Sensitivity* depends on many factors but is primarily related to how accurately the fingertips can pick up small small vibrations and small changes in force and position. Thus sensitivity is a function of grasp isotropy (if the fingers can impart forces with accuracy then they can also measure forces with accuracy) and stiffness (a compliant grasp is more sensitive to small changes in force).

- *Precision* is a measure of how accurately the fingers can impart small motions or forces to the object. Thus, precision requires light grasp forces, full manipulability and isotropy.

- *Dexterity* is similar to precision but implies that larger motions can be imparted to the object. Thus dexterity depends both on manipulability and the kinematic workspace of the hand (and often the wrist).

- *Stability* includes both the definition in Table 1.1, in which a stable grasp will return to its nominal position after being disturbed and the ability of the grasp to resist external forces without slipping.

- *Security* is related to stability, but is most closely associated with resistance to slipping.

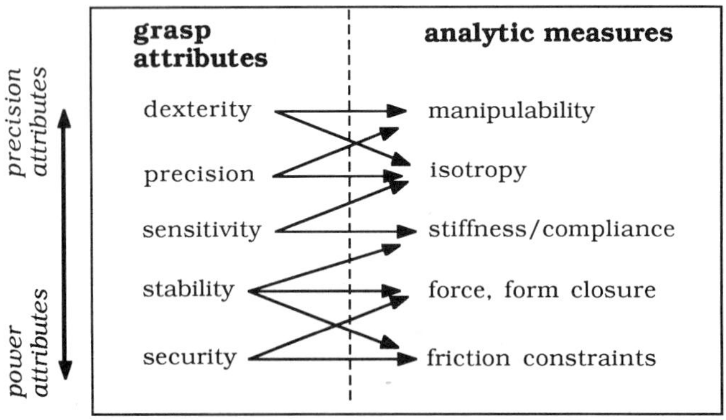

Figure 1.6: Human grasp attributes in terms of analytic grasp measures (©IEEE, Cutkosky 1989).

1.4.2 Grasp choice issues

The analytic measures are useful for comparing grasps, but the resulting constrained optimization approach to grasp choice is impractical for generating candidate postures. This approach requires specification of exact finger placements, and so requires search of a very large space of possible contact positions [e.g., Jameson and Leifer 1987].

Several workers have suggested that grasp choice is best approached as a two stage process: first a basic hand configuration or posture is selected, then the exact placement of the fingers is chosen [Ji 1987; Iberall *et al* 1988; Stansfield 1988]. Our experience with Grasp-Exp indicates that a knowledge-based approach using descriptions of the geometry and forces of the task and object can be used to generate candidate hand postures. From the taxonomy of Figure 1.1 we also see that there are a finite number of fundamentally different hand postures. If the search is done in this space it is a simple problem, and a relatively small rule-based system is capable of choosing a grasp based on object and task requirements. Stansfield [1988] and Cutkosky [1989] both suggest that about 100 rules are sufficient for this job. Ji's work [Ji 1987] outlines how heuristics might be combined with analytic concepts to generate grasps. In Section 1.5 we discuss questions of object representation for grasp planning, which can greatly simplify the generation process.

There is some question about which to consider first, the geometry of the hand and object, or the task forces. It seems that neither one should necessarily come first. Rather, these are simultaneous constraints and either one may prune off larger pieces of the search space initially. A good strategy is to first apply those constraints that reduce the search space fastest.

Once candidate grasps have been generated, the quality measures can be used for deciding between candidate grasps or optimizing the exact placement of the fingers. A rule-based system could also decide which of the analytical measures to use in assessing a grasp. This could be based on considerations about which of the simplifying assumptions (such as point versus soft-finger contacts) apply to a particular grasp. Note that the taxonomy is helpful in this, since choice of the correct approximations can be based on both the generic type of grasp and the particulars of the object and task. Alternatively, the system could decide which of the measures to treat as constraints and which to optimize, based on the circumstances.

Other work [Tomovic *et al* 1987; Iberall 1987b] suggests that once the basic posture is known, tactile sensing must be used to optimize placement of the fingers as the grasp is executed. Fearing [1987a] also shows that tactile sensing is crucial to robotic manipulation with large motion.

1.4.3 Virtual fingers, oppositions and internal forces

An interesting sub-topic that surfaces in both the analytic grasp work and in observations of human grasp is the dual nature of forces applied in grasping and manipulation. Forces can be decomposed into external forces which equilibrate the forces and moments of a task and produce motions of the grasped object, and internal forces which stabilize the grasp.

In the work of Iberall, external and internal forces appear in the context of virtual fingers and oppositions. While we recognize the usefulness of virtual fingers for generalizing the taxonomy of Figure 1.1, our own interpretation is slightly different and therefore we have added "virtual finger" numbers to the revised taxonomy of Figure 1.1. The Opposed-Thumb (Grasps 6-9) and Lateral Pinch (Grasp 16) are two-fingered grasps since there are two independently controllable gripping surfaces. Even the Opposed Thumb-4 Finger grasp is basically a two-fingered grasp since the four fingers act in unison. At the other end of the spectrum, power grasps 1-3 and 11 are difficult to describe in terms of virtual fingers since they completely envelope the part with something approaching uniform radial symmetry, but have no independent contact areas. Finally, the non-clamping grasps (almost non-grasps) such as the Platform and Hook grasps have one virtual finger.

The relationships between virtual fingers and oppositions and the analytic concepts of internal and external grasp forces becomes clear with a simple example such as the Tripod grasp, as seen in Figure 1.7. Iberall classifies the Tripod grasp as a two-virtual-finger grasp with one opposition. In the past we have argued that because there are three independently controllable contact locations it is a three-virtual-finger grasp. However, from a practical standpoint, there are just two oppositions, both involving the thumb. The human hand has some coupling between the index and third fingers, and the magnitude of the internal grasp force between these

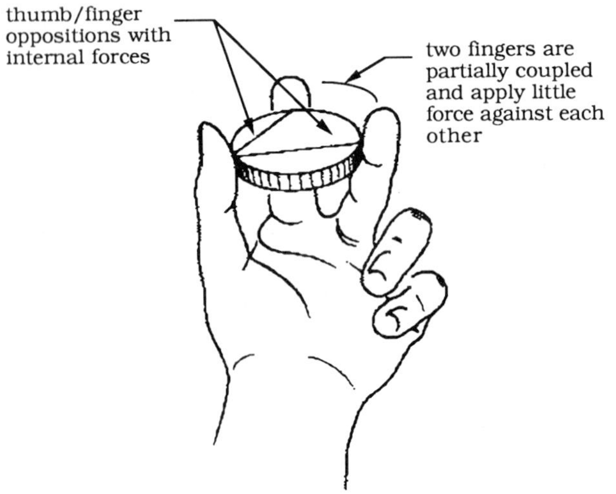

Figure 1.7: The Tripod grasp involves three independently controllable fingers but is dominated by two thumb/finger internal forces.

fingers is never large. The thumb is not just another finger with a special orientation, it is opposed to the fingers and thus can exert much larger forces against them. Moreover, if one examines the kinds of manipulations commonly done with a tripod grip it will be seen that the dominant axes of object motion are defined by moving the thumb up-and-down or side-to-side with respect to the fingers. In contrast, for the Stanford/JPL hand [Mason and Salisbury 1985] the tripod grasp becomes a 3-virtual-finger grasp with *three* opposition pairs, since all fingers are independent and nearly equal forces can be applied between each pair of fingers.

This suggests that virtual fingers correspond to independently controlled contact sites, and oppositions correspond to internal grasp forces. A tripod grasp may therefore have either two or three virtual fingers depending on the amount of coupling between the index and third fingers. In other words, the number of contact sites which *could* be independently controlled may be larger than the number actually used. (Of course, the number of contact sites which can be independently manipulated is also limited by the rigid body constraint of the object.)

The distinction between the number of virtual fingers and the number of opposition pairs (for those grasps where the numbers are different) speaks to the number of degrees of freedom required in the task and the amount of coupling among the fingers. This coupling is expressed analytically as off-diagonal terms in the grasp stiffness matrix [Cutkosky and Kao 1989].

1.5 Lessons for future analyses and for hand design and control

1.5.1 OBJECT REPRESENTATION AND GRASPING FEATURES

As mentioned earlier in describing the grasp taxonomy, approximate geometric descriptions of the objects (long, thin, disk-shaped, etc.) are too vague. A perspicuous representation would make reasoning about grasping easier and more effective. One key aspect is the designation of grasping "features." Features should not be neutral descriptions of the part geometry (e.g., cubes, cylinders) but should emphasize elements of the geometry that are important for grasping and task execution. Thus a hammer would be described largely in terms of its handle and the contact point on its head. With a feature-based description of objects, Grasp-Exp would ultimately resemble rule-based planning systems for setup and fixturing of machined parts, such as GARI [Descotte and Latombe 1984]. Feature-based descriptions of parts have also been explored for automatic robot grasp planning [Lozano-Pérez *et al* 1987; Popplestone 1987].

A proper object representation for grasping can be used not only in working with well-characterized tools and parts, but also with unknown objects in an unstructured environment. Stansfield [1988] has considered this problem in the context of grasping incompletely described objects whose shapes are determined by sensing. She uses a "spatial polyhedron" representation, which facilitates incorporation of new geometric information as it is sensed.

One important type of feature is possible grasp location. Following an earlier idea by Jameson [1987], Ji [1987] suggests algorithms for finding "double normals" on the surface of an object. Double normals are basically faces with opposed, collinear normals. Ji shows that double normals are good places for exerting and controlling the internal forces to maintain stability. In other cases, it is advantageous to position fingers symmetrically with respect to one of the ends of a double normal. Ji gives some heuristics for choosing possible grasp locations: for objects with parallel, planar faces position the contacts so that the contact normals are antiparallel; otherwise, position the contacts so that the inward pointing normals meet at a point. These double normals can also be thought of as candidate locations for the oppositions described by Iberall.

There are many important issues here which have yet to be explored. For example, how do we accommodate multiple task axes for a single tool? A screwdriver can be used to pry or stab in addition to turning a screw, and each task specifies a different principal axis. Also, objects can have several levels of detail, and it is not always clear which are unimportant. Thus a screwdriver handle is basically a cylinder, but has minor features such as ridges and knurls, which are important for applying torque.

The level of detail needed to describe an object may determine the fundamental approach to geometry. If the space of all object shapes can be

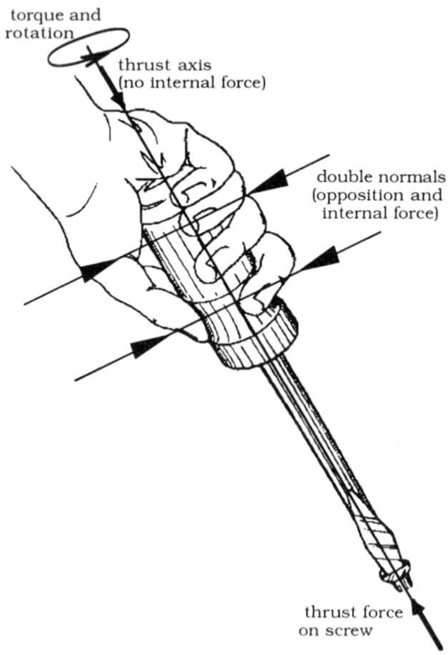

Figure 1.8: Principal task axes and double normals for turning a rusted screw.

represented by a small catalog of prototypes (or built from combinations of a few primitives) then it may be possible to use a rule-based approach. The question of object description also relates to the tactile sensing-based approach to finger placement. If exact finger placements can be determined on an ad hoc basis while grasping, then the grasp planner need only work with a rough prototype shape.

1.5.2 Relating object and task axes

When an object is used to perform certain tasks (e.g., turning a screw with a screwdriver) certain axes and features (e.g., an axis of rotational symmetry or a pair of opposed faces) tend to become most important. In this way, object geometry, in terms of features, can be correlated with the needs of a task.

Consider the screwdriver in Figure 1.8. The principal task forces include a moment about and thrust along the central axis, while the principal motion is a twist about the central axis. The magnitudes of the forces and motions on these axes dictate the choice of grasp. The tip is the action feature. The handle is the graspable feature. Following the grasp taxonomy, or the rules in Grasp-Exp, we would conclude that because the thrust force and torque are large, we need a power grasp. Because the handle is a thin cylinder, we choose a wrap. Finally, since the ability to apply twisting motions with the wrist is important (more important than tactile sensitivity, or the ability to apply radial forces or moments perpendicular to the tool axis) we choose Grasp 4.

From a more analytic perspective, we see that the tool handle is approximately axisymmetric and therefore has an infinite number of double normals in the radial direction. Applying internal forces across these double normals, we can resist thrust force and torque through friction. But if the required thrust force is large, as in trying to free a rusted screw, then friction on the handle will not suffice. In this case, we might look for a pair of double normals that involves the handle's end. But of course, there is no such pair. When the palm is used to apply a thrust force to the end of the handle the opposition is with the screw itself. Thus, there is no ability to independently control internal forces along the central axis; this is acceptable as long as the thrust force is predictable (e.g., the screw never suddenly pulls on the screwdriver). Looking at the same axis from a kinematic point of view, we observe that we do not have form closure along it.

As for the task motion, although we do not actually impart the twist with our fingers, we adopt a grasp that maximizes our ability to twist with the wrist. By contrast, if we use the same screwdriver for prying open the lid of a paint can we focus on an axis perpendicular to the central axis.

1.5.3 Hand design issues

In retrospect, the most useful contribution of the study of human grasps, from the standpoint of designing and controlling robot hands, has been a better appreciation of how task requirements and object geometry combine to dictate grasp choice. The study has resulted in a grasp taxonomy, which makes it possible to identify particular grasps and to trace how they derive from generic grasp types. The fact that both task requirements and geometry are important is clear from everyday experience. The grasp we use for picking up a pencil is entirely different from the one we use for writing, although the object geometry remains the same. On the other hand, if we consider the task of filing a machined part, the grasp we use for a flat file is different from the grasp we use for a round one, although the forces and motions are the same.

From the standpoint of hand design, we find that although the expert system contains a great deal more information than can be represented in a taxonomy, the taxonomy remains useful as a design aid since it allows one to see very quickly where a particular grasp resides in the space of possible grasps.

One fact that became clear in studying grasping in a manufacturing environment is that many grasps provide immobilization, while manipulation is performed by the wrist and/or arm. All of the power grasps fall into this category. Because the human arm has redundant degrees of freedom, the range of manipulation possible using the wrist and arm is quite large.

For these grasps the principal function of the fingers is to provide a large workspace for immobilization of the grasped object. Furthermore, the availability of this capability must influence grasp choice to a great extent. This observation leads to the question of what the proper role of the robot hand should be.

Unlike human arms, robot arms have excellent positional accuracy. But they also have comparatively few degrees of freedom and a large mass which limits their ability to control small forces at the hand. However, new manipulator designs can incorporate light weight wrists with redundant degrees of freedom. Does this mean that a passive gripper that can grasp a wide variety of workpieces is the solution for manufacturing environments? Perhaps not. We are successful in grasping and manipulation largely through our ability to reorient objects and modify grasps in response to shifting task requirements. This would obviously not be possible for a gripper capable of only immobilization grasps.

Another interesting aspect of the human grasping environment is the ubiquitousness of cylindrical handles, which are found on virtually every tool and many other objects as well. Our hands are clearly very good at grasping this shape, although this begs the question "Given the choice of any hand design, would another shape work as well?" A long, thin shape allows us to exert internal forces along a considerable axial distance, which permits us to resist torques about an axis perpendicular to the handle. Such moments are produced when forces are applied to the tool at some distance from the grasp region (e.g., in using a tennis racket). Many tasks produce just these sorts of moments in the hand. Hand surgeons have also noted that for a workman, the loss of the little finger can be more of a handicap than the loss of the ring finger, because of the decreased ability to resist such moments.

It is also possible to examine industrial gripper design in light of the taxonomy in Figure 1.1. For the most part, today's commercial grippers achieve particular instances of the power grasps on the left hand side of Figure 1.1. Increasingly, however, general-purpose grippers are inadequate for the variety of part shapes and tasks encountered in flexible manufacturing systems. A common solution is to provide an array of special-purpose grippers for each part style. The taxonomy in Figure 1.1 suggests, however, that if several grippers are to be used, they should be designed for classes of grasps and tasks – not for different part styles. To design a gripper for a part style is to design a tool, not a hand. Thus, like a phillips-head screwdriver which can only be used with phillips-head screws, the gripper is a special-purpose device.

A better approach is to start with basic task requirements and let those requirements dictate the design. For example, one might construct a gripper for precision grasps with opposed fingers and a second gripper for power wrap grasps. Another possibility is to construct a hand for two types of tasks with a single object. For example, a manufacturing hand used for picking up small power tools and then working with them could shift between the Opposed Thumb-4 Finger grasp, Grasp 6, and the Light Tool grasp, Grasp 5. Such generic designs can be adjusted to fit a variety of part shapes and finger adaptors may be used for specific constraints encountered with exceptional parts.

It is also unnecessary to achieve all of the different grasps in the taxonomy. For example, pulling on a wrench as shown in Figure 1.5 could easily be achieved with a wrap grasp. While it suits the machinist with his human hand to use a full repertoire of grasps, this Hook grasp may be unnecessary for a robot. In designing a robot hand and controller for a particular environment, the set of necessary and sufficient grasps can be determined from the taxonomy, and the capabilities of the robot gripper system can be matched to this set. In this way the grasp taxonomy can streamline hand design, construction, and control. Thus, in a form-follows-function sense, robotic hands can be made competent but not over-designed and hence overly expensive and difficult to use.

Acknowledgements: We thank Paul Wright at NYU and John Jourdain and Mark Nagurka at CMU for their assistance and suggestions on the grasp taxonomy and Kos Ishii at OSU for his help with Grasp-Exp. Mary-Jo Dowling of the Robotics Institute at CMU and Francis L'Esperance at Stanford provided drawings of hands and grasps. This work has been supported by the National Science Foundation under grants DMC8552691 and DMC8602847.

2

Opposition Space and Human Prehension

Thea Iberall[1]
Christine L. MacKenzie[2]

> *"Our hands become extensions of the intellect, for by hand movements the dumb converse, with the specialized fingertips the blind read; and through the written word we learn from the past and transmit to the future."*
>
> –Sterling Bunnell, 1944

ABSTRACT A problem that has plagued both motor psychologists in studying human behavior, and robot designers in reproducing it, is how to quantify that behavior. It has been argued that kinematic and dynamic models are inadequate for explaining human movement unless they also include both the performance constraints and the objectives that affect the neural and neuromuscular inputs. With a dextrous, multi-fingered hand, multiple grasping solutions are possible. This chapter addresses the question faced by the controller, that of how best to use features of the hand to achieve the task goals, given anticipated object properties and predictable interaction outcomes. The opposition space model takes into account the hand's ability to apply task-related forces while gathering sensory information, in terms of its precision and power capabilities. By separating implementation details from functional goals, the study of human hand functionality can lead to the design of better dextrous robot hands and their controllers.

[1]Department of Computer Science, University of Southern California, Los Angeles, California 90089-0782

[2]Department of Kinesiology, University of Waterloo, Waterloo, Ontario N2L 3G1

2.1 Introduction

Sterling Bunnell, the founding father of hand surgery, developed surgical procedures stressing the restoration of function over simple reconstruction. One such function of the human hand is to grasp a wide variety of objects under highly complex task requirements. As noted by Napier [1956], task requirements involve aspects of strength (power) and fine control (precision). Researchers have developed various schemes for matching hand functionality to task requirements, searching for ways to capture the hand's versatility in prehensile performance [Napier 1956; Lyons 1985; Iberall, Bingham and Arbib 1986; Cutkosky and Howe, this volume]. Napier, for example, suggested that the hand can form a power grasp or a precision grasp, thus creating a hand posture that matches the power and precision task requirements. Although a variety of prehensile postures have been suggested (e.g., lateral pinch, pulp pinch, enclose grasp, hook grasp, etc.), few clues are given as to what is a posture's essence or how it can be related to the wealth of analytic tools used by roboticists. We define prehension as the application of functionally effective forces to an object within a task, given numerous constraints. This definition stresses the fact that the human hand is both an <u>input</u> and <u>output</u> device. As an output device, it can apply forces both to match the anticipated forces in the task (stable grasp) as well as to impart motion to the object as necessary (manipulate). As an input device, it can gather sensory information about the state of the interaction with the object during the task in order to ensure grasping and manipulative stability. In a <u>precision</u> grasp, it can comply with small forces, impart small motions, and gather precise sensory information to match manipulative requirements of the task; in a <u>power</u> grasp, it can match larger anticipated forces while still ensuring a stable grasp through sensory information, using the arm and wrist to provide grosser motions. The words 'functionally effective' are used to highlight the fact that the forces must be applied within the functional constraints of the task; i.e., while forces can be used to effect a stable grasp and impart motions as needed in a task, there are functionally specific demands on how this is accomplished, such as their resolution of control, their precision requirements, the security needed, etc. The human hand has a variety of ways to grasp objects stably; the decision that the central nervous system (CNS) must make in choosing one of these depends ultimately on balancing the functional requirements of the task with the functional abilities of the human hand and body. Our goal, by making explicit the functional issues of prehension, is to address the question of how the hand can, under the direction of the CNS, achieve versatility in light of the multiple constraints and goals of dextrous manipulation.

As seen in other chapters in this volume, roboticists have a variety of analytic tools for controlling robot hands. At the same time, kinesiologists try to understand the CNS by studying human movement. In contrast to

Group	Level of Analysis	Examples
High Level	Social/Cultural	don't stick out elbows; stick out little finger
	Motivational	thirst; anger
	Informational	convey affection, anger
	Functional	don't drop object; manipulate object; 'move as quickly and as accurately as possible'
Physical	Object Properties	intrinsic (texture, surface length, weight, etc); extrinsic (location, distance, environment, etc)
	Biomechanical/ Mechanical	kinematics; dynamics; limitations on force generation due to bones, muscles, tendons, ligaments, skin; effect and use of pads
Sensori-motor	Neural	temporal and spatial limitations on CNS; pyramidal tract needed for fractionated finger movements; sensory info needed to sustain movement; sensory information needed to preshape hand; tonic vibration reflex
	Perceptual	types, locations, and response characteristics of receptors; numerous tactile receptors in pulps with small receptive fields
	Anatomical/ Physiological	structural limitations on movements, directions, and extents; length of phalanges; additional muscles in index and little finger; pads; anatomical couplings
	Evolutionary/ Developmental	evolutionary pressures; five fingers; pyramidal tract develops in about eighth month

Table 2.1: Sources of constraints on grasping. Movement constraints and goals act to limit the way the human hand can be controlled in prehension.

roboticists, they can only make hypotheses about the underlying control mechanisms by measuring key behavioral components (in essence, they are reverse engineering the controller). To do this, experimental conditions are manipulated and their effects are quantified through spatial and temporal characteristics of prehensile movements. These characteristics involve the emergence of hand postures (the preshaping of the hand) during transport to the target [Jeannerod 1984] and the timing of movement features [Jeannerod 1984; Marteniuk *et al* 1987]. Varying the experimental conditions allows the use of information from various sensory modalities to be evaluated [Jeannerod 1984, 1986], as well as the effect of object properties [Jeannerod 1984] and intentions [Klatzky and Lederman, this volume; Marteniuk *et al* 1987]. All of these appear to be important in defining the rules for effective prehension.

While roboticists and experimentalists seem to be looking at the same problem, they use different techniques and vocabularies to do it. It has been argued [Nelson 1983; MacKenzie and Marteniuk 1985; Marteniuk *et al* 1987] that interactions are occurring among multiple performance constraints. Nelson [1983] argued that dynamic models, such as [Hollerbach 1980], are inadequate for explaining human movement unless they also include both the performance constraints and the objectives that affect the neural and neuromuscular inputs. MacKenzie and Marteniuk [1985] describe two types of constraints: those variables that constrain the use of feedback and those due to structural limitations that affect the preparation and execution of goal directed movement. Under the first category are included issues such as how fast feedback can be used, the types of feedback that can be used, and the experience of the performer in learning sensorimotor representations. In the second category are limitations from anatomical and neurological structure, from the structure of information communicated, the structure of environmental information, and constraints due to context and intentions. Marteniuk *et al* [1987] emphasize the interaction between knowledge (which includes a person's experience and movement objectives) and the biological structure. Rewording the list of MacKenzie and Marteniuk [1985], Marteniuk *et al* [1987] make even more explicit the notions of task constraints and conditions of speed and accuracy. By taking a task-specific view of movement planning and control, they define a task as the interaction of the performer with the environment under given movement goals. This view predicts that movement is optimized and specific to the unique requirements that arise from an individual motorically interacting with the environment.

The dextrous and versatile performance of human prehension can be viewed as emerging from a large multidimensional constraint space. In order to study the complex interaction between movement goals, object properties, environmental characteristics, and the performer's biological structure and experience, we argue that there are different levels of analysis within this constraint space. These constraints, or levels of analysis, are grouped into three broad categories, as seen in Table 2.1.

This list expands and reorganizes the constraints listed in [MacKenzie and Marteniuk 1985; Marteniuk *et al* 1987]. The key point is that it brings in the notion of higher level goals working together with harder constraints, as well as suggests ways that a roboticist's view of prehension can be integrated with a kinesiologist's view. For example, when a roboticist studies what constitutes a stable grasp, he/she is detailing the constraints imposed by an *implicit* fact that a functional goal is 'do not drop the object.' On the other hand, when a kinesiologist tells a subject to 'move as quickly and accurately as possible', he/she is imposing constraints on how to plan and effect the movement, given the person's biomechanics and sensorimotor constraints. Even physicians, in evaluating impairment of hand function, make distinctions between physical and functional evaluations [Swanson, Hagert and Scanson 1983]. By making explicit the influences of the multiple constraints acting on these systems, we argue that roboticists can develop higher level robot languages for control of more sophisticated dextrous robot hands, and kinesiologists can further refine their models of human performance using the tools provided by roboticists.

In the next section, this table is discussed in terms of the three levels of analysis. In the third section, a model is suggested for capturing human prehensile versatility. The final section summarizes this chapter while at the same time points out open research questions relevant both to roboticists and experimentalists.

2.2 Constraints on Human Prehension

2.2.1 Physical constraints

The laws of physics, captured in a variety of equations, create limitations on the planning and control of prehension. Whether the CNS solves these equations directly or not, they detail the meaning of movement within inertial reference frames and the effects of making contact with the environment. In the example where the functional constraint is to 'not drop the object', a posture must be chosen that effects a stable grasp. The three requirements for a stable grasp are [Fearing 1986b]: 1) the object must be in equilibrium (no net forces and torques); 2) the direction of the applied forces must be within the cone of friction, which is twice the angle between the arc-tangent of the coefficient of static friction and a normal to the surface; and 3) it should be possible to increase the grasping force's magnitude to prevent any displacement due to an arbitrary applied force. Creating a stable grasp means taking into account active forces and torques as well as passive ones caused by the frictional components of the skin contacting the object surface. In addition, counteracting forces must be separated into their grasping and manipulation components [Yoshikawa and Nagai, this volume]. If the applied force is too powerful, there is an unnecessary waste of energy; however, what is being optimized in human systems is a question [Nelson 1983].

Although useful, sensory information is not a necessary condition for effecting a stable grasp. Mason [Mason and Salisbury 1982] has shown that there are predictable results when two fingers are about to grasp an object even though there is some object position uncertainty. For example, in picking up a glass with the right hand, if the index finger makes contact before the thumb, the glass will rotate out of the grasp if the direction of the line of pushing (a line parallel to the motion of the finger, drawn through the contact point) is to the near side of the center of friction between the glass and the table (a system of frictional forces arising at a contact area reduces to a single force acting through the center of friction). If, instead, the line of pushing is through the center of friction or to the far side, the glass will translate towards or rotate into the thumb, respectively.

Hand postures afford different ways to apply forces. Biomaterials such as bones, muscles, tendons, ligaments, and skin create limitations on static and

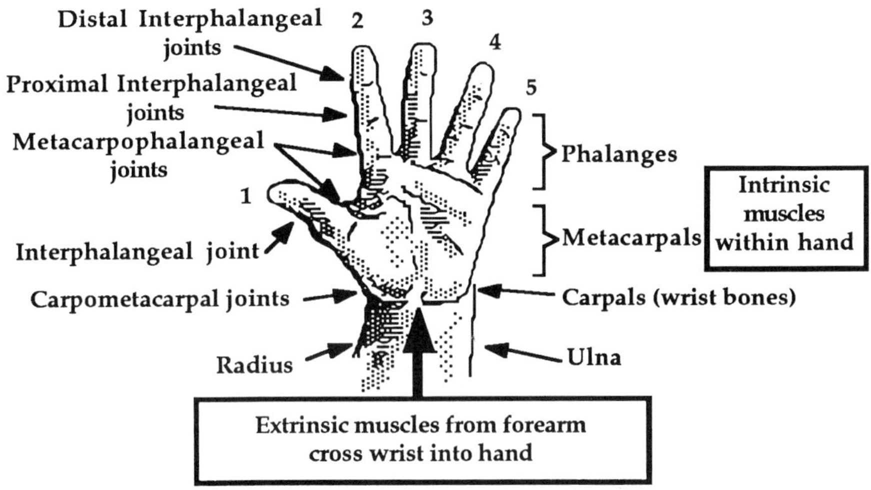

Figure 2.1: Bones and joints of the human hand. Note distinctions in types of muscles.

dynamic force generation. Chao, Opgrande and Axmear [1976] argue that typically used postures are ones where joints are stabilized and where forces can be applied optimally without undue stress and strain on the ligaments, joint capsules, tendons, and muscles. An example would be where articular surfaces are coadapted, thus permitting increased force generation while decreasing stress on ligaments and muscles. Results from cadaver studies have been used to develop and verify computer models of how muscles and tendons generate force across the various joints for different isometric hand postures [Chao *et al* 1976; Cooney and Chao 1977; Wells, Ranney and Keeler 1985; An *et al* 1985]. Extrinsic finger flexors (as seen in Figure 2.1, forearm muscles that insert into the phalanges by sending long tendons through the wrist and hand) have a greater mechanical advantage than the extrinsic extensors [Tubiana 1981]. Within the hand, the intrinsic muscles of the index finger produce more force in a lateral pinch than in an enclosing grasp [An *et al* 1985]. The fingers and thumb have an average greater strength in a power grasp than in a pulp pinch [Cooney and Chao 1977; An *et al* 1985], particularly since multiple points of contact are applying a force against the object. However, the size of the finger separation in the power grasp influences the gripping forces. In a task involving the isometric grasp of different size cylinders, Amis [1987] determined that the distal phalanx exerts the largest gripping (normal) force in all fingers, with a smaller force produced by the middle and proximal phalanges. Summing up these forces within a finger, the total normal force was largest for the smallest objects, and then decreased as the object size increased. Shearing forces at the distal and proximal phalanges for the index, middle, and ring fingers tended to pull the object into the grasp for smaller objects. As the object got larger, shearing forces on the middle and proximal phalanges tended to zero out, while at the distal phalanges, they tended to push the object out of the grasp. Other studies have analyzed the effect of wrist position on available force in precision and power grasps [Hazelton *et al* 1975]. Because the extrinsic muscles send tendons through the wrist, the amount of force that the fingers are able to apply is greater in wrist extension and/or ulnar deviation than in other positions. Hazelton *et al* [1975] noted that the relative amount of force available at each finger during a hook grip remains constant across wrist positions (25% at the index, 33.5% at the long finger, 25% at the ring finger, and 16.5% at the little finger). Amis [1987] found the mean contributions to be 30%, 30%, 22%, and 18% to the overall grasp force. The lesson here (and further detailed in Section 2.2.2) is that there is a differential in the individual finger's potential contribution to force application, thus a factor in choice of postures.

Skin has properties that affect the frictional component of force generation. Typically, the law of static friction (Amonton's law) states that the tangential force F_T of friction is constrained to be no greater than the product of the normal force N with the coefficient of static friction μ_s, or $F_T \leq \mu_s$ N (thus creating the cone of friction mentioned above). The coefficient of static friction, measured by various researchers [Comaish and Bottoms 1971; Westling and Johansson 1984; Buchholz *et al*.1988], involves an interaction between the constraints of the skin surface and object surface (see below for discussion of object properties). Comaish and Bottoms [1971] found that the hand's palmar surface has a higher coefficient of static friction than does the back of the hand. Buchholz, Frederick and Armstrong [1988] show that the coefficient of static friction increases when moisture is present, a situation that particularly occurs when the sweat glands of the palm are activated. Comaish and Bottoms also found that with some surfaces, the coefficient of static friction increased as the contact area increased (which would imply that Amonton's law is not the best model to use here, since it says the coefficient of static friction is dependent on the magnitude of the forces, not the amount of contacting area). Interestingly, the coefficient of static friction decreases with increasing load, reducing the effectiveness of friction for heavier objects [Comaish and Bottoms 1971; Buchholz *et al* 1988].

The frictional component of force generation is further enhanced when the pulps on the distal palmar surface of the fingers are used. These are highly specialized for prehension, in that they provide friction, due to the epidermal ridges and to the sticky self-lubricating excretions through the tops of the ridges, and that they to comply with (instead of being displaced by) objects touched in the environment [Quilliam 1978; Glicenstein and Dardour 1981; Thomine 1981]. While their existence is more an aspect of the hand's anatomy (see Section 2.2.2), their use in a prehensile posture is such that they can automatically supply a force sufficient to counteract small amplitude perturbations of the object, useful in stable grasping (along the lines of a 'soft finger' as a contact type that resists moments about the contact normal [Salisbury and Craig 1982]). The lubricant is greasy, having good adhesive qualities at low shear velocities, enhanced by the hills and valleys of the ridged surface which extends the total shearing surface area [Moore 1975]. At high shear velocities, friction is reduced, thus minimizing wear and tear of the hand surface. With ridges that are concentric in orientation, there will always be some perpendicular to the force exerted (as opposed to in the palm where that is not true). However, due to biomechanical properties, skin responds in non-linear ways to loads [Wilkes, Brown, and Wildnauer 1973]. As the load increases, collagen fibers in the dermis reorient from an irregularly arranged pattern to an orientated one in the direction of the applied load. As more fibers become aligned and thus resist extension along their length, the skin becomes more stiff, reducing its ability to comply with the object.

Object properties, and the object's relationship to the environment, can affect the chosen posture. Jeannerod [1981] made a distinction between intrinsic and extrinsic properties of objects. Extrinsic properties are those which are relational with respect to the person (i.e., location, distance, angle, etc.) and would include support constraints, such as a table supporting a surface, and other obstacle constraints (forcing one to reach around another object). Lederman and Klatzky [1987] have shown how subjects can extract intrinsic object properties relevant to the task. These include surface properties (texture, hardness, and temperature), structural properties (shape, volume, weight), and functional properties (part motion, specific function). Before contacting an object, many object properties can be perceived visually (see [Klatzky and Lederman, this volume]). In a study where subjects had to use a precision grip to grasp one of ten different sized wooden disks, Marteniuk *et al* [in press] show that the maximum aperture reached during preshaping between the index finger and thumb closely correlated with disk size. In terms of the constraints that object properties impose on prehension, examples are numerous: object shape can delimit grasp postures and the number of fingers potentially contacting a surface; the weight of the object can determine the type and strength of grip; and the availability of an object surface will constrain the orientation of the reach and potential contact locations.

Other object properties have an effect on force generation as well. Cochran and Riley [1986] found that handle shape affects the force exerted: higher forces were seen with irregularly shaped handles (rectangular and triangular) than with uniformly shaped ones (circular). Westling and Johansson [1984] look at precision grasping of objects with different textures and different weights. If the object's surface texture is sandpaper, then there is a higher frictional component to the interactive forces than if the surface texture is sandpaper or silk. They look at the ratio between the grip force (force normal to the object's surface) and the load force (force tangential to the object surface in line with gravitational force). The value for this ratio below which the object will slip out of the fingers is called the slip ratio. Johansson and Westling find evidence that subjects maintain a safety margin above this location, and even use one of two strategies for maintaining a safety margin: either the safety margin is a constant fraction or else a constant absolute value of the grip force. While the slip ratio is a physical constraint, maintaining such a safety margin above it is a functional constraint, imposed by some goal of the performer (in fact, for Johansson and Westling to measure the slip ratio, they must request their subjects to 'slowly separate the thumb and index finger until the object drops', a difficult task to perform).

How the CNS actually obtains task relevant object information is a function of the interaction with the object. Klatzky and Lederman [this volume] have shown that subjects use their hands in particular ways to extract the sensory information needed in a task. For example, lateral motion is used to extract texture information and unsupported holding is used for extracting weight information. Even though a procedure is specialized for just one dimension, it is still informative along others. For example, they show that contour following is used specifically to extract exact shape and volume, but it can as well extract sensory information for all the object properties they measure. While a slower process than the more specialized ones, contour following nevertheless is a procedure typically observed in prehension. In an informal mug grasping study [Arbib, Iberall and Lyons 1985], the subject's fingers were seen to move along the inside of the handle while capturing the object into the grasp, thus perhaps providing the CNS additional knowledge about the texture, hardness, temperature, and weight of the mug.

In summary, prehensile behaviors are subject to the constraints imposed by the laws of physics. A dextrous, multi-fingered hand is not constrained at this level of analysis to unique prehensile postures for a given object. We turn next to a discussion of constraints explicitly imposed by the sensorimotor system.

2.2.2 Sensorimotor constraints

The hand of *Homo sapiens sapiens* represents millions of years of evolutionary pressures, changing it from a tree-dwellers tool into what it is today. Holder [1983] notes, after studying 145 limb skeletons from amphibians, reptiles, birds, and mammals, that the overall pattern of skeleton and muscles has varied little in 350 million years, even though the limbs have become adapted for different functions. LeGros-Clark [1959] noted the development of the pentadactyl (five-fingered) hand, the replacement of claws by nails, and the specialization of the finger pads. Other signs of refined functionality have been noted due to changes in tendon insertions [Abbott

1970], the remodeling of joints [Lewis 1977; Marzke 1983], and intradigital proportions between phalanges [Musgrave 1971]. A functional axis of a primate hand can be distinguished by various criteria [Rabischong 1981] such as the longest finger or the insertion of the dorsal interossei muscles (intrinsic hand muscles that abduct the fingers). In the human hand, two dorsal interossei insert into the middle finger making it the functional axis of the hand. However, it is interesting to note that the head of the 4th metacarpal (see Figure 2.1), unlike its counterparts in the other fingers, is symmetrical, suggesting instead that the ring finger may be the functional axis of the hand [Dubousset 1981]. These, and other characteristics of hand structure, including why we even have five fingers, are evolutionary constraints in the sense that they have modified how prehension has been organized over the eons (see [LeGros 1959; Napier 1962] for further reading). Today, as one moment in time, they have left the human hand with a variety of highly asymmetrical anatomical and physiological features and constraints.

The anatomy and physiology of the hand and arm create structural limitations on possible joint configurations, movements, directions and extents. In the robotics literature, the directions and extents of movements are referred to as the 'workspace'. In the human literature, the region of extrapersonal space where prehension may occur has been termed 'grasping space' [Grusser 1983]. Grasping space is a function of motion about the shoulder, elbow and wrist joints of the arm, given that the body is stationary.

Although the hand has over 25 degrees of freedom, many are coupled by the nature of the ligamentous structure and location of tendon insertions [Kapandji 1982]. For example, although each finger has four degrees of freedom, they are not individually controllable. The flexors tend to work together [Tubiana 1981]: when the index finger begins to close in toward the thumb in a precision grip, the flexion of the proximal interphalangeal joint (see Figure 2.1) releases tension on the retinacular ligament, allowing flexion of the distal interphalangeal joint. Flexion of the proximal interphalangeal joint puts the intrinsic muscles under tension, initiating flexion of the metacarpophalangeal joint. In another example, the intrinsic lumbrical muscle runs from the extrinsic flexor tendon to the extensor tendon and plays multiple roles in flexion and extension [Ranney and Wells 1988]. It is primarily responsible for interphalangeal joint extension, which it does by decreasing the flexor force. Because of its attachment to tendons moving in opposite directions and the great number of muscle spindles (sensors), Ranney and Wells suggest that the lumbrical muscle monitors the rate of hand closing during grasp.

Other aspects of the muscular system show distinguishing workspace characteristics. An obvious feature is the increased mobility of thumb and its placement on the palm, both of which enable its opposability to the fingers. While the tendency is to think only of the thumb as being unique, there are distinguishing features among the fingers besides their differential size and contribution to force generation. While the extrinsic extensors going to all four fingers have a common origin in the arm, additional extrinsic muscles to the index and 5th fingers give these fingers more independence of movement [Kapandji 1982]. The 5th carpometacarpal joint, with a mobility of about 25 degrees, has a unique intrinsic muscle that allows the palm's shape to be modified for complying to a variety of object shapes. The differential action of the fingers is seen in the composite distal palmar crease [Bunnell 1944]: the distal of these creases (i.e, the 'heart' line in palm reading) is related to the action of the middle, ring, and little fingers in opposition to the thumb; the more proximal crease (the 'head' line) is related to the action of the index, middle, and ring fingers. Bunnell also notes that the palm, with webbing between the fingers, is broader and longer than the back of the hand.

A final feature of the hand's anatomy is the skin and the underlying fatty pads. Skin consists of multiple layers and tissues, notably the outer epidermis and inner dermis [Thomine 1981]. Palmar skin, which is thick, tough, resistant to pressure, and good for complying with objects, can be contrasted with dorsal skin, which is fine, supple, and does not impede mobility in flexion. The palmar skin, unlike the dorsal skin, is covered with epidermal ridges and exocrine sweat glands, and contains malleable 'fat pads'. At the distal ends of the fingers and thumb are the specialized pulps [Glicenstein and Dardour 1981]. Shearing forces at the pulps are resisted by strong interdigitating folds between the epidermis and dermis, where interpapillary ridges prevent the epidermis from sliding over the dermis [Quilliam 1978]. The dermis is further anchored to the connective tissue around the bone of the distal phalanges with fibrous connections. It is particularly noteworthy that the trajectory followed by the fingers during thumb opposition brings the finger pulps into the same plane as the thumb pulp [Kapandji 1982]. In perturbation studies of rapid pinch movements, Cole and Abbs [1987] observed that subjects consistently brought the finger pulps into contact although that was not part of the task instructions. For subjects to do this in response to the thumb perturbation, it required the reciprocal adjustments at two index finger joints, an adjustment more complex than a single joint one. The reason for this could be due to a higher level goal (see Section 2.2.3).

In terms of the nervous system, its anatomy and physiology create temporal and spatial physical limitations. The types of receptors (both cutaneous and proprioceptive), and their spatial and temporal response characteristics serve as constraints in the control process. Studies analyzing the motor response to tactile sensory information show low level interactions. Torebjörk *et al* [1978] present evidence that a tonic vibration reflex can cause the fingers to flex. In this study, they placed a small motor against the finger, which caused the finger to vibrate. They found that all subjects would increase their finger flexion force against a small plate, in a frequency-dependent way. Torebjörk *et al* argue that the signals from particular skin mechanoreceptors could be involved in such a motor response. In the finger pulps, numerous tactile receptors are found, more so than most other parts of the body, thus giving the CNS much information about the object with which they come into contact [Vallbo and Johansson 1984]. These mechanoreceptors are classified by their adaptation response to sustained skin deformation and the structure of their receptive fields. Receptors with small and well-defined receptive fields are especially dense in the finger pulps. Westling and Johansson [1987] observe that at higher grip forces (when the skin is less compliant) these receptors are less responsive. In terms of proprioception, receptors in the muscles, tendons, and joint capsules signal the CNS about the current state of the limb [McCloskey 1978]. An interesting result relevant to prehension shows altered muscle spindle sensitivity (i.e., dynamic changes in 1a afferent firing patterns) with flexion and extension of the hand [Edin 1988]. This could possibly play an important role in hand preshaping.

Constraints on the speed/timing of sensorimotor integration exist. Sensory information appears to adjust ongoing movements in less time than previous estimates of visual reaction time, estimates in the range of 190-260 ms [Keele 1968] or 120-200 ms (see [Schmidt 1982]). Cole and Abbs [1988] have shown responses in grip force by the index and thumb with a latency of 60-90 msec after onset of perturbing increases or decreases in load force. The changes were dependent on the size and velocity of the perturbation force and independent of the original grip force level. The latency is longer than a monosynaptic reflex but shorter than the reaction time estimates, reflecting a rapid, automatic mechanism responding specifically to unexpected load force changes and not maintaining a preferred safety margin. They showed that this depends on cutaneous stimulation such as increased shear forces at the digital pulps due to object slips. One possible explanation is that sensory information is being used here for triggering a response, instead of as feedback for modulating a response.

As well, availability of a sensory modality can affect movement. Movements occur differently depending on whether the sensory information is available. An example is seen in a patient with a lesion in the parietal area of the cerebral cortex who could not preshape her hand while reaching to grasp an object, since she lacked somatosensory information [Jeannerod 1986]. Once the hand was within view, the patient could under visual guidance open her hand and grasp the object, although she was still unable to apply the correct grasping forces. Peripheral nerve damage can influence the behavior as well [Rothwell *et al* 1982]. The patient in this study could initiate grasping tasks, but a stable grasp could not be maintained. He could not perform automatic reflex corrections in voluntary movements, nor could he sustain levels of muscle contraction without the use of vision. Johansson and Westling [1987] have shown that as human subjects pick up objects, microslips occur beneath the conscious awareness of the subject; that is, the object slips slightly in the grasp as the object is lifted. Recording from the median nerve, Johansson and Westling demonstrate that sensory information about the object's state is being transmitted into the nervous system. The microslips are reduced to zero as the force level is adjusted and balanced to the frictional conditions, stabilizing the object in the grip. The Rothwell *et al* patient, having lost access to microslip information, may have lost the ability to make low level fine adjustments. These data suggest physiological limitations on prehensile movements, specifically on the underlying sensory and motor pathways.

Other types of damage to different neural substrates constrains available grasping behaviors. It has been shown that cutting the pyramidal tract, a direct pathway between various cortical areas and the motoneurons in the spinal cord, results in the loss of fractionated finger movements [Lawrence and Kuypers 1968]. Primates can still perform collective hand grasping movements because less direct pathways are still intact, but are unable to oppose thumb and fingers. One other aspect of the nervous system is the issue of sensorimotor representations (see [Iwamura *et al* 1981] as an example). While it is known that the motor and somatosensory cortical representation for the hand is quite large, it is still not known what kind of processing is going on.

In summary, human prehensile behavior is further constrained by anatomical and physiological aspects of the body and nervous system, just as robots and their computer controllers are constrained by their construction and interface. However, without a 'reason' for the movement, which in itself can constrain the formation and modulation of the movement, there would be no need for this complex machinery. Therefore, we turn to a discussion of higher level goals.

2.2.3 High level constraints

At the highest level, we group together four classes of constraints in living systems. These constraints are imposed for a variety of reasons. The importance here is that there are probably large individual differences. For our purposes, we have labeled them: semiotic/informational, motivational, social/cultural, and functional/intentional.

Semiotic, symbolic, or emotional information may be conveyed in the way an object is grasped [Nespoulous and Lecours 1986]. For example, one might hold a cigarette in a suggestive manner. Anxiety or anger may be communicated by gripping an object tightly. Motivations, such as needs being satiated, may add an urgency to the movement. As well, sociocultural constraints create boundaries on what is acceptable behavior. Refined behavioral interactions with objects in one culture might be taboo in another. For example, sticking out the little finger as one grasps a teacup may be a sign of good upbringing in a culture (see Section 2.2.2 for a possible anatomical reason). On the other hand, sticking out one's elbow as one reaches might be an invasion of someone else's extrapersonal space.

While this is only anecdotal evidence for these various effects on prehension, kinesiologists and psychologists have begun to explore the functional issues. Movement constraints act within and around the intentions of the performer, which delineate goals for a movement. An example would be the 'sherry-glass response' [Traub, Rothwell and Marsden 1980]. Subjects were asked to maintain their thumb and index finger a few millimeters from the rim of a full glass of sherry. A perturbation was made to the wrist that could cause the hand to hit the glass and knock it over. A short-latency grab response (50 msec) occurred in the thumb muscle in response to the perturbation, saving the glass from being knocked over. However, the response was observed only when the subject's intent was to prevent the glass from falling over. Traub *et al* argue that this suggests a grab reflex, where the digits respond within a time too short to be voluntary, flexing in order to maintain contact but only if that is the person's intent.

In a formal empirical situation, intent emerges from the experimenter's request to the subject to perform a certain task. In MacKenzie *et al* [1987] for example, the subjects were asked to point with a stylus 'as quickly and as accurately as possible' to a target of varying size and at varying distances. The question being asked was whether there was a reliable kinematic measure of the speed and accuracy requirements of the task. In this case, Fitts' Law [Fitts 1954] was used, which states that movement time (MT) is directly proportional to the index of difficulty (ID) of the task:

$$MT = a + b \times ID \tag{2.1}$$

where the ID=$\log_2$(2A/W), A=amplitude of movement (an extrinsic object property), and W=width of target, or target tolerance (an intrinsic object property). When plotting MT against ID, a linear relationship is seen. MacKenzie *et al* [1987] measured the MT of the tip of the stylus, its time to peak resultant velocity, and the percentage of movement time after peak resultant velocity. They found a differential effect of target size and amplitude on these parameters. As seen in Figure 2.2(a), in plotting ID vs the time before the peak resultant velocity (the acceleration time), there is a noticeable effect of amplitude. For each target size, the acceleration time increased as the ID (amplitude) increased. For each amplitude, there is no effect of target size on the acceleration time. In Figure 2.2(b), the data are normalized to 100 points to examine the percentage of MT after peak resultant velocity (the deceleration phase). For each amplitude, this measure increased as the ID increased (target size decreased). These results indicate that the resultant velocity profile is not symmetrical. Fitts' Law argues that the MT will increase as the target size decreases; here, we see that the reason that the MT is increasing is because of a relatively longer deceleration phase. As well, if the movement is further, there is a longer acceleration phase. The results show that the time spent in the deceleration phase was predicted by ID as well or better than MT was. This was not

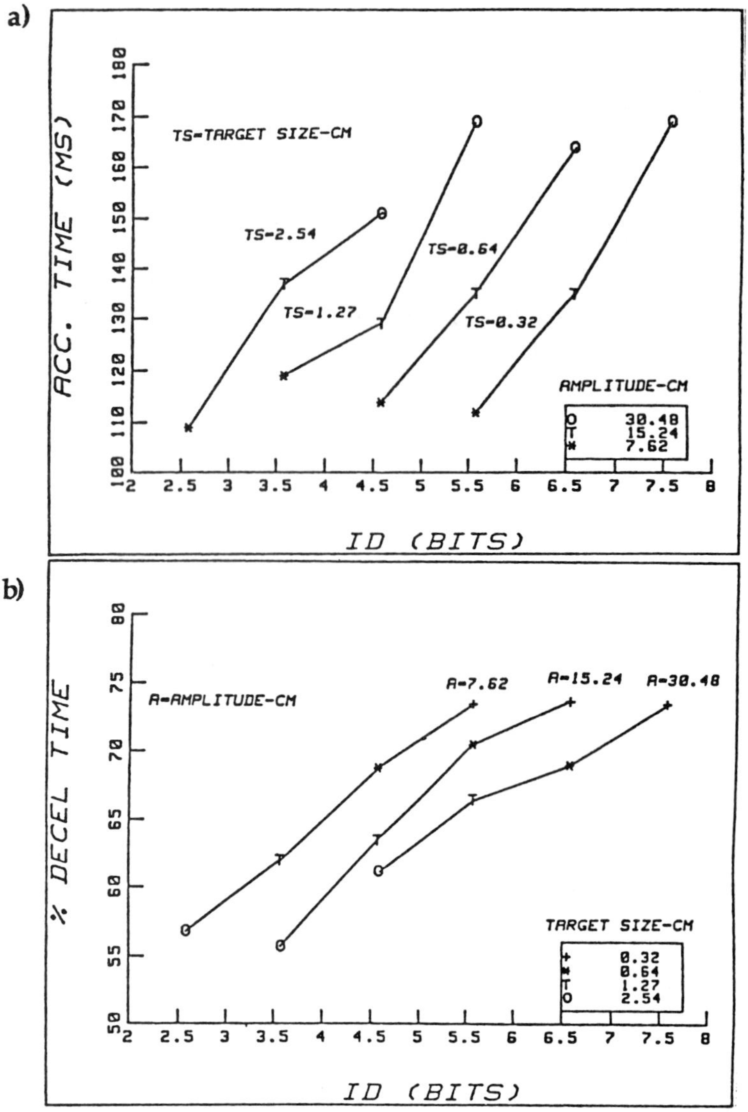

Figure 2.2: (a) Acceleration time (msec) as a function of ID, for each of the target sizes. Time to peak speed was a function of amplitude, not target size. (b) Percentage of MT spent in deceleration as a function of ID, for each of the amplitudes. More time was spent in the deceleration phase as target size decreased. Amplitude had no significant effect. (From C. L. MacKenzie *et al* (1987): *Q Jour Exp Psych*, 39A: 629-647. Reprinted with permission from the Experimental Psychology Society.)

Conditions	Task	W	A	% Time in Deceleration Phase
{Point to / Grasp} disk  2 Widths, 2 Amplitudes	Point	2cm	20cm	34% ± 10.2
			40	34% ± 12.7
		4	20	33% ± 5.1
			40	30% ± 11.2
	Grasp	2	20	58% ± 2.3
			40	50% ± 4.6
		4	20	52% ± 1.8
			40	51% ± 2.6
Grasp {tennis ball / light bulb} 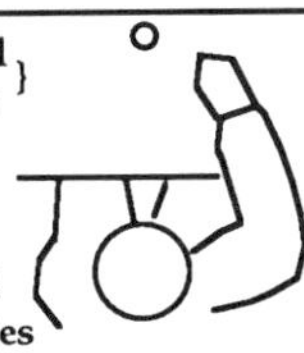 Width fixed, Amplitude fixed, 2 Object properties	Grasp ball	6	30	48% ± 3.8
	Grasp light bulb	6	30	51% ± 4.5
Grasp, then {throw / fit} 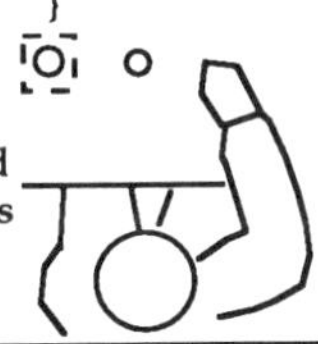 Width fixed, Amplitude fixed, Object properties fixed, 2 Contexts	Grasp and throw	4	30	46% ± 4.4
	Grasp and fit	4	30	52% ± 5.4

Table 2.2: Comparison of percentage of time in deceleration phase for three task choices, data from [Marteniuk *et al* 1987]. Note that as task gets more complex, percentage of time in the deceleration phase increases.

the case for acceleration time. Only movement amplitude affected the time to peak velocity. Thus, amplitude and target size effects were dissociable in that the shape of the tangential velocity profile was a function of target size (accuracy), and the peak speed along the path of the trajectories was scaled according to movement amplitude.

In grasping and aiming tasks [Marteniuk *et al* 1987], precision requirements were varied, showing how intention, context, and object properties affect timing parameters of prehensile movements. Table 2.2 summarizes the experimental conditions and results. The first experiment varied goal (to point to a target or to grasp a disk); the second varied object properties (grasp compliant tennis ball or fragile light bulb); the third varied movement intent (to place or to throw an object; note that MT is measured from start of movement just to the grasp). Using Fitts' Law to equalize the ID's, it was observed that the grasping took longer than pointing. Table 2.2 shows that the percentage of time in the deceleration phase was longer for grasping than pointing, grasping the light bulb than grasping the tennis ball, and placing than throwing. They argue that all these effects could be due to the requirement for "precision". Fitts' Law predicts that MT increases with aiming precision requirements, but they demonstrate it is due to the duration of the deceleration phase being lengthened disproportionately to the rest of the movement. As well, they show other influences, such as object properties and task intent. Less variability between conditions was seen in the early or acceleration phase of the movement, and more variability during the deceleration phase. They argue that the early part of the movement is more likely to be directly influenced by central stereotyped movement planning or programming, while the later part of the movement, during the deceleration phase, is using more sensorimotor adjustments for controlling the movement, causing more variability. Increasing precision requirements of a task may induce subjects to use more sensory information, particularly in the 'homing in' part of the task. This relates to [Jeannerod 1984], where it was argued that an initial ballistic phase gets the hand into the vicinity of the target, and a second phase using feedback guides the hand to the target. It is interesting to note that in pointing tasks, where contact is not being made with the environment, velocity profiles have been shown to be symmetrical [Atkeson and Hollerbach 1985].

Another dimension to task requirements, besides precision, is the anticipated forces acting in the task. In [Iberall *et al* 1986], an informal study was performed to observe the effect of anticipated forces on the posture chosen to grasp a vertically standing dowel. Subjects were asked to pick up the dowel to either place it on a marker or else to shake it. It was observed that in the less forceful, more precise placing task, the subjects used precision grasps. In the more forceful, less precise shaking task, the subjects initially used a precision grasp to lift the cylinder, and then adjusted their posture into a more powerful posture once the cylinder was lifted. The posture chosen seemed to match the forces currently acting in the task. During the lifting (whether for the placing or shaking task), only the light weight of the cylinder was acting, and therefore the subject used a precision grasp. The subjects did not use the more powerful posture before it was necessary.

In summary, people grasp objects only if they have reasons to do so. Once these are established, be they to achieve some functional goal, convey some information, or satisfy some internal motivations, the movement can be carried out within socio-cultural guidelines. While goals in the strictest sense are not constraints (people do drop objects), their effect on human prehensile behavior can be seen in the selection of movement parameters and the choice of prehensile postures. Since multiple grasping solutions are possible, the controller is faced with the question of how best to use hand features, in relation to the anticipated object properties and predicted interaction outcome, in order to achieve the goals of the task.

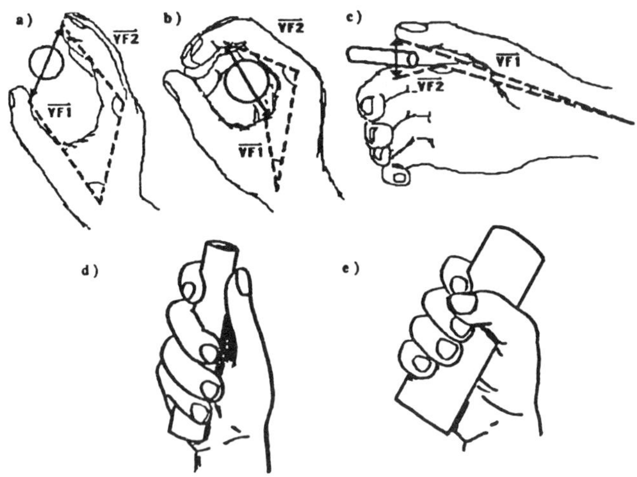

Figure 2.3: (a) Pad opposition along axis generally parallel to palm (b) Palm opposition along axis generally normal to palm. (c) Side opposition along axis generally transverse to palm. (d) Napier's power grasp (combination palm/side opposition). (e) Napier's coal hammer grip (palm opposition only). Solid line: opposition axis, dashed lines: VF vectors. (a-c ©IEEE, Iberall 1987; d-e based on [Napier 1956]).

2.3 Opposition Space: Mapping into human and robot hands

In [Arbib *et al* 1985], the *virtual finger* concept was introduced as a functional unit, replacing the analysis of the mechanical degrees of freedom of individual fingers by the analysis of the functional roles of forces being applied in carrying out some task. Real fingers group together into a virtual finger (VF) to apply an opposing force against other VFs or against task torques. Each VF is represented by the magnitude and orientation of a vector relative to the palm (see Figure 2.3), with values that change as the fingers and thumb flex and extend. A method for setting up a pair of oppositional forces is called an *opposition* [Iberall *et al* 1986]; it has been shown that prehensile postures involve combinations of three basic oppositions [Iberall 1987b]. These include pad opposition (Figure 2.3(a)), used in precision tasks, palm opposition (Figure 2.3(b)) with its more powerful capabilities, and side opposition (Figure 2.3(c)), a compromise between the two. What gives this language of opposition spaces its expressive power is that the hand is not limited to one opposition at a time. For example, the posture that Napier called a power grasp (Figure 2.3(d)) is really a combination of a palm opposition for applying strong forces and a side opposition for directing the shaft of the tool. By bringing the thumb onto the back of the fingers and creating a 'coal hammer' grip (Figure 2.3(e)), the posture reduces to palm opposition alone, where "the thumb has ceased to contribute any element of precision to the operation" [Napier 1956, p. 909].

Using combinations of these functions, various prehensile classifications can be captured with this framework [Iberall 1987b]. An ***opposition space*** describes a hand posture as a collection of VFs able to apply functionally effective forces against an object for a task. It provides a theory of hand movement which explains preshaping, rather than merely describing it as others have done using symbolic classifications. As a parameterized view of human prehension, it provides components for modeling a biological subsystem for the control of natural, prosthetic, and robotic hands. For example, as the hand reaches out to grasp an object, it preshapes into a posture that closely resembles the posture taken on in contacting the object [Marteniuk *et al*, in press]. This 'preshape posture' consists of combinations of oppositions and virtual to real finger mappings that are going to be used in the task. Each VF has geometric parameters that describe it: its length, orientation to the palm, the orientation of a normal to the surface of its distal endpoint (the anticipated contact point), and the size of the contact area. For a VF in pad opposition, this contact point is the finger pulps. For palm opposition, it is the centroid of the pressure distribution of the multiple contact points. In addition, each VF has sensory parameters at the contact point: spatial resolution, type of sensory information available, and sensitivity (sensory threshold). Finally, each VF has force and motion parameters: amount of force available and its direction of application at the contact point, motor resolution, motor accuracy, and range of motion relative to the other VF in the opposition. Using the opposition space concept, the preshaping phase sets up critical parameters for a selected opposition space, and the enclosing and manipulation involves sensations and the application of forces appropriate for the chosen opposition space.

There are, however, multiple constraints acting upon these opposition space parameters. Marr [1981] suggested three levels for analyzing systems that perform complex information processing: task level, representation level, and implementation level. Using Marr's approach as a guiding principle, we suggest that the Opposition Space level can be re-represented or mapped into a Biomechanical level (Figure 2.4). At this level, opposition space parameters are re-represented in terms of the forces and torques acting at the joints and contact points. Inverse kinematic and dynamic equations translate VF parameters into real joint angles and torques. For example, if one objective of the movement is to 'not drop the object', the opposition space chosen for the grasp must effect a stable grasp, taking into account the active and passive components of force generation by the muscles, tendons, and skin surfaces. In addition, the chosen posture must relate to stress and strain on biomaterials and take into account mechanical advantages. Using a pad opposition, the enhanced frictional component of

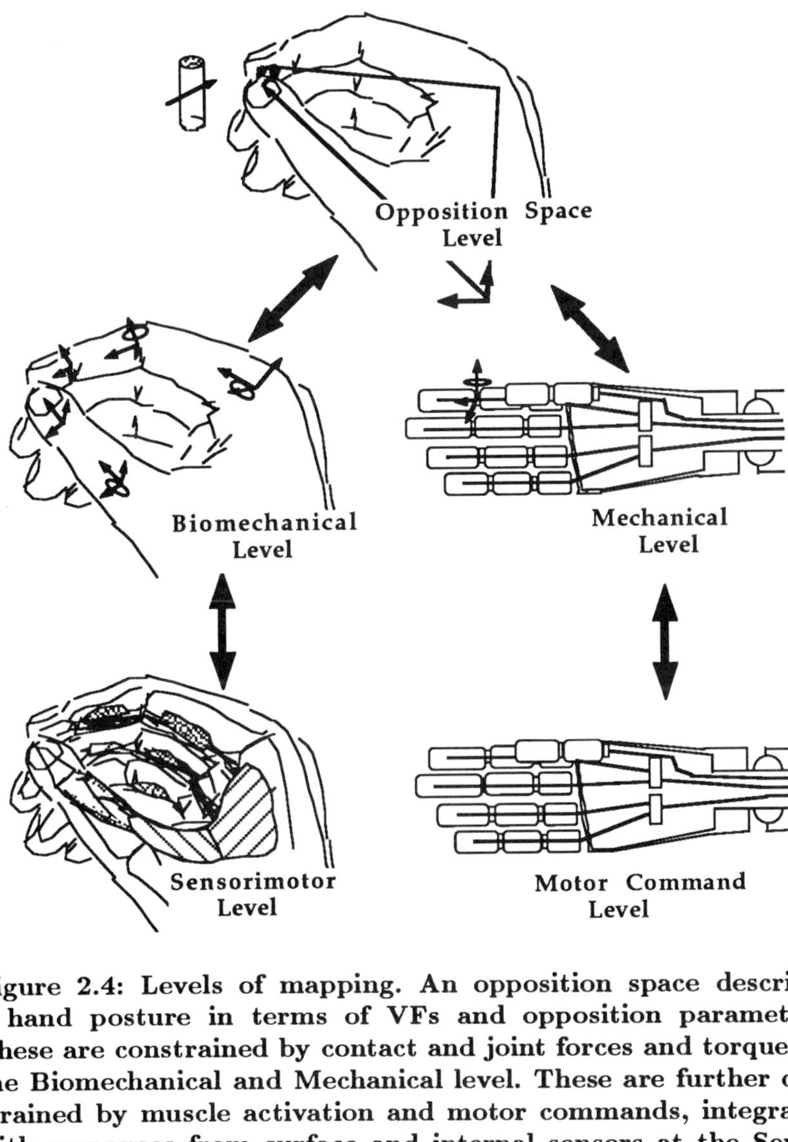

Figure 2.4: Levels of mapping. An opposition space describes a hand posture in terms of VFs and opposition parameters. These are constrained by contact and joint forces and torques at the Biomechanical and Mechanical level. These are further constrained by muscle activation and motor commands, integrated with responses from surface and internal sensors at the Sensorimotor and Motor Command level. A controller chooses an opposition space that both satisfies the requirements of the task as well as the constraints imposed by the hand.

the finger pulps helps in reducing the amount of active force needed. However, if the object is heavy and/or if there is a low coefficient of friction between the skin and object surface, the posture involves a different opposition space, one where active available forces are larger. A palm opposition, with its multiple phalangeal contact points and greater strength, would be used. As well, the palm, as a large virtual finger, brings in a torquing component to a prehensile posture, if so needed in a task. The tradeoff is in not having the finger pulps in contact with the object; a compromise would be to use a combination of side and palm opposition, where the thumb pad contacts the object and applies a force transverse to the palm. If the task goal is to 'move as quickly and as accurately as possible', timing parameters are chosen consistent with Fitts' Law while also consistent with kinematic and dynamic constraints acting on the hand and arm. Not only must the anticipated forces be matched, an opposition space must be chosen that allows perceptual systems access to the information necessary to ensure the accuracy requirements. Accuracy suggests pad opposition, but only if the forces are not too great. With greater forces, one compromise is to use more fingers in the VF opposing the thumb.

The final mapping from the Opposition Space level (Figure 2.4) is the Sensorimotor level. Opposition space parameters are re-represented at this level in terms of the activation level of the muscles and receptors acting on the fingers, which in effect is the implementation level. Anatomical constraints, for example, limit the number of ways real fingers can be mapped into VFs. The task goal of 'not dropping the object', expanded by the ways that forces can be generated to achieve the goal, is further translated into a posture that judicially uses the skin mechanoreceptors, optimally placing the pulps, with all their receptors, against the object. The chosen VFs must have adequate sensory resolution and sensitivity at the contact points. With large task forces, pad opposition would not be effective due to loss of skin sensitivity and lack of muscle strength. Being able to anticipate much of the object's behavior as contact is being made allows part of the choice to be influenced by what VF parameters will correctly position task-specific sensors for triggering the next action. If the object starts moving, the tonic vibration reflex will force the fingers to close tighter. Perhaps in contour following, where the fingers are moving instead of the object, similar principles apply. The advantage of performing contour following is that one gains additional knowledge about the object's various properties, and the fingers can perhaps be better placed for effecting a stable grasp. If palm opposition is used in order to account for the larger forces, the posture still gives information about the object state, particularly because the palm can adapt to the object's shape and therefore place more skin surface in contact with the object. In terms of constraints on the mapping from virtual to real fingers, there are significant distinctions in how the individual fingers contribute to prehension (number of muscles, types of articular surfaces, lengths and widths of phalanges, type of support at wrist, etc.).

Affecting the choice of an opposition space, intrinsic and extrinsic muscles make differential contributions to movement, while muscle, tendon, and joint receptors provide information about results of those movements. We note the existence of low level sensorimotor features – such as the tonic vibration reflex, pad alignment in opposition, coupled degrees of freedom in the fingers, ligaments being used for passive control, and rapid grip force adjustments – and higher level neural control, such as active control of muscles and the pyramidal tract for fractionated finger movements.

Opposition space as a model takes into account the hand's ability to be both an input and output device, dealing with applying task forces while gathering sensory information. It addresses the hand in terms of its precision and power capabilities, providing combinations of oppositions and VF mappings that match the requirements of the task. We see two important benefits to using a Marr type view for prehension. On one side, it separates implementation details from a task description, a trend occurring in programming languages in general, and robot programming languages in particular. This will allow the study of human hand functionality to be carried over to dextrous robot hands. The mapping from opposition space into the human hand, with its particular physical/biomechanical and its sensorimotor constraints, can be replaced with a robot hand that has its own mechanical and motor constraints (Figure 2.4, right side). The mapping changes without changing the overall high level description of hand functionality. Of course, until a robot has a reason to grasp an object, other than it was told to do so, the only high level effects would be functional ones.

The other issue is the goal to understand the versatility of human prehension in general. With a large repertoire of available movements and the potential use of many prehensile strategies at one's disposal, the choice for using one strategy over another can be based on anything from physical laws to whimsy. Roboticists and experimentalists seem to currently be at odds in terms of what are important variables to be examining in the quantification of motor control. By identifying and separating hard, physical constraints from softer, functional constraints, we hope to find ways to integrate the various views of motor behaviors.

2.4 Conclusion

The versatile performance seen in human prehension emerges from within a large multidimensional constraint space. By making explicit the influence of various constraints on the environment/performer, we address the need for a common vocabulary between experimentalists and roboticists. The systems of interest, while seemingly different (made of different materials, have different characteristic workspaces, etc), must obey many of the same constraints (laws of physics, achieve goals). With a dextrous, multifingered hand, multiple grasping solutions are possible, and only additional constraints will guide a controller in making a choice. It is faced with the question of how best to use hand features in relation to the anticipated object properties and predicted interaction outcome, in order to achieve task goals. By separating out levels of analysis, the study of human hand functionality can be related to dextrous robot hand functionality.

Evolutionary specializations have enhanced the human hand's ability (through control by the CNS) to perform a wide variety of prehensile tasks, equipping the hand as both a precision and power device. Within the complex interaction of physical constraints, active and passive sensorimotor features, and task goals and motivations, prehensile postures are chosen that will apply functionally effective forces to an object. We use the term 'functionally effective' in order to make explicit the notion of task goals (i.e., applying forces, imparting motions, gathering sensory information) creating constraints on how a human hand is shaped into a force-applying and sensory-gathering posture. The opposition space model focuses on the question of how to discuss human hand functionality in terms of task requirements. By separating anatomy from function, it provides a way to study the complexity of motor control, as well as suggests ways to design and control versatile dextrous robot hands. Open research questions include: 1) how the CNS works within this multidimensional constraint space, 2) to what extent do humans rely on sensory information, and 3) whether high level knowledge may be considered separate from the biological (or mechanical) implementation.

Acknowledgements: Preparation of this chapter was supported in part by funds provided by NSF grant DMC-8719579, a gift from Rockwell International to the USC Center for Neural Engineering, and NSERC grant OGP0008303.

3

Coordination in Normal and Prosthetic Reaching

Alan M. Wing[1]

3.1 Thirsty?

If you want to take a tumbler of water in order to drink from it, you will probably reach for it from the side with the forearm midway between pronation and supination (so that the palm of the hand is turned towards body midline). What does the reaching movement comprise? At the end of an initial, rapid, distance-covering phase of movement that leaves your hand close to its target, thumb and fingers will have opened sufficiently to allow the tumbler to be encompassed. The second phase is made more slowly and leads upto contact of the hand with the tumbler. In this phase, coordination between transport of the hand by the arm and changes in the distance between thumb and finger (hand aperture) becomes critical if the impact of the collision on contact is to be minimized - especially if the tumbler is full to the brim!

Since transport of the hand and hand aperture are subserved by anatomically distinct elements that are capable of being moved separately under voluntary control, the nature of coordination in reaching is of some interest. How does the movement control architecture of the brain specify the linkage between hand transport and aperture? The neurophysiologist might look for an answer in terms of the nervous system pathways that mediate movement and the patterning of activity that determines the appropriate contraction of groups of muscles [cf. Greer 1984]. In contrast I present an approach, based in experimental psychology, that takes a higher-level view in which people are seen as adaptive systems that pick up and process information from the world and respond appropriately to it [Smyth and Wing 1984]. After describing a paradigm for studying reaching I summarize empirical data from normal subjects and then turn to the performance of a proficient user of an artificial hand. In both cases the goal is to understand the strategic basis for action because this points to the mental representation of movements of the digits, both in relation to the body (that is to movements of the arm) and in relation to the external object.

[1]MRC Applied Psychology Unit, 15 Chaucer Rd, Cambridge CB2 2EF, UK.

And, I would like to suggest, it is this representation that determines the functional pattern of coordination seen in reach and grasp movements.

3.2 Reaching for an unstable object

In studying reaching behavior it is important to have a standardized task that is capable of yielding results that are reliable yet valid. For our research we have developed a task which, like many real-world reaching tasks, emphasizes accuracy. A thin cylinder made, for example, of wood dowel is placed on end some distance in front of the subject (see Figure 3.1). Starting with the hand close to the body, the subject is required to pick up the cylinder from the side, then keep the cylinder upright while moving it some distance before putting it down.

Because the cylinders used are relatively lightweight and their diameters are much smaller than their length, errors in positioning the hand or in adjusting the hand's aperture when picking up the cylinder are of consequence in that it is easily knocked over. Good coordination of hand transport and aperture is therefore important, particularly if a degree of positional uncertainty is introduced by varying the location of the cylinder from trial to trial. The task may be given a further dimension by using cylinders of various diameters. With objects of varying size it is found that the maximum aperture attained by the hand, which occurs around the end of the initial phase of reaching, is adjusted in proportion to their size [Jeannerod 1981].

3.3 Strategic adaptation of hand aperture to movement conditions

To illustrate the coordination in this reaching task I have taken data from a study of a group of normal subjects by [Wing, Turton and Fraser 1986]. Of interest were the effects of the availability of visual information about transport on aperture. Figure 3.2 shows illustrative trajectories obtained from video recordings. Examination of the changes in thumb and finger positions orthogonal to the approach line show a clear maximum in hand aperture. From the group average data in the two outside columns of the figure it may be seen that maximum aperture of the hand was greater with the eyes closed than with the eyes open. This increase in aperture probably served to provide greater tolerance for errors in hand transport which, under these conditions, cannot be corrected on the basis of visual checks of the position of the hand in relation to the object.

A widening of maximum aperture was also observed when vision was not completely removed but just limited in usability because the movement was carried out under time pressure. If a reaching movement is carried out very

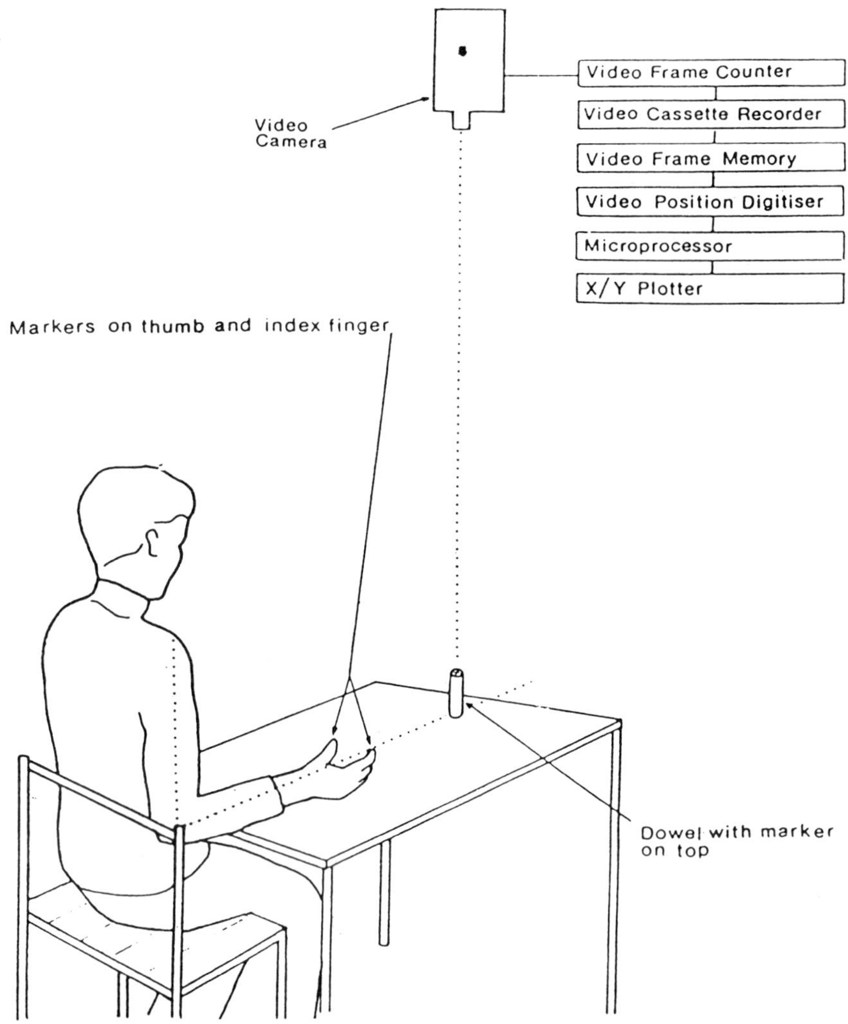

Figure 3.1: Prototypical pick-up paradigm: the task is to pick up the unstable cylinder between the pads of the thumb and index finger keeping the forearm midway between pronation and supination. Confining the movements to the horizontal plane simplifies subsequent digitization of trajectories from the video recording. (From A. M. Wing, A. Turton and C. Fraser (1986): *J Motor Behavior*, 18(3): 245-260. Reprinted with permission.)

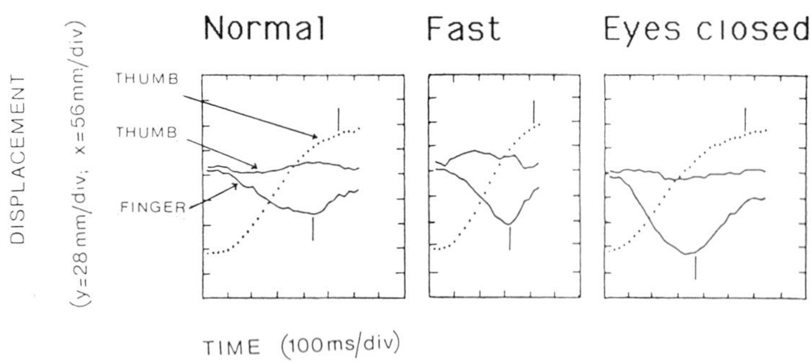

	Normal	Fast	Eyes closed
Max aperture	80	93	116
Pre-contact SD	3	5	9

Figure 3.2: Illustrative trajectories from a single subject (above) showing the thumb's approach (dotted line) to a target with contact being made at the point indicated by the vertical line. The solid lines show the position of the thumb (upper trace) and finger (lower trace) orthogonal to the approach line (maximum aperture shown by the vertical line below the finger position). The values below are the group averages for the standard deviation of thumb position (orthogonal to the approach line) at maximum aperture and one frame before contact (in mm). Mean data from 10 subjects, 6 trials. (From A. M. Wing, A. Turton and C. Fraser (1986): ***J Motor Behavior*,** **18(3): 245-260. Reprinted with permission.)**

rapidly, there is little time for visual feedback to be processed in order to improve the accuracy of the end-point of the hand transport movement. As may be seen in the middle column of Figure 3.2, when the subjects were asked to make rapid reaching movements, hand apertures were observed to be larger than normal. In this experiment movement velocity and time to process visual feedback were confounded. It is thus worth noting that Wallace and Weeks [1988] have subsequently shown that the effect is not due to the change in velocity *per se* but indeed results from the lack of feedback processing time before movement completion.

Thus maximum hand aperture is related, not only to the 'external' constraint provided by the size of the object, but also to an 'internal' factor, namely, the manner of hand transport. It has been suggested that reaching for objects in the environment involves a visuo-motor channel linking the perception of intrinsic object attributes such as shape to a control process governing hand shape (including aperture size) and a separate channel links the extrinsic attribute of position to a transport controller governing position [Arbib 1981; Iberall 1987a]. Within this framework the effects of the experimental manipulations described above suggest the existence of a cross-link between the two channels.

A direct demonstration of such cross-linking is provided by a paradigm described in [Athenes and Wing 1988]. When reaching for an object, a perturbation of the spatial relation between hand and target object was introduced on randomly selected trials, either by deflecting the arm or by moving the object. With displacements on these trials of equal size but opposite in direction, the hand's sideways displacement relative to the object was the same and required similar corrective action by the subject.

In this latter experiment the issue was whether subjects, knowing that there might be a perturbation, would use a strategy to reduce its likely effects if it should actually occur. In particular would the maximum aperture of the hand, which the earlier work had shown can compensate for uncertainty about position, be larger in blocks of trials where perturbations might occur than in blocks where the subject was informed that perturbations would not occur?

Because the interest was in anticipatory compensation rather than with the way in which subjects correct ongoing movement [cf. Abbs, Graco and Cole 1984], measurements were only taken from trials on which there was no perturbation. On those trials when a perturbation could have, but did not, take place, maximum hand aperture was observed to be of the order of 10% wider than in the blocks of control trials without perturbation. This strategic adaptation to the perturbation conditions would tend to improve the chances of the hand still encompassing the target had a perturbation actually occurred. Thus, in functional terms, control of grasp and transport are interrelated.

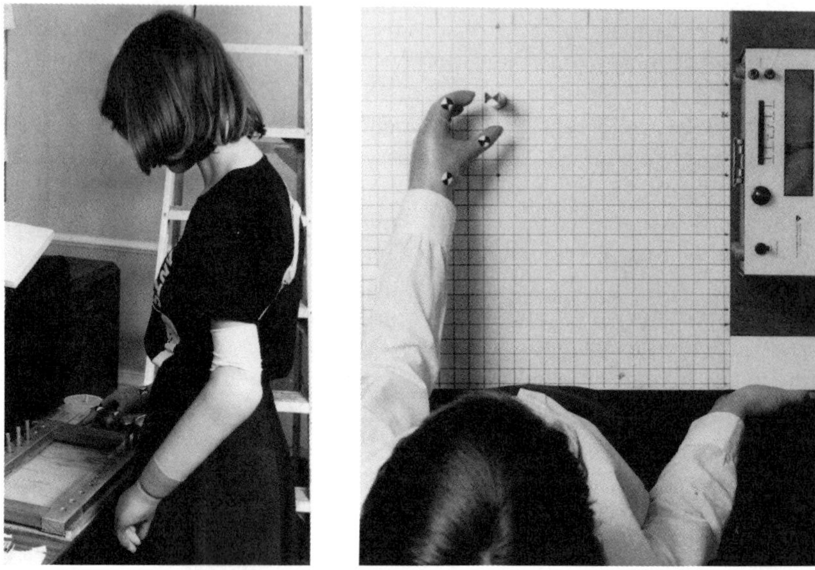

Figure 3.3: (Left) CY transferring pegs using her natural hand, leaving the opening cord of the artificial hand (which runs from the right shoulder, across the back to the left elbow) untensioned so that the prosthesis is held closed by its internal spring. (Right) The camera view of the reach and grasp task. Note the 20 mm calibration grid (not present during data collection) and the markers on thumb, finger and wrist of CY's artificial left hand (used to aid subsequent digitization of position).

3.4 Proficient prosthetic prehension

Another study of reaching behavior involved a single-case study of a young girl with an artificial hand [Frazer and Wing 1981; Wing and Fraser 1983]. The aim of the research was to find out whether the aperture adjustments that are observed during reaching with the natural hand would also obtain to the artificial hand. Since the artificial hand movements were controlled by very different muscles, such similarities would be evidence for a level of movement representation above that of a motor program specific to a particular group of muscles.

Thirteen years old at the time of the study, the subject CY had, from the age of two, used a below-elbow prosthesis that provided a form of precision grip with a strong spring keeping the hand normally closed. Opening the hand involved tensioning a cable which ran to the elbow and across the back to a harness around the contralateral, right shoulder (see Figure 3.3). Forward flexion of the right shoulder girdle increased the tension and opened the hand against the action of the spring. The extended period of wearing

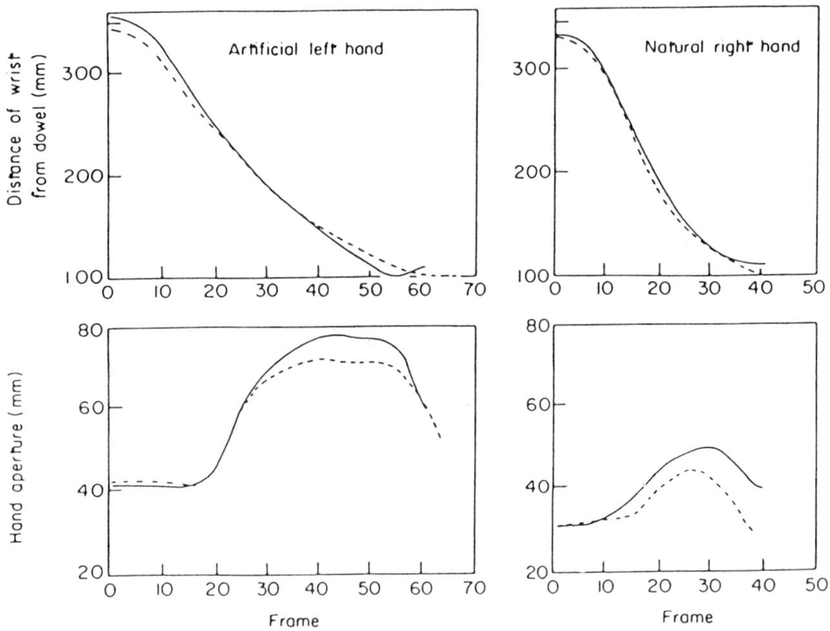

Figure 3.4: Illustrative trajectories for CY reaching and grasping 12 mm (dashed) and 22 mm (solid line) cylinders. Hand position (above) and aperture (below) are plotted as a function of time when using artificial (left) and natural (right) hands. (From A. M. Wing and C. Fraser (1983): *Q Jour Experimental Psychology*, 35A: 297-309. Reprinted with permission.)

this appliance from an early age gave CY a notable degree of confidence in its use in carrying out everyday activities. Despite CY's proficiency in using her artificial hand, it will be appreciated that it lacked the attributes of compliance and tactile sensibility which contribute to the final contact and grip phase of picking up an object with the normal hand. Thus, for CY, vision was particularly critical in successfully picking up our unstable cylinder.

Figure 3.4 shows the transport and grasp components of CY's natural and artificial hand movements as a function of time. It will be noted that the modulation of aperture with cylinder diameter seen in the natural hand is also evident in the artificial hand. However, there is a delay in closure of the artificial hand. Consequently, its closing takes place with less change in the distance between the hand (or the wrist to be precise) and the target and this is illustrated in Figure 3.5.

Although CY's artificial hand was quite natural in appearance, when it was opened or closed the design introduced approximately equal and opposite degrees of rotation of the thumb and finger. Despite this, the recordings

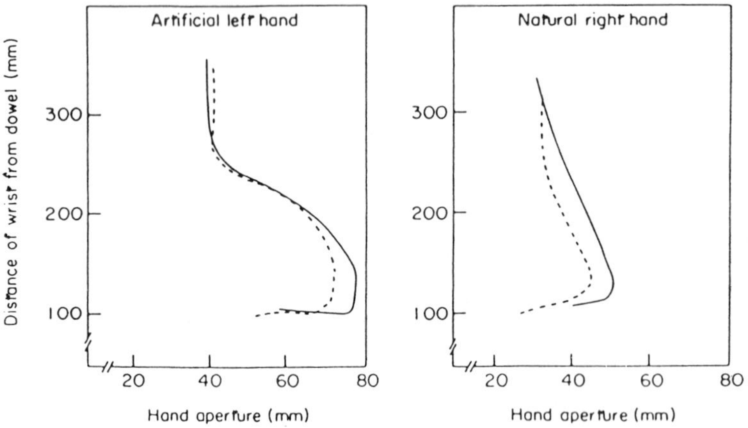

Figure 3.5: Illustrative trajectories for CY showing artificial and natural hand aperture as a function of distance of the wrist from the cylinder. (From A. M. Wing and C. Fraser (1983): ***Q Jour Experimental Psychology*****, 35A: 297-309. Reprinted with permission.)**

of reaching movements showed (see Figure 3.6) a stability of the thumb relative to the line taken by the hand in approaching the target object. Further examination of the figure shows the relative stability of the thumb in the natural hand. Indeed it seems to be a general characteristic being also evident in the experiment of Wing, Turton and Fraser (see Figure 3.3). To achieve this stability in her artificial hand, CY used forearm rotation in a horizontal plane around an axis near the wrist. The rotation tended to cancel movement of the thumb relative to the object while amplifying finger movement.

3.5 Internal representation of the hand in movement

Why had CY evolved a strategy of stabilizing thumb position in a manner not envisaged in the mechanical design of the hand? With the absence of tactile feedback from the prosthetic hand, vision is particularly important in guiding the final approach so that the hand encompasses the target without knocking it over. The maintenance of relative stability of the thumb as the hand approaches the target object, with the finger providing the changes in aperture, might then confer simplicity for feedback control. The forearm rotation geared to contralateral shoulder function may thus represent a low-level, automatic process that developed as a result of experience in order to allow the separation of visual feedback about hand position and

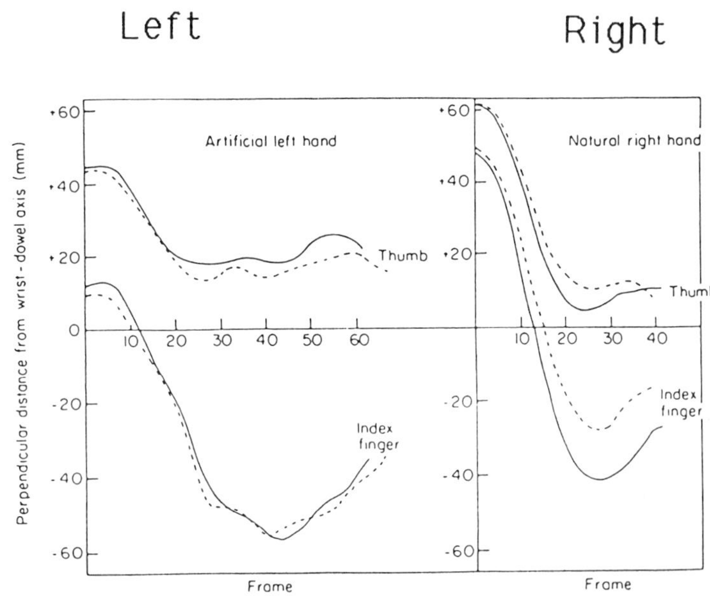

	Left	Right
Thumb	+4	−3
Finger	+26	+15

Figure 3.6: Illustrative trajectories from 2 single reaches by CY with the artificial hand (left half) and the natural hand (right half) showing change of thumb and finger positions relative to wrist-dowel axis. The average changes in thumb and finger position (from peak aperture to contact) for both narrow and wide objects over 4 trials are shown below (in mm). (From A. M. Wing and C. Fraser (1983): *Q Jour Experimental Psychology*, 35A: 297-309. Reprinted with permission.)

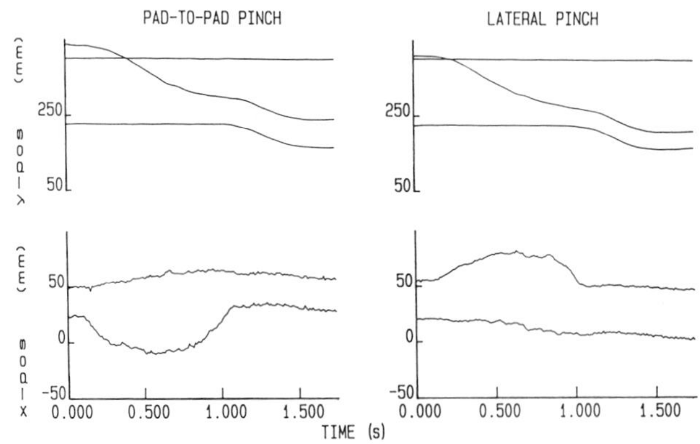

Figure 3.7: Two illustrative normal hand trajectories illustrating single-sided opening with palm facing midline (left) so there is pad opposition of index finger and thumb or with palm facing down (right) so the thumb meets the lateral aspect of the index finger. The upper three, y-position curves show the home position, movement of the wrist, and movement of the target (which is stationary until it is picked up after approx 1.5 seconds). The lower two curves show, relative to the wrist marker, the x-position of the thumb (above) and, either the tip (left) or the proximal interphalangeal joint (right) of the index finger.

aperture. But is there anything significant about the allocation of the roles for the thumb and finger?

The design of CY's prosthetic hand results in symmetrical movements of the two opposition surfaces contributing to precision prehension [Iberall, Bingham and Arbib 1986]. Thus it might be thought that the choice of relative stability of the thumb was arbitrary. A differently structured task might have revealed a different pattern of relative stability. Some support for this speculation comes from the casual observation that people do seem to be able to make effective reach and grasp movements with either the thumb or index finger stabilized. For example, in picking up a small object, such as a pin, from the surface of a table people may use a lateral precision pinch in which the side of the index finger provides a stable surface to which the thumb moves, sweeping the pin up as it does so. The relative stability of thumb or index finger is thus dependent on the position of the forearm, as illustrated in Figure 3.7. With the forearm pronated (so that the palm of the hand faces down) most of the movement occurs in the thumb. When the forearm is semi-pronated (such that the palm of the hand becomes visible) most of the change in the hand's aperture is achieved with the finger.

It might be argued that the contrasting contributions of thumb and finger movements in the development of aperture in the natural hand should be interpreted on purely anatomical grounds. With the forearm pronated, the range of thumb movement in producing opposition between thumb pad and side of the index finger is much greater than the lateral movement of the finger. In contrast, with pad opposition, the range of finger motion is much greater, even exceeding that of the thumb. But while these anatomical constraints on movement exist, the grasp organization seen in prosthetic prehension further suggests the form of coordination between thumb and index finger reflects the functional representation of the hand during reach and grasp movements. The intriguing question that remains to be answered is whether relative stability of one of a pair of opposition surfaces arises as a matter of trial-and-error. Alternatively, it might be a matter of transfer of a higher-level control structure that developed with use of the natural hand [Hay 1984].

Since these data on proficient prosthetic prehension are based on a single case, albeit an in-depth study, it will be appreciated that any conclusions will necessarily be tentative if the reader's interest is in applying them to questions of design. At the present time there are no published data on individual differences in reaching behavior in normal subjects, let alone in groups of patients with motor disabilities. Since the latter can exhibit widely varying degrees and styles of adaptation to disability, it is quite possible that the variability between different users of artificial hands would be more than normal. Thus it would be wise to view the present data as a basis for working hypotheses about the way people think about their manipulative devices. If the results suggest something about the mental models that people hold of their hands during movement, and these in turn suggest avenues for further behavioral research, we may perhaps begin to approach principled solutions for human-machine interface issues in robotics and teleoperation.

4

Intelligent Exploration by the Human Hand

Roberta L. Klatzky[1]
Susan Lederman[2]

ABSTRACT Humans are remarkably successful at identifying and learning about objects haptically. During the course of haptic object identification, purposive "exploratory procedures" are executed. A variety of factors control the course of haptic object exploration by constraining and influencing the selection of the next exploratory procedure to use. These factors include stored information about objects and their perceptual attributes, associations between exploratory procedures and attributes, and constraints arising from the nature of exploratory procedures themselves. A conceptual model describes such factors and their interactions in object processing.

4.1 Introduction

For the past several years, we have been engaged in a research program to study the competencies of the human haptic system for object recognition. Haptics includes cutaneous sensors in the skin, providing pressure, vibratory, and thermal information, and mechanoreceptors in muscles, tendons, and joints, providing position and movement information. Thus the system comprises touch, position, and movement. Following Gibson [1962], we stress the function of haptics as a purposive exploratory system rather than a passive receiver.

This chapter describes our research by focusing on a particular issue: What determines the sequence of exploratory activities when objects are haptically apprehended (i.e., their properties are encoded) and categorized? Or, to put it more simply, what determines the next hand movement?

We will first describe background studies that led us to this issue. Then we will present a conceptual framework for viewing haptic exploratory activity. This framework leads us to consider a wide variety of influences on

---

[1]Department of Psychology, University of California, Santa Barbara, CA 93106.

[2]Department of Psychology, Queen's University, Kingston, Ontario, Canada K7L 3N6.

the course of exploration; those influences are the principal concern of the chapter.

4.2 Initial Findings

Our research program began with two seminal findings. First, we documented that the haptic system is extremely effective at object recognition [Klatzky, Lederman and Metzger 1985]. In a study in which subjects were given 100 common manipulable objects and asked to name them, the success rate was 96%, and if near misses were accepted, it was 99%. The modal (most frequent) response latency was 1-2 sec, indicating that not only was recognition accurate, it was fast. This first finding, then, led us to ask, what enables the haptic system to recognize objects so well?

Our second finding tells a more complex story, but one with an equally simple point. The point is that the haptic system capitalizes on the motor capabilities of the hand, in order to enhance its sensing capabilities. The result is a set of specialized "exploratory procedures," or stereotyped movements for apprehending object properties [Lederman and Klatzky 1987].

Much of our work is based on the documentation of exploratory procedures, and we will therefore go into them in some detail. Figure 4.1 indicates the procedures we have studied. Six of them are directed to encoding basic properties of the substance or structure of objects: texture, hardness, apparent temperature, weight, size, and global and exact shape. We consider these six properties to virtually exhaust the directly sensed, fundamental properties of objects, although others are arguable (for example, instead of "hardness," one might distinguish among rigidity, brittleness, and elasticity). Two other properties we considered are higher-order attributes – the nature of part motion (e.g., linear vs. rotary), and the function of the object as inferred from its global or local structure (e.g., container; noisemaker).

The figure indicates that each property can be paired with an exploratory procedure that elicits it. An exploratory procedure is a way of interacting with an object that has certain invariant and typical features. Lateral motion, for example, is the procedure used to extract texture. Its invariant property is lateral movement between the skin and object. Its typical properties are rapidity, repetition, and occurrence on a homogeneous portion of the object surface rather than along an edge or at a junction of different materials. To go on with our list of procedures, there is pressure for encoding hardness; static contact for thermal sensing; unsupported holding for weight; enclosing for volume and envelope contour; and contour following, which is used to extract precise contour information as well as global shape. We have also found specialized performative procedures for encoding our two higher-level object properties, function and part motion. These procedures essentially execute the function or motion; for example, placing the hand into a container instantiates its function of containing.

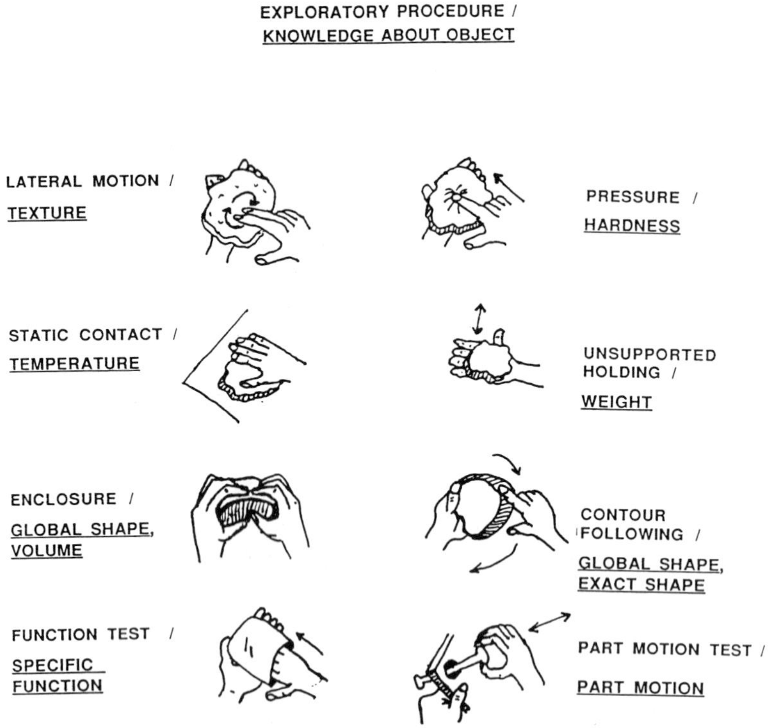

Figure 4.1: Haptic exploratory procedures and the object attribute(s) with which each is associated (©Academic Press, Lederman and Klatzky 1987.)

How do we know these procedures are used? We asked a group of subjects to take part in a match-to-sample task. On each trial, they were given a sample object, followed by three candidates for a match. They were to pick the candidate that best matched the sample. Most importantly, on each trial some object property was targeted for matching. For example, subjects were asked to match the objects on their envelope shape – roughly spherical, rectangular, or whatever – regardless of texture, weight, size, and so on. The results showed clearly that the distribution of exploratory procedures shifted with the targeted property. Specifically, use of the procedure(s) relevant to a given property increased sharply when that property was targeted, despite the multidimensional nature of each object.

4.3 Influences on Exploration: A Conceptual Framework

Up to this point, we have presented the background to the main point of this chapter. Our concern here is with the factors that control the course of haptic object exploration, or "the next hand movement." We can express this concern in the context of a conceptual framework for haptic object processing, as described in Figure 4.2. The framework identifies three distinct levels of information representation. The top level represents information about objects – their parts, their surfaces, their component materials, and so on. The middle level represents information about perceptual attributes such as roughness, divorced from particular objects. The bottom level represents information about exploratory procedures. A recognition model making these distinctions was implemented successfully by Stansfield [1987] with a robotic arm and end effector.

We view these three levels as modules within a parallel interactive system, in which representations are activated, exploration is effected, and objects are apprehended and recognized. An important aspect of the system is that exploratory procedures cannot all be executed in tandem. Elements within the system compete for control of exploration, and a series of decisions is made about which exploratory procedure to execute next. Influences on this decision come from the flow of activation (positive or negative) among representations, as well as other constraints such as those within the motor system. These influences can be associated with different levels within our framework.

Some directives for exploration are clearly knowledge driven, or "top down," and some are clearly data driven, or "bottom up." Others fit less clearly into the data-driven/knowledge-driven distinction. At the object level, acting top-down, are direct "calls" for object attributes, based on hypotheses; this level may also call directly for a test of higher-level attributes such as function. The attributes level works bottom-up to activate

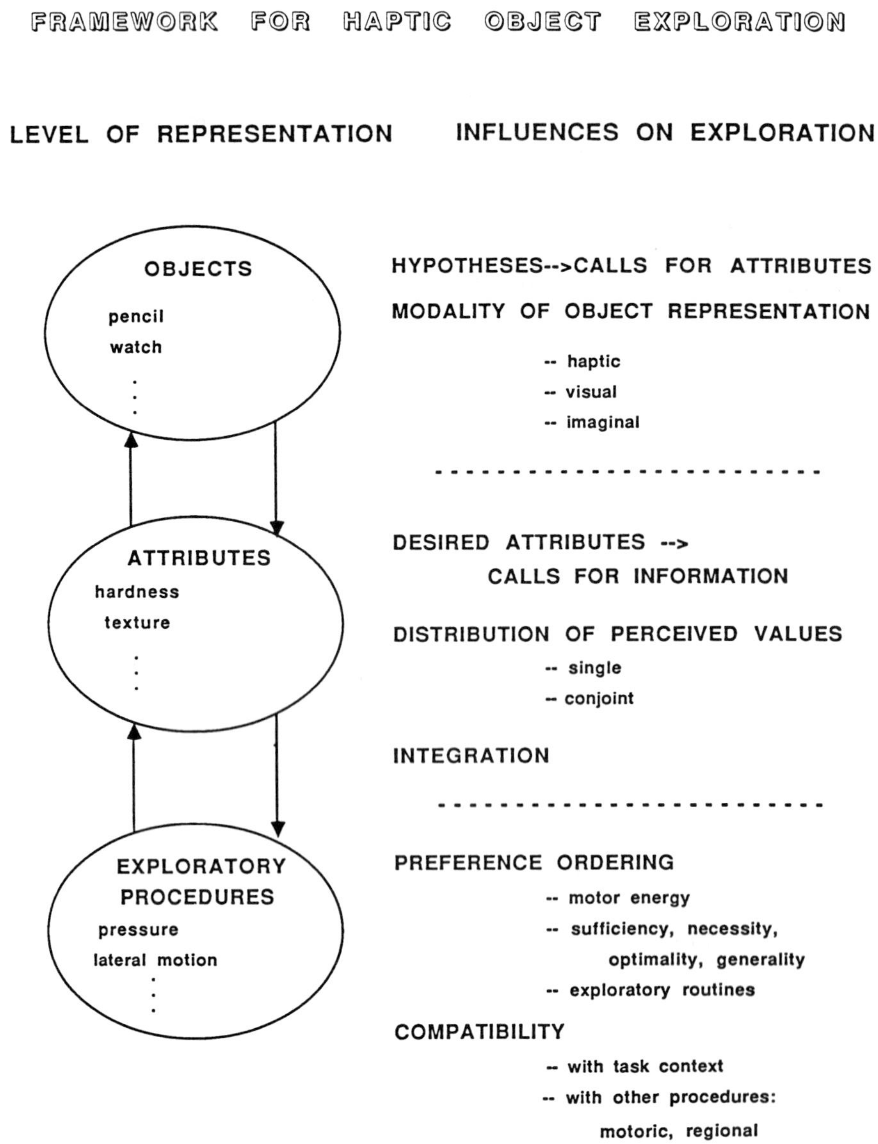

Figure 4.2: A general framework describing influences on the course of haptic exploration.

object information; it works downward to call specific procedures. Each attribute sends a call for the exploratory procedure that is optimally associated with it; for example, the texture attribute would call the lateral motion procedure. Conversely, each exploratory procedure, when executed, sends information about the values on the corresponding attribute dimension, acting bottom-up.

In addition to these direct pathways between representational levels, there are other elements guiding exploration. One such influence, which we have associated with the top level in our system, is the modality of representation that is currently being constructed. Within a robotic or human system, an object might be described in terms of visual, haptic, or abstract conceptual primitives. Each of these constitutes a "modality," in present terms. Psychologists distinguish among modalities not only at the perceptual level, but also at higher cognitive levels. There is substantial evidence that the cognitive representation of an object can be abstract and related to its meaning or function, or alternatively, it can be more concrete and analogous to physical media. One much-studied representational medium is visuospatial imagery, which has been described as activation of an internal pixel-like depiction from a more abstract description in memory [Kosslyn 1980]. Such a representation will convey different attributes than haptically evoked imagery or than a direct representation of an object's function.

Another influence on exploration arises from the middle or attributes level in our system. It is the distribution of previously perceived values on various attribute dimensions. Consider, for example, the attribute of hardness. In the world of common objects, most are rigid. A very soft object represents an extreme on the distribution of hardness values. Its softness is unusual, which might initiate further exploration. We must consider not only the distributions of values on single dimensions but also conjoint distributions, which might have extremes. An object that is small but that seems heavy might induce further exploration for weight, for example.

Also associated with the attributes level is the extent to which object properties are integrated, or treated together in processing. Psychological theories have proposed that certain properties are integrated involuntarily during perception; an example from color vision is hue and chroma [Garner 1974]. Other attributes may be processed jointly at later, more voluntary stages, for example, during categorization decisions. Whether attributes were integrated involuntarily or selectively, we might expect exploration for one to be accompanied by exploration for the other.

Several influences on exploration arise from the bottom level, because of the nature of exploratory procedures. There are factors which might impose a general preference for the order in which exploratory procedures are used. These would include the energy required for the motor act. The extent to which a procedure is broadly sufficient for apprehending object properties, as opposed to being specialized, is also a likely influence on exploration, with general procedures being preferred for early, open-ended encoding.

UNBIASED HAPTICS (BLINDFOLDED)
"SORT SO THAT OBJECTS THAT 'GO TOGETHER'
ARE PLACED IN THE SAME BIN."

HAPTICALLY BIASED (BLINDFOLDED)
"SORT SO THAT OBJECTS THAT 'FEEL SIMILAR'..."

HAPTICS PLUS VISUAL IMAGERY (BLINDFOLDED)
"SORT SO THAT OBJECTS THAT
HAVE SIMILAR 'VISUAL IMAGES'..."

HAPTICS PLUS VISION (SIGHT ALLOWED)
"SORT SO THAT OBJECTS THAT 'GO TOGETHER'..."

Figure 4.3: Instructions used in the study of representational modalities and their influence on exploration.

Other influences at the bottom level occur because procedures are motor acts, operating under motor constraints. Whether a procedure is executed is likely to depend on other ongoing actions. For example, some procedures may be motorically compatible, so that one can be initiated while another is being executed. We will provide evidence that this is the case for lateral motion and pressure, which can be evoked together by a hybrid action that applies both lateral and normal force. Another motor constraint is manipulatory – whether a procedure is executed may depend on its compatibility with manipulatory activities of the exploring or the free hand. The pressure procedure, for example, cannot be executed without object stabilization, either by holding it while squeezing or by using the other hand.

To summarize, we have proposed a substantial list of influences and constraints that should play a role in directing haptic exploration at all levels. Our research to date has exposed and illuminated a number of these influences. In the following sections, we will review that research and point out its relevance to the issue of exploratory control.

4.4 Influences on Exploration: Empirical Findings

4.4.1 The Object Level

Influence of Representational Modality on Exploration.

The first influence on exploration to be described is at the top level in our system – it is the modality of the object representation being constructed. Our work has contrasted, in particular, representations in the modalities of visual perception, visual imagery, and haptic perception. We predicted that structural attributes, particularly contour information, would predominate in vision and visual imagery, whereas the haptic modality would be likely to emphasize attributes related to substance, such as texture and hardness.

To address the representation issue, we had people explore objects freely, either haptically, or haptically and visually [Klatzky, Lederman and Reed 1987]. Different task instructions, along with these perceptual variations, were intended to lead to four different types of mental representation. These instructions are shown in Figure 4.3. The cover task for all groups was a similarity judgment. Participants were asked to sort a set of objects into bins, so that objects that were similar were placed into a common bin. Under unbiased instructions, with or without vision, participants were not given any particular definition of similarity. These conditions should lead to representations directly based on haptic, or haptic plus visual, perception. To bias toward a visual imagery representation, we told participants to think of similarity in terms of the object's visual image (remember that exploration was haptic without vision). Finally, to induce a haptically biased representation, we told them to think of similarity in terms of how the objects "felt."

The objects in this study were created from all combinations of three values on each of four dimensions: hardness, shape, roughness, and size. Thus, for example, each object was rigid, slightly compliant, or soft; it was oval, hourglass-shaped, or clover-shaped; and similarly, it had one of three possible roughness values and three possible sizes. Each object was of homogeneous material and planar, that is, it had two-dimensional contour, and its third dimension was thin and invariant.

From the participants' sorting behavior, we created a measure of how vivid, or "salient," each dimension was in the object representation. A dimension was salient to the extent that its levels were segregated in different bins. Thus for example, shape would be maximally salient if all the ovals were in one bin, all the hourglass shapes in another, and all the clovers in a third. In this case, each other dimension would have to have zero salience, because its levels would be mixed within each bin. The bin of oval shapes, for example, would have to include some of each possible size.

Finally, in addition to the salience score of each dimension, we had a measure of how frequently each of four exploratory procedures occurred. Those procedures were lateral motion – related to texture; pressure – related to hardness; contour following – related to shape; and enclosure – related to shape and size. Our concern was not only with salience, but with its relationship to the pattern of object exploration.

Our results are summarized in Figure 4.4. It shows data for each dimension in a separate panel. Within each panel, we can see how groups of subjects with different instructions fared. There are two dependent measures of interest – the salience of the dimension, and the frequency of occurrence of the directly related exploratory procedure(s). The salience scores

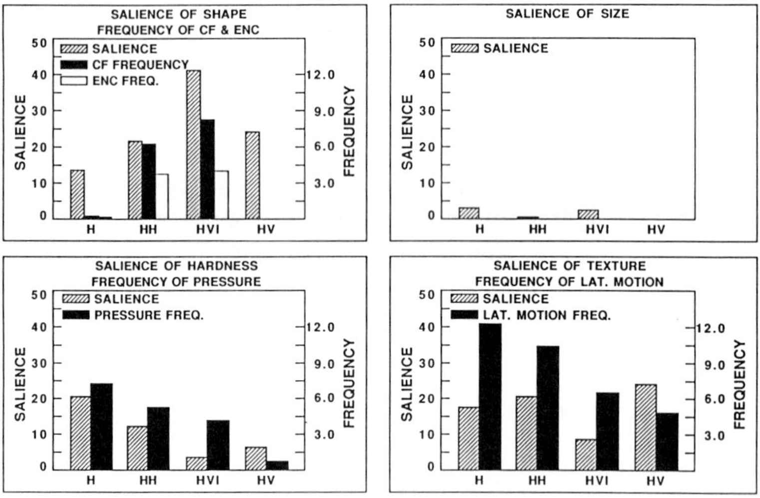

Figure 4.4: Salience scores for each of four dimensions, by group (H = unbiased haptics; HH = haptically biased haptics; HVI = haptics plus visual imagery; HV = haptics plus vision), together with the frequency of the relevant exploratory procedures for the dimension (CF = contour following; ENC = enclosure; LAT = lateral). Note that no salience score for size differs significantly from zero and therefore no procedure is displayed. (©American Psychological Assn, Klatzky, Lederman and Reed 1987).

indicate differences in how the objects were represented, depending on the perceptual condition and instructions. Consider the dimension of shape: This had relatively low salience for groups that were exploring haptically without any particular bias, or that were instructed to think about how objects felt. But these same groups showed relatively high salience for the substance dimensions of texture and hardness. The visual imagery group showed strong salience for the shape dimension. And vision plus touch acted like a moderator: Those who could see the objects as well as feel them found shape and substance, particularly texture, somewhat salient. We note that size was not found salient by any group, possibly because of the range of sizes we used – all within the span of the hand, allowing relatively little variation. (Also, this design pits dimensions against one another, so that size cannot be found salient when other dimensions dominate.)

Next consider the patterns of exploration, as indicated by the percentages of occurrence of targeted exploratory procedures. For the three groups who were denied vision, there is a direct relationship between the salience of a dimension, and the extent to which relevant exploratory procedures are performed. When shape is salient, there is exploration for shape – contour following and enclosure. When texture is salient, lateral motion tends to occur; and similarly for hardness and the occurrence of pressure. Not surprisingly, vision strongly reduces the amount of haptic exploration.

These data show a clear relationship between the salience of an object property to a haptic explorer, and the pattern of exploration. Further, this pattern of results clearly reflects an influence on exploration of the desired modality of representation. Consider, for example, someone who wants to form a visual image. The external envelope of the object is the most important property in such a representation. Accordingly, a call is more likely to be sent for shape information, leading to enclosure and contour following. Similarly, instructions to consider what objects feel like are more likely to lead to calls for texture and hardness than for shape, and hence to the appropriate procedures for exploration.

On the other hand, the relationship between salience and exploration may also reflect influences from the bottom level of our system. We have previously found that contour following is relatively slow and subject to considerable error with complex contours. Lateral motion and pressure, in contrast, can be executed very quickly. In the absence of instructions to form a particular representation, people may choose to execute these latter procedures because of their low demands on motor energy. As a result, the representation may emphasize the corresponding dimensions of texture and hardness.

In short, either representational salience may invoke patterns of exploration, or patterns of exploration may determine representational salience. When salience invokes exploration, the influence acts top down. When exploration determines salience, the influence is bottom up.

4.4.2 The Attributes Level

Influence of Attribute Distributions.

We next consider what people know about object attributes through experience with haptic perception, and how that knowledge influences exploration. They are likely to know the general range of values that objects take on a dimension such as hardness. They are also likely to know which values on different dimensions tend to co-occur. We call such co-occurring values "natural correlations." For example, we know that large objects tend to be heavy; a large light object such as a balloon is an anomaly. How might such knowledge of distributions affect exploration? One strong possibility is that an anomalous observation leads to further exploration, for purposes of verification. Another possibility is that knowledge of natural correlations can be used to "prune" the tree of potential exploratory movements, eliminating those that are likely to encode redundant features of objects.

We are currently planning a program of research to determine which attributes of objects tend to be correlated, and how correlations affect exploration. Some preliminary data on this point derive from another study [Lederman and Klatzky, in press], in which people were asked to rank attributes according to their importance in categorizing objects. For example, weight might be critical in categorizing an object as a cast-iron frying pan. We found that certain attributes tended to be correlated, in that if one was highly ranked, the other was as well. The most strongly related pair, in this sense, was size and shape. Texture and hardness were also highly correlated. The general pattern was to find strong correlations between structural categories, or between substance categories, and weaker ones across these boundaries.

4.4.3 THE EXPLORATORY PROCEDURE LEVEL

Earlier, in conjunction with the study of mental representations, we raised the possibility that the effort or energy involved in an exploratory procedure may determine whether it is used. This influence is associated with the bottom level of our system; that is, it is due to the nature of exploratory procedures themselves. We now consider a number of influences associated with that level.

Influence of the Sufficiency, Optimality, and Generality of Exploratory Procedures.

The first factor to be considered is the breadth of attribute information that an exploratory procedure provides. We distinguish among exploratory procedures according to their sufficiency, necessity, optimality, and generality for encoding haptic attributes [Lederman and Klatzky 1987].

By our definition, an exploratory procedure is *sufficient* to encode an object attribute if it permits one to discriminate between objects along that particular dimension. For example, enclosing an object is not the best way to encode its texture, but it might be sufficient to do so, because of small-scale lateral motion that occurs during the act of enclosing.

An exploratory procedure is more than sufficient to encode some dimension – it is *optimal* – if it provides better discrimination performance, in terms of accuracy (or speed, where there is a tie in accuracy), than other exploratory procedures. And if it is the only procedure to be sufficient to encode that dimension, it is termed *necessary*.

Sufficiency, optimality, and necessity are judged of each procedure relative to each dimension. By looking across haptically encoded dimensions, we can ask whether an exploratory procedure is relatively specialized, or in contrast, more general. A procedure is specialized to the extent that it discriminates one attribute well, and the others much less well. It is general to the extent that it encodes multiple attributes at about a constant level

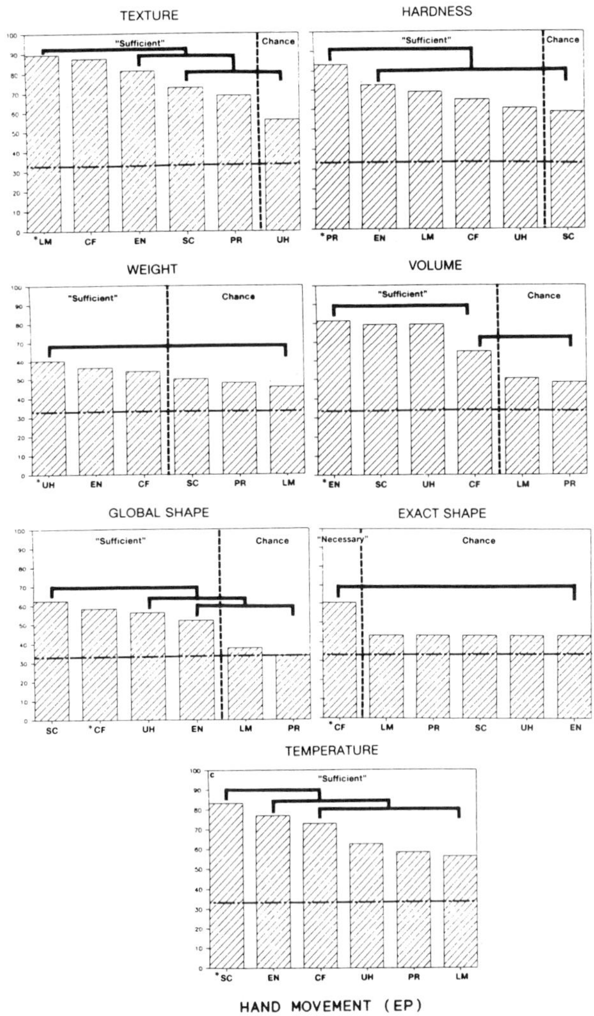

Figure 4.5: Performance in 'matching objects on specific dimensions' task, by dimension and exploratory procedure (PR=pressure; EN=enclosure; LM=lateral motion; CF=contour following; UH=unsupported holding; SC=static contact). Procedures that did not attain performance significantly above chance are shown to right of dashed vertical line. Asterisk indicates procedure used most often during spontaneous matching on same dimension. Heavy brackets at top join procedures that did not differ statistically. (©Academic Press, Lederman and Klatzky 1987.)

of discriminability.

The importance of sufficiency, optimality, and generality for patterns of object exploration should be clear. Why waste time on a nonoptimal procedure, if there is only one attribute of an object that is to be encoded? Alternatively, if many attributes are of interest, a nonspecialized, highly general procedure is of greater value. Thus the task context, together with the known performance of an exploratory procedure, is likely to strongly influence which procedure is executed.

We have tested the sufficiency, optimality, and generality of exploratory procedures in a variant of the match-to-sample task. In this case, participants were constrained to explore in a particular way, and to match objects on a particular attribute. Most combinations of exploratory procedures and attributes described earlier were tested. For example, the lateral motion procedure was used to match not only texture, but weight, size, and so on.

Figure 4.5 shows the results of this study. We can see that in general, the procedure that was spontaneously used in the unconstrained match-to-sample task (marked with an asterisk) is also the optimal one, in terms of accuracy (objective, or determined by consensus of observers in the original task), or in the case of a tie, speed. Thus, for example, lateral motion gave the highest accuracy for texture matching. In the case of matching precise contour, contour following was not only optimal but necessary.

Another interesting outcome of this study was that procedures differed in their generality of application, or conversely, their specialization. Pressure was the most specialized procedure, in that it yielded much higher accuracy for hardness than for any other dimension. Enclosure was the least specialized: It was grossly sufficient, producing above-chance accuracy on matching most attributes of objects, but it was generally not optimal. If enclosure with some pressure were accompanied by lifting (unsupported holding), that is, if a simple grasp were performed, it is clear that we could find out a great deal about an object very rapidly. For this reason, grasping would be ideal for initial contact.

Exploratory Routines

In fact, grasping does appear to be our favored way of contacting unknown objects. In this sense it constitutes a habitual "routine" for exploration. The existence of such general routines may be an important determiner of exploration, at least in its early stages.

We examined exploratory routines as part of an ongoing study of object categorization [Lederman and Klatzky, in press]. Blindfolded subjects were given an object in their upraised palms and asked if it was in a particular category. Of interest here is that they tended to follow highly routinized patterns of exploration for the first few moments. Although there were variations, the most general pattern was an enclosure of the object in one or both hands, followed by unsupported holding. This routine is likely to be very successful for recognizing objects, especially at the "basic level" – the level of common naming, such as jar or pencil. Enclosure, we know, is sufficient to grossly discriminate values on many attributes. Accompanied by lifting and even slight pressure, other attributes are added, and a snapshot of the object can be obtained with minimal exploratory effort.

Influence of Compatibility of Exploration and Task Context.

Another bottom-level factor likely to be critical to the selection of an exploratory movement is its compatibility with other aspects of the task environment. Compatibility takes various forms. A procedure may be compatible or incompatible with other exploratory procedures that are also under execution. Or its compatibility may be judged relative to general manipulatory constraints in the task context, for example, whether the object is fixed or moveable or whether one or two hands may be used.

We have found clear evidence of compatibility effects of this latter sort in the bin-sorting task described previously [Klatzky *et al* 1987]. Recall that objects varying in size, shape, hardness, and texture were sorted and the most salient dimension(s) was determined. In one condition, we constrained participants to sort the objects either with one or both hands. This was intended to affect their ability to stabilize the object and to reorient it for exploration in different regions.

If two hands are available, one can serve the stabilizing and orienting function (which we have called "task maintenance"), and the other can serve the exploratory function. Given one-handed exploration, the two functions can still be observed: Part of the hand typically stabilizes and part explores. For example, the fingers may hold the object while it is surveyed by the thumb. This limits the ease and extent of exploration, of course. Procedures such as lateral motion and pressure, which are performed in a small region of the object, can easily be performed along with stabilization by the same hand. Contour following is more difficult under one-handed conditions. It tends to occur in a broader region, and it requires not only stabilization but also periodic reorientation to bring new parts of the contour into focus. If contour following is limited by one-handed exploration, we should see an effect on the salience of the corresponding object property – shape.

The results of this manipulation showed a clear effect. In the condition where shape tended to be salient (visual imagery), it was far more so under two-handed exploration. In fact, the salience score was twice as great as for the one-handed condition. Evidently, this general constraint on the manipulatory context influenced the course of exploration and apprehension of object properties.

Influence of Compatibility Between Exploratory Procedures.

As noted above, we can also speak of compatibility between exploratory procedures. Some procedures can effectively be produced together, whereas others may interfere. Hence if one procedure is selected, it will influence what others are chosen.

We have found compatibility of exploratory procedures to affect performance in a classification task, using the same multiattribute objects described previously [Klatzky, Lederman and Reed 1989]. Participants repeatedly classified 9 stimuli into three categories. The categories could be defined by one dimension only; for example, all "A"s are hard, all "B"s medium hard, and all "C"s soft. Alternatively, the categories could be defined by two dimensions redundantly: "A"s are hard and oval, "B"s medium hard and two-lobed, and "C"s soft and three-lobed. We also included a condition where three dimensions – texture, hardness, and shape – redundantly classified the objects. (Size was not used as a redundant dimension in this study.) As each object was classified, we measured the response latency, defined as the time between initial contact with the object and the vocal response. Our interest was in redundancy effects – reductions in response time due to additional redundant cues to classification – and also in the exploratory procedures executed during classification.

We did find that redundant dimensions speeded classification, but only to a point. Two redundant dimensions led to faster classification than one, but a third redundant dimension did not further reduce response times. Examination of exploratory procedures revealed why: Given the redundant dimensions of texture and hardness, subjects tended to execute both. When contour was provided as a further cue to classification, subjects largely behaved as if it were not present. They tended to produce the same exploration as for texture and hardness, with relatively little contour following.

A variety of compatibility effects may underlie this preference to execute the procedures for texture and hardness. One is motoric. The relevant exploratory procedures can be produced together, in the form of a hybrid movement with both lateral and normal force. In fact, this was frequently observed. Further, the extraction of hardness competes with the extraction of shape information, in that the appropriate degree of normal force is quite different. Applying pressure in order to determine hardness may actually deform the shape of a compliant object. Another aspect of compatibility concerns the appropriate region of the object for exploration. For thin planar objects, contour following is movement along the object's edge. In contrast, the preferred position for lateral motion and hardness is within a larger homogeneous region; the edge may actually interfere with their encoding.

Several converging operations in these studies supported the idea that texture and hardness are jointly encoded more than either dimension is joined with planar contour. In one study, subjects were instructed to classify objects on the basis of one dimension, such as texture. At the same time, another dimension varied redundantly with the explicitly mentioned one. After 100 or so classification trials under these conditions, the implicit redundant dimension was withdrawn. That is, the stimuli being classified were changed so that the redundant dimension was now invariant. Of course, the explicit classification rule still applied.

Elimination of the implicit redundant dimension greatly interfered with classification, when the two dimensions were texture and hardness. When either dimension was paired with shape, however, the effects were minimal. It appears that texture and hardness were both considered, even under instructions to attend to just one. The exploratory procedures supported this assumption, showing overt exploration for both dimensions.

These results suggest that exploratory compatibility is important in determining whether dimensions are processed together in tasks such as object identification. On the other hand, as was mentioned above, influences at the attributes level are also likely. That is, there may be perceptual or cognitive preferences for integrating certain dimensional combinations, which would call for the relevant exploratory procedures together.

4.5 Conclusion

We have now discussed evidence, from our ongoing work on human haptics, for a variety of influences governing patterns of exploration. The determination of "the next hand movement" is indeed complex. To understand patterns of haptic exploration, it is essential not only to discern what factors influence it, but also to determine their weights in various contexts. Given a fuller understanding, our initial conceptual framework may be developed as a model that more precisely predicts exploratory sequences for apprehension and identification.

As dextrous robotic effectors with multiple sensors are developed, the issue of how to direct exploration and analysis of sensory data will become increasingly critical. Given exploratory procedures that optimize the extraction of particular object properties, the coordination and sequencing of the procedures must be determined. The general domains of influence on this process that we have described can be applied to robotic systems as well as to the biological system we study. Factors that potentially affect the course of robotic exploration include knowledge about objects and their properties, the history of previous exploration, and constraints imposed by the sensors, the effectors, and the procedures.

Part II

Dextrous Hand Control Architectures

Subramanian T. Venkataraman

The notion of a system architecture has only recently taken over the field of robotics. The view that intelligent systems cannot be developed through ad-hoc combination of small components, each developed independently, is gaining strength. Albeit this realization, most existing systems are relatively small, and represent engineering solutions to specific sets of problems in a given problem domain. The main reason for this is the sheer imagination that is required to develop an architecture that sends strong philosophical messages without neglecting details that are acutely necessary for implementations.

An architecture typically lends structure to the solution to a given problem. It describes computational elements and information flow. Pictorially, an architecture resembles a three dimensional connected graph, and may span across several abstraction layers in a hierarchy of control or information representation.

A control architecture lends structure to control algorithms. Typically, control algorithms contain mathematical relationships that are necessary for the generation of actions required to correct discrepencies between desired and actual behavior. For example, during grasping, if vibration sensors placed upon fingertips detect slip, then the corresponding grasp reflex control algorithm provides the mathematical relationships necessary to generate appropriate squeeze forces.

Computational architectures, on the other hand, describe how information is processed during the execution of a corresponding control algorithm. They define how computations are performed, and how knowledge- or sensor-based information is processed. In the previous example, a grasp reflex computational architecture would describe how vibration sensor information may be processed to obtain *measures* that indicates the *extent of slip*, and also define how this information (along with the knowledge-based information describing nominal grasp conditions) must be processed to obtain control signals.

And finally, a corresponding hardware architecture describes how a control algorithm may be implemented on a physical system. In the grasping example used above, details about the end effector and actuator hardware, associated sensing apparatus, and I/O devices. In some cases, issues that pertain to low-level computational needs are included within the hardware architecture description (such as processor hardware, and bus schemes), while in others, details of the robotic, interface, or sensing hardware may be included within details that outlay low-level computational architecture details.

Chapters in this part of the book focus primarily upon architectures for dextrous manipulation. Most robotic architectures typically deal with the hierarchical decomposition of abstract task commands into actuator signals for successful task execution. And, while their top-down decomposition of abstract commands is well defined, due consideration has not been placed upon the development of appropriate feedback mechanisms. Examples include [Albus *et al* 1983; Saridis 1983; Lozano-Pérez *et al* 1987; Venkataraman and Lyons, this volume]. On the other hand, architectures such as the ones in [Brooks 1986; Klatzky and Lederman, this volume] differ in their philosophy, as well as their functionality. Their architectural designs are based on sensing and feedback pathways, with primitive control actions defined within this context. There is often not even the notion of a global plan. For example, Brooks [1986] implements intelligence through an aggregation of simple reactive primitive actions.

The exact computational nature of most robotic architectures is yet unknown. Except for Lozano-Pérez *et al* 1987 and Brooks [1986], most of them have not even been implemented completely. The road from control architecture to a computational architecture design is often rocky, since it demands that its designer be an expert in several technology areas. Many researchers have, therefore, taken an alternate route towards the development of computational architectures. Instead of a computational architecture that maps a given control architecture appropriately, they developed one that would contain the general suite of functionalities that would be required from a computational architecture for prototypical robotic implementations. The intent here was to develop an architecture on which a large spectrum of control architectures could be mapped onto. Examples of such architectures include [Chen *et al* 1986; Raibert 1986; Narasimhan, Siegel and Hollerbach, this volume].

Any robot hardware architecture on which a control architecture is to be implemented must naturally conform to the suite of tasks in question. A poor design or choice will significantly complicate implementations. For example, the application of a Cincinnati Milacron robot to watch making would be as ineffective as the application of the Stanford/JPL hand to pressure weld two metal parts. The design of the Stanford/JPL for grasp stability, the design of the Belgrade/USC hand for automatic shape adaptation in [Bekey, Tomovic and Zeljkovic, this volume], the arm design for isotropic mass distribution in the arm's operational space in [Khatib 1989], and the arm design for structural modularity in Robotics Research arms all illustrate the inherent coupling between tasks and hardware design.

Dextrous robot hands is the main focus in the subsequent parts of this book. Therefore, a preliminary working definition for dexterity is provided. A dextrous robot may be defined as a robotic mechanism that has the ability to impart arbitrary motions and forces upon *its* environment. A dextrous robot hand is an end effector that has the abovementioned ability. Note that many other constraints such as energy minimization, grace, or stability in posture may be added to the description of dexterity (defined here based on motions and forces). While the Belgrade/USC robot hand, as of today, does not conform to this definition, we view it as a bottom-up approach towards dexterity.

Chapter 5 deals with the integration of dextrous end effectors into flexible assembly cells. Venkataraman and Lyons propose an architecture that structures the control of dextrous hand in a task-oriented manner. To develop this architecture, they first examine their domain of application, extract certain functional invariances out of its structure, and describe the decomposition of abstract assembly operations into a sequence of low-level task *primitives*. They also consider resource allocation (and reallocation) of tasks to physical resources and present a vocabulary of task primitives, formally developed from the physics of mechanical interactions at the contacts between a robot and its environment, and describe how tasks may be executed using these primitives.

In the next chapter, Narasimhan, Siegal and Hollerbach discuss the development and implementation of the *CONDOR* multiprocessor hardware/software architecture for the real-time control of the Utah/MIT hand. They present a summary of existing processor options, discusses design tradeoffs, and justify the *CONDOR* system design. The system uses Motorola 68020 processors, and a SUN workstation provides the tools and utilities necessary for a clean, yet flexible environment to design and develop hand control programs.

The last chapter in this section describes the design philosophy, the structural details, and servo-control of the Belgrade/USC hand. Bekey, Tomovic and Zeljkovic discuss the design goal of this five-fingered device, which was to obtain a hand suitable for stably grasping objects of varied geometry. Conforming the hand to the geometry of each object is achieved by *autonomous shape adaptation* with a very simple control architecture that implements reflex control.

Chapters in this part of the book reflect predominant architectural approaches to dextrous manipulation. Like their counterparts for multiple arms or mobile robots, vital pieces of architectural information are still being hard hard-coded, based on the deep engineering insights of their designers. These architectures also lack a measurable performance evaluation criteria to test and verify how *good* it really is.

In order to develop performance benchmarks and compare one approach with another, a more formal approach to the design, development and implementation of architectures will need to be taken. If, some day, robot hands will be at a technological horizon where they may be effectively applied in prosthetics, medicine, industrial assembly, and planetary exploration, serious attention must be paid now to system architectures. Rigorous work will also be needed in understanding how to combine a top-down design from a problem domain specification with a bottom-up design from a set of available actuators and sensors. Then, it might be possible to develop an architecture that represents the *functioning* of an integrated dextrous hand system, and not just the dextrous hand, its sensors, and its computers independently.

5

A Task-Oriented Dextrous Manipulation Architecture

Subramanian T. Venkataraman[1]
Damian M. Lyons[2]

ABSTRACT Much of the previous work on dextrous manipulation has concerned itself primarily with hand construction, sensor development, servocontrol or grasp planning. Efforts towards the development of an overall manipulation system have either been restricted to a particular set of implementations, or been far too general to be practical. In this chapter, we take the position that a dextrous hand can be studied most efficiently when embedded within the context of an *application domain*. The main focus of this work is to develop a control architecture for a flexible assembly cell equipped with dextrous hands.

The central idea is to use constraints from the description of the task to be executed to select an appropriate abstract model and physical robot hand-arm system with which to execute the task plan. With this idea as target, we describe how allocation can be performed on the basis of a *task criterion* for a task plan. We also introduce a set of control models that can be allocated and interfaced to real hand-arm systems for the actual execution of task plans.

5.1 Introduction

In this chapter, we develop a control architecture for a flexible assembly cell equipped with dextrous hands. The controller accepts a high-level assembly operation as input and decomposes it into more primitive operations taking into account robot resource capabilities and availabilities in the cell and executes these operations. A unique aspect of this chapter is that it describes a first step at integrating dextrous hands into industrial assembly. The main thrust of this chapter is to bring out the constraints that result during resource allocation. The applicability of this approach is demonstrated through the integration of control models for the execution of task

[1]Tele/Autonomous Systems Group, MS 198-330, Jet Propulsion Laboratory, 4800 Oak Grove Drive, Pasadena, CA 91109

[2]Manufacturing Systems Research Department, Philips Labs, 345 Scarborough Road, Briarcliff Manor, NY 10589

plans within the architecture.

A flexible assembly cell consists of a number of robot systems, along with the tools used during assembly by them. Each system is composed of an arm, some number of end-effectors, and associated sensory and actuation equipment. In particular, some of the end-effectors may be ***dextrous hands***, commonly defined as devices that are capable of imparting arbitrary small motions and forces upon the object that they hold. These assembly cells are typically required to handle small and mixed batch assembly. To make this technology viable, and progress towards 'one of a kind' assembly [Bourne 1989], flexible assembly cells must function with little or no setup/changeover time. They must also be robust. Note that a set of robot systems, each of which may be computationally or structurally different, can be used to handle unexpected events or error conditions that might cause problems for a single robot or piece of fixed automation.

Dextrous hand research, thus far, has concerned itself primarily with hand construction and servo-control (a survey of related articles may be found in [Venkataraman 1988]), sensor development (an overview of related articles may be found in [McCammon and Jacobsen, this volume]), grasp planning [Lyons 1985; Ahmad and Feddema 1987; Nguyen 1988], or with manipulation [Yoshikawa and Nagai, this volume]. We believe that dextrous hand research is at a point where giant strides can be made only after certain issues that impede its effective use have been resolved. This chapter attempts to resolve such issues by integrating dextrous hands into the automated assembly system, and describing in detail *how* a hand may be effectively used.

The design and construction of a controller for a flexible assembly cell is a complex problem involving issues in task decomposition, resource allocation, control, and sensing. Our work in this chapter focuses on task-directed usage of dextrous hands, and on the development of a framework for allocating and utilizing robot resources (such as dextrous hands) in assembly.

Input to the control architecture is an object-level assembly operation. This input could be generated by a human programmer or by an automatic task planning program [Vijaykumar and Arbib 1987]. Our view of an assembly operation is based on our previous work in [Vijaykumar *et al* 1987; Lyons 1988]; the main extension to that view in this work will be the concept of a *task criterion* associated with a task plan (i.e., assembly operation). Each task criterion specifies demands made on robot resources by the task plan to enable successful task execution.

The controller decomposes a high-level assembly operation into more primitive actions which can be carried out on appropriate robot arms and end-effectors. The user need have no knowledge of the arms and tools available to the cell; this mapping is dynamically evaluated by the controller to suit the task and the state of the workcell. This is what we mean when we call our architecture ***task-oriented***: The choice of a physical manipulator as well as the dynamic and control models of that manipulator are determined by the requirements of the task to be executed. The controller also deals with the issue of reallocating resources when the current allocation is no longer appropriate (e.g., if an error occurs). We find that some problems, such as regrasping an object, can be naturally expressed by the concept of *reallocating resources* for a task plan.

Once a robot resource has been allocated, an interface procedure is called up to translate commands coming from the task plan into a form on which the robot can act. A key step in this will be the introduction of a vocabulary of *task primitives* to program *actions* performed by each finger, and constructing hand task primitives, using finger task primitives. Note that a wide selection of grasps, such as those in [Lyons 1985], can be built using these hand primitives. Note also that manipulation tasks involving a dextrous hand (small manipulations), or involving multiple arms (gross manipulations) can be programmed using these primitives.

The remainder of the chapter is laid out as follows: In Section 5.2 we describe the design process for the architecture. Section 5.3 introduces an assembly operation example that reappears throughout the more detailed sections of the chapter. The architecture is described in detail in Section 5.4. In Section 5.5 we discuss a useful format for the task criterion, and describe the mechanism of robot resource allocation. In Sections 5.6- 5.10 we describe an appropriate set of task primitives for dextrous hand usage and discuss the mechanism for interfacing to physical resources. We conclude in Section 5.11 by outlining the current state of implementation of this system and future directions.

5.2 Design

In this section, we motivate the design of the assembly cell architecture. The central design tenet is to base the architecture on the structure of assembly operations.

5.2.1 Assembly Operation Structure

An *assembly operation* may be considered to consist of four phases:

1. *Reaching* for the component.

2. *Acquiring* the component.

3. *Transporting* the component to the vicinity of the appropriate sub-assembly, and finally,

4. *Parts-mating* the component with the sub-assembly.

Although distinct, these phases are strongly linked. For example, a component should be grasped to enable the subsequent parts-mating operation

operation [Lyons 1985; Ahmad and Feddema 1987; Li, Hsu and Sastry 1987] and to facilitate finding a path to the vicinity of the parts-mating operation [Lozano-Pérez and Brooks 1985]. The transportation phase must bring the component into a suitable position from which to start the parts-mating operation. However, different strategies for the parts-mating operation may generate different sets of feasible grasps and transportation paths. Thus, although the phases of an assembly operation are distinct, they can not be decoupled.

5.2.2 Robot Information

Two types of information about robots in a workcell will be made use of: their capabilities (*allocation* information), and how to command them (*interface* information). An important extension is that the allocation and interface information for physical arms and end-effectors, as well as control models, such as a grasp configuration for a dextrous hand, or a compliant control model for an arm will be provided. We will assume that allocating a model entails allocating a physical robot on which to implement the model[3]. This combination of a model and a physical robot will be called a *virtual robot*; the reason why this concept is necessary and what form it should have is discussed in detail in Section 5.5.

5.2.3 Candidate Architectures

We build our architecture based upon the structure of assembly operations. By this we mean that if a number of robot systems are available to the controller, then, broadly, the choice of an arm is made for the reaching/transportation phase, while choice of an end-effector is made for the acquisition/manipulation phase. In Figure 5.1, the three major candidate architectures are presented. In each case, the shaded boxes represent allocation functions (programs to allocate a task plan to an appropriate robot system based on given task criteria), and the non-shaded boxes, the interface function (programs to send commands to the allocated robot). In option A, allocation and interfacing of the end-effector and arm are performed *together*; in B, the arm allocation precedes the end-effector's; in C, the end-effector allocation precedes the arm's.

5.2.4 Evaluation of Candidate Architectures

To evaluate these candidates, we need to introduce some terminology. Let A be the set of all available arms in the cell, and H the set of all end-

[3]We assume this for technical convenience: Allocating the model and then choosing a robot to match the needs of that model could be phrased within framework we present – it's a case of 'more of the same.'

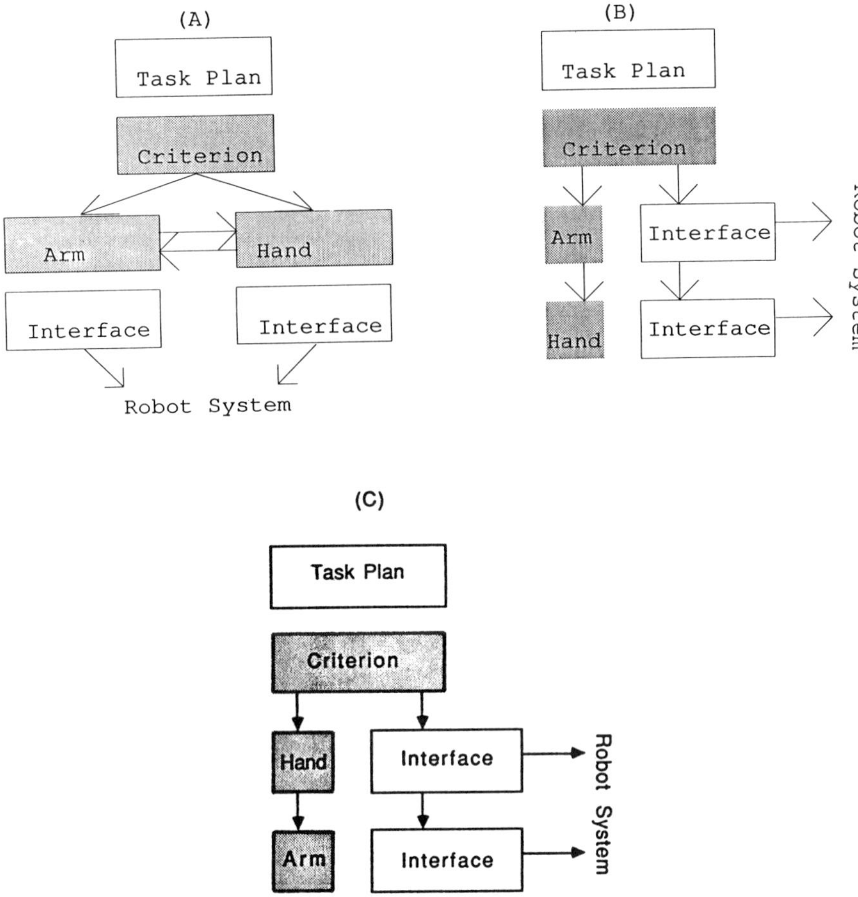

Figure 5.1: The Three Candidate Hierarchical Architectures

effectors in the cell. The library functions $LA(a), a \in A$ and $LH(h), h \in H$ describe the characteristics of the arms and hands in the cell. The functions $IA(a), a \in A$ and $IH(h), h \in H$ describe the way in which robots in the cell can be interfaced. (The construction and use of LA, LH and IA, IH will be discussed in detail in Sections 5.5 and 5.6 respectively).

We will assume that we can construct $H_a \subset H, a \in A$ the subset of end-effectors on, or available to, arm a. For example, for an arm with a gripper exchange mechanism, H_a would list all the end-effectors that could be used on the arm. For an arm with a dextrous hand, H_a will include all the *different grasps* that can be done with the hand (since we have included allocation and interface information for control models as well as for physical manipulators). Similarly, $A_h \subset A, h \in H$ describes what arms can be used with any one end-effector. When A (or H) is subscripted with a set, we mean the union of all A_h (H_a respectively) for h in that set.

Let χ, the *task criterion* for a task plan, be the statement of what robot resources are necessary for the successful completion of that task plan. (We will expand χ and describe the information contained within it presently.) χ will be used in conjunction with the library functions LA and LH to allocate the execution of task plans to available robot systems. We will assume that we can break χ down into χ_a and χ_h the criteria for selection of arm and hand respectively.

We evaluate these candidate architectures based on how well they model the structure of the assembly operation, how well they facilitate allocation, interfacing and reallocating, and how well they suit the assembly cell scenario we postulate.

1. **Assembly operation structure:** In scheme B, the reach phase is scheduled first. In this, the scheme reflects the relationship between allocation and the assembly operation structure: Unless the object can be acquired, the assembly operation cannot even be started. Scheme C, on the other hand, places allocation for the manipulation phase at highest priority, while A does not represent the above relationships at all.

2. **Allocation, interfacing and reallocation:** The table given in the following page outlines allocation and interfacing for the three candidates: Scheme A is the most computationally complex scheme. The other two differ only in the order in which they allocate resources types. Both these schemes facilitate reallocation: the allocation scheme is hierarchical, and thus lower-level choices can be replaced without affecting the higher levels. Computationally, the complexity will depend on the types of arms and hands in the workcell: if there are many complex arms, the set $H_{\hat{A}}$ will probably be large and thus the complexity in scheme B will approach that of scheme A. In similar fashion, if there are many complex hands, then $A_{\hat{H}}$ will be large and the complexity of scheme C will approach that of A.

	Allocation	Interface
(A)	ARM: scan all $a \in A$ for χ_a. HAND: scan all $h \in H$ for χ_h.	Arm and hand commands must be split by task plan.
(B)	ARM: scan all $a \in A$ for χ_a and produce $\hat{A}$ and $H_{\hat{A}}$. HAND: scan all h in $H_{\hat{A}}$ for χ_h.	ARM: Filter out gross motions. HAND: Manipulative motions only.
(C)	HAND: scan all $h \in H$ for χ_h and produce $\hat{H}$ and $A_{\hat{H}}$. ARM: scan all a in $A_{\hat{H}}$ for χ_a.	HAND: Filter out manipulative motions. ARM: Gross motions only.

3. **Assembly cell scenario:** In reallocation, B allows choice of hand to be switched while using the same arm (e.g. a new grasp chosen, or a replacement end-effector), while C allows choice of arm to be changed while maintaining the same end-effector (a difficult feat to implement). Both schemes allow reallocation of arm and hand (forcing either a hand-off situation, or temporary placement of the object). Our workcell description emphasizes dextrous hands: In general, we expect the hands in the workcell to be more complex than the arms. Thus, in our assembly cell scenario, B is the most useful alternative.

Our criteria support architecture B as the best choice for our workcell controller.

5.3 Running Example: Compliant Placement

We introduce an example assembly operation, compliant placement, which we will use to illustrate the more detailed parts of this chapter.

The actual choice of assembly operation is not that important, since the majority of our architectural components are similar no matter what operation is involved. Our example involves a place operation for which a high level assembly task planner (such as the one described in [Vijaykumar and Arbib 1987]) has prescribed a compliant control strategy. The example task may be formally defined as follows (see [Vijaykumar *et al* 1987] for

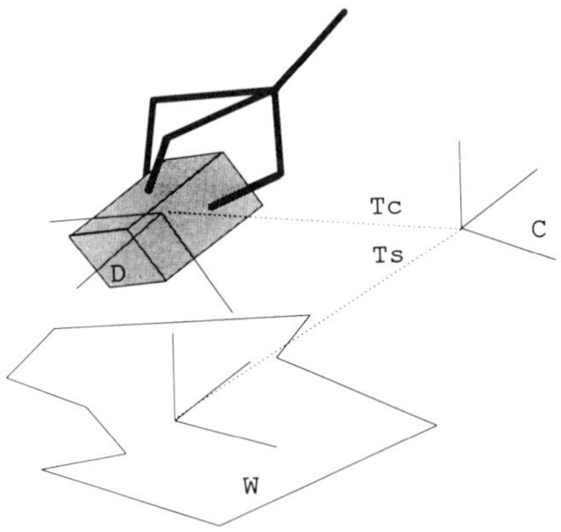

Figure 5.2: Compliant Place Strategy

similar examples):

- Goal, G_p: Object surface D against surface W.
- Strategy, S_p: Move coarsely into position over the destination surface W (Figure 5.2). Move towards W complying to torques on x and y (in object coordinates) until some threshold force f_z is reached.
- Invariant, N_p: There should be no intervening objects between D and W.

We will introduce more and more detail into this example as our exposition proceeds.

5.4 Architecture

5.4.1 Command Input and Task Decomposition

The input to the controller is an object-level assembly command, e.g. insert A into B, place C to the left of D, etc . The command must directly reference a task plan that carrys out the operation. The task-plan contains information about ***goal states***, ***sensory pre- & post-conditions***, ***control strategies*** and a ***task invariant condition*** [Vijaykumar ***et al*** 1987]. The hierarchical process of expanding a command into its components is called ***task decomposition.*** The details of task decomposition are not covered in this chapter, for this and more complex planning techniques see [Vijaykumar and Arbib 1987; Vijaykumar *et al* 1987; Lyons 1988].

The control strategy describes one *particular* way to achieve the goal. For example, for a peg-in-hole-insertion assembly, at the object-level, a control strategy would be to be to comply to reaction forces from the environment along all but the insertion axis. Note that a poor choice of control strategy may result in its failure to achieve its goal; in the compliant place example, if there are objects between the end-effector and the target surface, then the strategy will fail.

5.4.2 Allocation

Allocation will be implemented hierarchically according to scheme B. At each level, a robot that exactly meets the task criterion is called the *ideal* robot. In general, such a robot will not physically exist, and a *real* robot plus an interface procedure will be used to implement the ideal robot. Note that decomposition of commands and allocation of resources can occur in parallel; in fact, to allow reallocation of resources during the execution of an assembly operation, they *must* occur in parallel.

5.4.3 Hierarchical Levels

Figure 5.3 describes the three levels of the controller architecture. The topmost level (denoted *ideal task level*) is the third level in our architecture. The input command is an object-level assembly instruction. The control strategy, goal condition and the sensory conditions are *operated upon* to derive a set of *actuation signals* that will nominally cause task execution to terminate only when the object reaches its goal state. Output from this level (denoted object commands) represents commands to the ideal robot.

At level 2 (denoted *ideal arm level*), we begin the allocation process. In keeping with the philosophy of allocation scheme B, we allocate the reach phase first. If this cannot be scheduled, then the task cannot be executed. If all robot systems having the required functionalities are not available (scheduled on other tasks), then this task must wait in a queue until one becomes available. Since dextrous hands have been chosen for end-effectors, the likelihood of an end-effector on the chosen arm performing the acquisition phase is high and so, it may be more important to perform the allocation for transportation immediately after reach. If this allocation fails, hand-off will be necessary. If successful, then one or more arms become potential candidates for the task.

Allocation for the parts-mating phase follows at level level 1 (denoted the *ideal hand level*). After complete allocation, control strategies, goal conditions and sensory conditions are used to generate task plans to directly command the arm or hand. The interfaces between the boxes denoted *real arm* at level 2 and *real hand* at level 1 and the hardware, as well as the

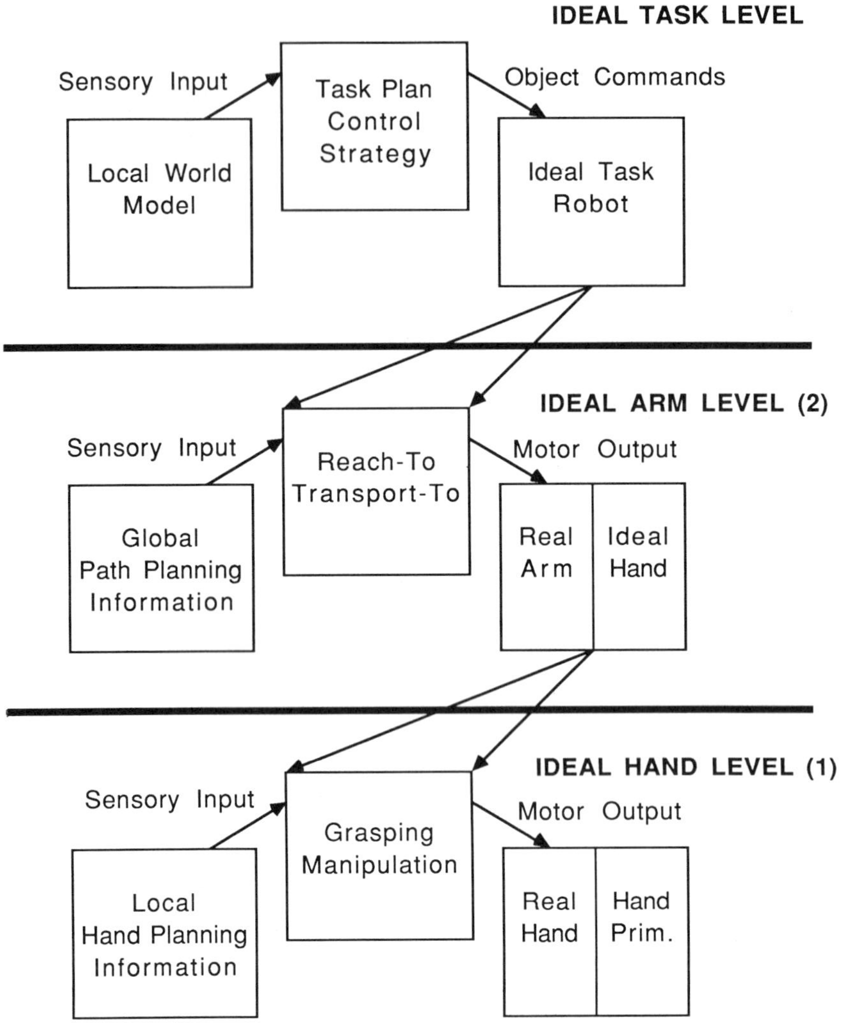

Figure 5.3: The Task-Oriented Architecture

details of the control models that can be allocated along with physical manipulators, are discussed in detailed in the Section 5.6. The following is a summary of what *control actions* imply at each level of the architecture:

- **Ideal Task Level:**

  The object-level assembly command is used to select a task-plan at this level. The task plan is decomposed into its components. The components, such as the control strategy, output commands to the next level. These commands assume that the next level can provide an ideal robot that can change the state of the object directly.

- **Ideal Arm Level:**

  At this level, the ideal task robot is decomposed into an ideal arm and an ideal hand, and the task is decomposed into a reach/transportation for the ideal arm and an acquisition/parts-mating for the hand. Upon successful allocation of the ideal arm tasks to a real arm and/or control model, an interface algorithm is chosen so that ideal arm commands can be transformed into real arm commands. This algorithm may need to add extra constraints to the incoming commands in order to make the translation possible. Ideal hand commands are passed to the next level. Global path planning information (e.g. from cameras or a built-in world model) is used to decompose and execute arm commands at this level.

- **Ideal Hand Level:**

  At this level, an real hand and/or control model (e.g. a grasp) is chosen to match the ideal hand. Upon successful allocation, an interface algorithm is chosen so that ideal hand commands can be transformed to real hand commands. A model of the grasped object (e.g., a geometric model which may be updated by sensory information) is used to decompose and execute hand commands at this level.

  Real hand commands describe changes in the states of the hand, and specify constraints induced by measures such as grasp stability or manipulability [Li and Sastry, this volume; Yoshikawa and Nagai, this volume], and the inequality constraints on contact forces. The actuation signals command the hand directly and have the ability to control constraint parameters.

5.4.4 REALLOCATION

Reallocation is performed subject to certain *consistency* constraints. For example, when hand-offs occur in scheme B, consistency demands either that the object be placed down for regrasping by the more appropriate arm-hand system, or the exchange happen through some dual-arm-hand coordination scheme. Allocation (and reallocation) and command decomposition occur in parallel in this architecture. The boxes labelled *global pathplanning information* at level 2 and *component models* at level 3 provide the necessary sensory information for command decomposition and execution (Figure 5.3).

5.5 Task Criteria and Robot Allocation

We now explain the notion of task criteria and describe how it helps in allocating a robot system for the execution of task plans.

5.5.1 Task Space

We define *task space* $\mathcal{C}_T$ to be the space in which a task plan may be completely defined. Let $\mathcal{C}_G$, and $\mathcal{C}_S$ represent the spaces in which goal conditions, and sensing conditions and task invariants are defined. Let $\mathcal{C}_C$ be the space in which the control strategy and actuation signals are defined. Then, for a task plan representation as discussed in [Vijaykumar *et al* 1987], we have:

$$\mathcal{C}_T = \mathcal{C}_G \cup \mathcal{C}_C \cup \mathcal{C}_S \quad (5.1)$$

Note that in contrast with [Vijaykumar *et al* 1987], actuation signals are explicit in the task plan in this chapter. We place no constraints on the structure of the task space, but for many assembly operations it will typically be changes in the kinematic and force states expressed in terms of object-centered coordinates.

For example, for the compliant placement operation, an element of the task space $(R \cup \#)^6$ is $(p_x, p_y, p_z, \theta_x, \theta_y, f_z, \tau_x, \tau_y)$ where the τs are torques in object coordinates (see Figure 5.2), f_z is the z component of force and the ps (θs) are position (orientation) components, again in object coordinates. The # element will denote we don't care what happens on that degree of freedom; allowing us to specify commands in a hybrid position force paradigm.

5.5.2 Task Criterion

A *task criterion* is a specification of what robot resources a task plan needs in order to execute successfully. Thus, the task criterion is strongly connected to the description of the task space. For example, Nakamura, Hanafusa and Yoshikawa [1987] deal with the issue of simultaneously executing several prioritized task plans on a single redundant robot. Only the configuration subspace, R^{m_i} that needed to be controlled for some task plan i, was considered as a definition of the task criterion. Li, Hsu and Sastry [1987], on the other hand, describe a more complex task criterion for grasping. It consists of two (grasped) object-centered task spaces, one with *wrench* and one with *twist* coordinates. The task criterion in each space is an ellipsoid inequality called the *task ellipsoid*. The hand-object configuration which maximizes the task-ellipsoid is the most appropriate one with which to execute the task plan. This method extends the task criterion to include *ranges* on the axes. A limitation is that this method is strongly linked to the twist and wrench space descriptions, and so, generality (i.e., arbitrary nature) of task space is lost.

One approach to extend the power of the task criterion is to continue to 'tack' components on as necessary. For example, defining the task criterion as including a task ellipsoid constraint, a payload constraint, etc. However, the danger is that the task criterion would soon become unwieldy. Lyons [1985] represents the task criterion for a dextrous hand using a *grasp-based feature space*. The feature space axes represent the amount of precision movement necessary and the amount of grasp security necessary to complete the task, and a set of coarse object properties. The space is segmented into three regions, each corresponding to a *grasp*, a stereotype hand-object configuration that includes an appropriate simplified hand model and object *acquisition* and *manipulation* information. The advantage of using a *set of models* of the robot (rather than a single 'hardware' model) is that it simplifies translations from a task criterion to a dextrous hand control model. Thus, the task criterion can remain simpler and more elegant. A disadvantage is that it forces task criteria to be expressed in terms of the stereotypical grasps (in the dextrous hand case) rather than for the task. This disadvantage can be overcome by choosing the set of models so that they are task-oriented; that is, that the models are chosen with some task domain in mind.

In summary, from this short survey of literature, it is clear that a task criterion should consist of at least a task space definition, but is made more versatile by considering aspects of the quality of control in the task space. The inherent complexity of the task criterion can be reduced by linking it with appropriate task-oriented models of the robot system. This is the reason that we extend the allocation and interface information in our architecture to include control models as well as physical manipulators.

5.5.3 Construction of Task Criterion

We now discuss the pieces of information that should be in the task criterion. Let the task space for task plan i be written as R^{m_i} and let its task criterion be χ_i. Let $R = A \cup H$ be the set of all available robot systems.

We postulate a function f_i which for any task i maps from the task space to a single 'common' space C. This function will be a necessary part of any task specification. We assume that a sufficiently complex C is chosen to support an interesting set of tasks[4]. For each robot $r \in R$, we postulate a mapping function f_r that maps from C to the configuration space of that robot. Clearly, many f_r will not be surjective. The essential criterion for determining if a robot r supports a task i is to see if the range of f_i is a subset of the domain of f_r: $ran(f_i) \subset dom(f_r)$. Given this, the interface from task commands to robot commands can then be implemented by composing the functions, $f_r \circ f_i$. Not all robots will support all tasks. For example, if the task space contains a force axis, then $f_r \circ f_i$ is undefined for a robot whose dynamic behavior is described using a standard Lagrangian [Venkataraman 1988]. Another example is when the task space may contain the three Cartesian position axes, $f_r \circ f_i$ is not defined on a 2-D robot. Clearly then, f_i should be part of χ_i, and f_r should be part of the library functions LH, LA that were discussed earlier.

Given that the task space can be mapped onto the configuration space of some robot, the next question is one of the necessary quality of control in the task space. We propose the following additional components for χ_i:

1. *Range.* The necessary controllable range on each axis of the task space is given by:
$$\rho : Z \longrightarrow R \times R \tag{5.2}$$
where $\rho(n) = [p, m] \quad n \in Z$ where $Z = 1, ..., m_i$ and $p, m \in R$. In words, this means that the map ρ takes an integer for each task axis and gives the interval on that axis that must be controllable. This allows us to set unilateral conditions (e.g. for pushing an object rather than grasping it and moving it) through either $p = 0$ or $m = 0$.

2. *Accuracy.* The accuracy required on each task space axis is given by:
$$\alpha : Z \longrightarrow R \tag{5.3}$$
A command of $r \in R^{m_i}$ to the robot, will result in the robot carrying out $r = (r_1 \pm \alpha(1), r_2 \pm \alpha(2), \ldots, r_m \pm \alpha_i(m))$.

It is interesting to note that these components can provide information to decide if it is necessary to grasp the object (for some unilateral ρ requests, acquisition may not be necessary). Note that the outputs of a grasp planner like the one described in [Nguyen 1988] are usually desired wrenches and twists. In the task space, accuracy measures may be combined with the task ellipsoid concept through specification of error ellipsoids in the wrench and twist spaces rather than ellipsoids of the actual wrench and twist values.

The task criterion also needs to have information on what ranges and payloads will be necessary for the operation. The following three quantities capture this:

---

[4] One example C might contain Cartesian positions, forces and torques

1. The position of the component to be acquired, specified as a homogeneous transformation matrix, T_c, with respect to some base coordinate frame B.

2. The location of the subassembly or parts-mating site, T_s, as a homogeneous transformation matrix relative to B.

3. The loads that the robot will experience while acquiring the part, W.

These pieces of information will be used to determine the minimum workspace envelope necessary for an arm to carry out the reach, acquisition, transportation and parts-mating phases. Note that allocation of a robot implies that information contained in the task criterion successfully matches the characteristics of the chosen robot. The library functions LA, LH discussed earlier should contain enough information to determine if a task criterion can be met. A useful solution is to have LA, LH store exactly the same types of information as χ but for a specific robot rather than for a task. So, for example, if χ_i is the task criterion for task i and $LA(r)$ is the library function for robot r, then

$$\begin{aligned} \chi_i &= (f_i, \rho_i, \alpha_i, T_{ci}, T_{si}, W_i) \\ LA(r) &= (f_r, \rho_r, \alpha_r, W_i) \end{aligned} \tag{5.4}$$

Based on the task description, we can outline the task criterion for the compliant place operation, χ_p, as follows:

- α_p: The strategy hinges on good force and torque control, but the position accuracy can be coarse.

- ρ_p: All our position ranges will be bidirectional, and of the order of the component size. The torque and force ranges should be small, and the z force range could be unidirectional.

- T_c, W and T_s are given by the particular situation.

5.5.4 Task Criteria in the Architecture

At the top level (ideal task), an input assembly command is expanded into its components (as described in Section 5.3), one of which is a task criterion in the format discussed in the previous section.

$$\chi = (f, \rho, \alpha, T_c, T_s, W) \tag{5.5}$$

where, ρ specifies the range required in the task space from the ideal robot, α specifies the accuracy required and, T_c and T_s specify the transformation matrices for the object centered coordinate system and the sub-assembly centered coordinate system respectively. W describes the payload imposed by the task.

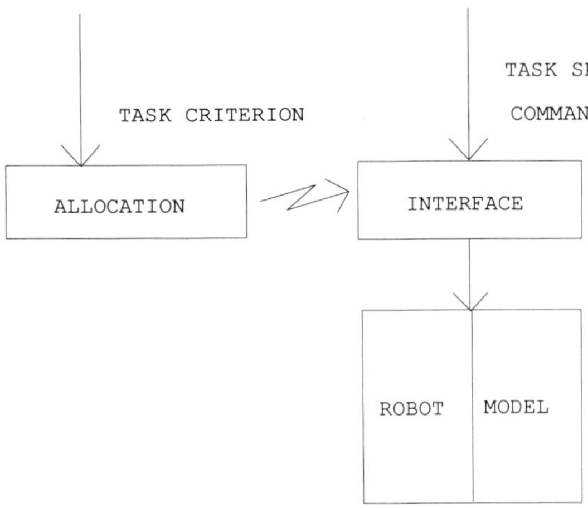

Figure 5.4: Allocation and Interfacing of Robot Model

At the next level (ideal arm), this criterion is broken up as follows: The workspace envelope for an appropriate robot arm is constructed as a function of the range ($RANGE(\cdot,\cdot)$) from the current position of the robot to the component, and from there to the sub-assembly. The purpose of this level is to carry out the acquisition and transportation phases of the assembly operation; to 'set the scene' for the parts-mating strategy.

$$\rho_a = Func(\ RANGE(T_r, T_c), RANGE(T_c, T_s)) \qquad (5.6)$$

where T_r is the current position of the robot r. The first range is crucial; you could ignore the second if you're willing to choose a new arm once the component has been acquired. These concerns can be represented in an implementation of $Func$.

At the next level (ideal hand), the choice of a particular hand model is subject to $\rho_h = \rho$ and $\alpha_h = \alpha$ being satisfied. Since we will consider a specific grasp on a dextrous hand as an example of a hand model, this allocation step ***subsumes*** grasp selection. Once a hand model has been chosen, commands issued to the ideal hand level are translated into the hand specific commands (Section 5.4) for a particular physical (possibly dextrous) hand; this mapping is $f_h \circ f$.

The basic building block in implementing this architecture is shown in Figure 5.4. At each level, the motor actions are issued to what might be called a ***virtual robot*** (by analogy with [Arbib, Iberall and Lyons 1985]); that is, a combination of a physical robot plus some interface and/or control model.

For the three levels of abstraction in our task-oriented architecture, we have:

Level 3: Ideal Task Level

- Allocation: $\chi = (\rho, \alpha, T_c, T_s, W)$
- Virtual Robot: Ideal Task Robot

Level 2: Ideal Arm Level

- Allocation: $\chi_a = (f_a, \rho_a, \alpha_a, T_{c_a}, T_{s_a}, W_a)$
- Interface: $f_r \circ f_a$ for some arm $r \in A$. (Reach-To, Transport-To in Figure 5.3).
- Virtual Robot: Real Arm, Ideal Hand

Level 1: Ideal Hand Level

- Allocation: $\chi_h = (f_h, \rho_h, \alpha_h, T_{c_h}, T_{s_h}, W_h)$
- Interface: $f_r \circ f_h$ for some hand $r \in H$. (Grasping, Manipulation in Figure 5.3).
- Virtual Robot: Real Hand

5.5.5 Allocation Constraints

We apply the following two constraints to the allocation process:

1. In the allocation of resources, if more than one robot meets the task requirements, then we should choose one that *minimally* meets the task requirements.

2. We should allow *dynamic reallocation* of resources to account for the possibility of the present allocation becoming *non-optimal*. Such reallocations must be subject to *consistency* constraints of the form:

   - Ideal Robot: Object must be placed down in workspace of new robot, which must now meet the full χ specification again.
   - Ideal Hand: Object may need to be placed down, but only χ_h needs to be satisfied again.

Note that the interface and allocation modules (Figure 5.4) are contained within the *Reach-to, Transport-to* box at level 2, and the *Grasping, Manipulation* box at level 1 (Figure 5.3).

5.6 Task Execution

Commands generated in the boxes (in Figure 5.3) denoted *Ideal Hand* or *Real Arm* represent low-level atomic tasks that must be executed on an allocated set of arm/hand hardware. Tasks, in general, describe *how* a specified goal must be reached. Atomic tasks do the same, except at a finer resolution of information. For example, a free motion task allocated on a 6 *dof* arm may be described using a sequence of Cartesian motions of its tool frame. Note that, henceforward, atomic tasks will be referred to as ***task primitives***.

Task primitives generated by the box denoted *Ideal Hand* in Figure 5.3 represent desired atomic changes in the state of the component (to be assembled). These task primitives must be decomposed successively into more and more primitive operations until the actuators that drive the hand can be commanded directly. The execution architecture in Figure 5.5 describes this decomposition process. In the sections that follow, we will discuss this architecture in detail, and describe a vocabulary of task primitives to represent actions performed by a dextrous hand at several levels of the architecture. We will also present appropriate mathematical representations for these task primitives.

The actual dynamic modeling and control methodologies for dextrous hand usage has been omitted from this work. Pertinent details are available in Chapters 4, 5 and 6 of [Venkataraman 1988]. Hand control architectures, as well controller design methodologies are described in Chapter 6 of [Venkataraman 1988]. Chapter 6 of [Venkataraman 1988] also analyzes task control of Stanford/JPL Hand [Mason and Salisbury 1985], using linearized hand dynamic models, approximated interaction force constraint models, and pole placement methods using multivariable feedback control techniques [Callier and Desoer 1982].

Subsequent sections are organized as follows: We begin with a description of the overall execution architecture in the next section, and develop a vocabulary of task primitives for the execution of ideal hand commands on an allocated dextrous hand and associated hardware. In the following sections, we describe the actual task decomposition process across levels of the hierarchy represented in Figure 5.5, provide with a mathematical representation for the task primitives, and describe their implications on task-based hand control.

5.7 Description of Execution Architecture

Figure 5.5 shows the task execution architecture for a dextrous hand. At level 2 in Figure 5.3, ***reach-to*** and ***transport-to*** phases are allocated on a robot arm. A corresponding set of ***ideal hand commands*** are generated at this level for ***acquisition*** and ***parts-mating*** operations. Such ideal hand

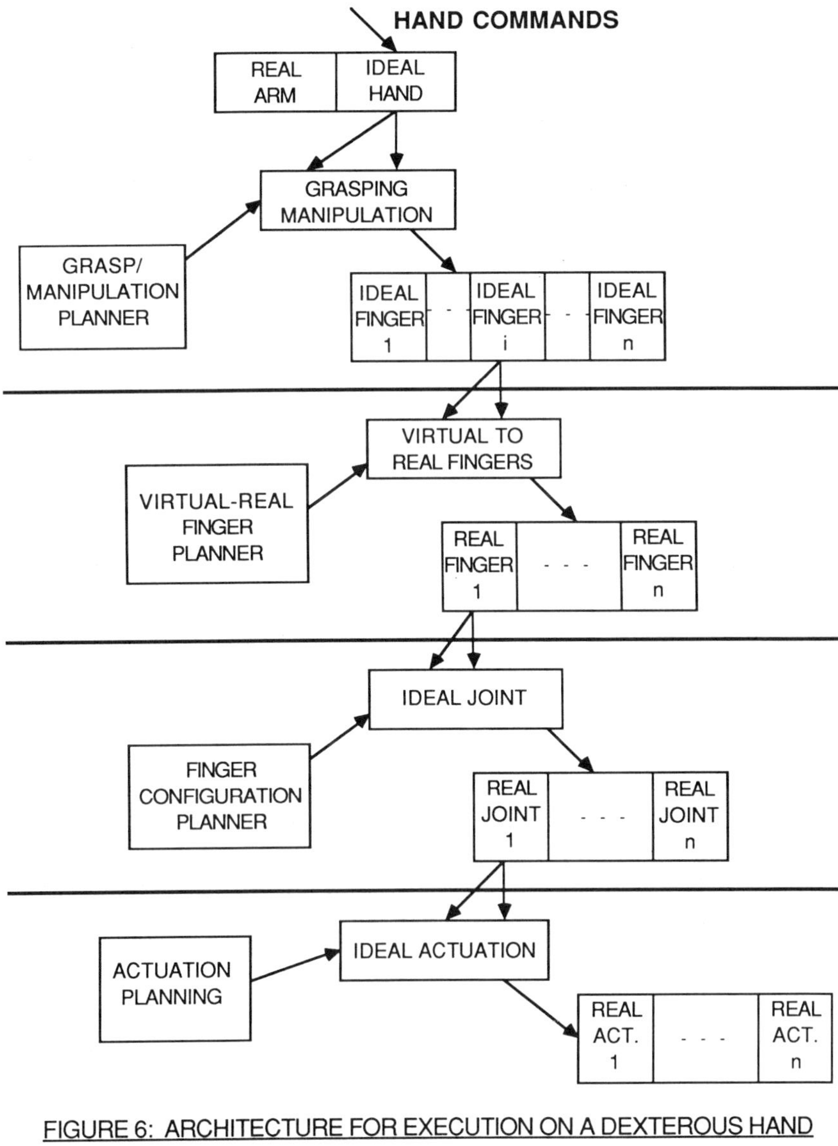

Figure 5.5: Task Execution Architecture

commands are considered input into the architecture. We will refer to these commands as ***object task primitives.***

Object task primitives are used by ***Grasp*** or ***Manipulation*** planning algorithms to generate ***hand task primitives,*** which command a real hand with *ideal fingers*. The number of ideal fingers that make up a real hand is usually the minimum number required for the task at hand. The Stanford/JPL Hand is an example where the number of real and ideal fingers are the same for grasping tasks. Note that ideal finger is used here as the 'desired' finger description, and 'virtual finger' as the machine that implements whatever actions are required, while in [Arbib *et al* 1985], this is assumed to be 'built in' to the task description. At the next level, ***finger task primitives*** represent commands input into the ***real fingers*** of a dextrous hand. The motivation behind this scheme of considering ideal fingers prior to real fingers, is to subdivide the problem of allocating the grasping or manipulation task into two steps: (i) allocate the task onto a hand that consists of a minimum number of fingers of variable dimensions, and (ii) allocate the ideal fingered hand onto a real hand.

Thereafter, each finger primitive, is decomposed into *joint task primitives*, and further, into *actuator task primitives*. Actuator task primitives are used to generate the actuation signals that drive the hand.

Note that structure of the execution architecture bears a strong resemblance to the overall task architecture. Note also that the execution architecture for a single arm will accept commands at the same abstraction level as finger task primitives and execute them.

5.8 Task Primitive Vocabulary

In this section, we list a vocabulary of task primitives for each level in the architecture shown in Figure 5.5. Our task analysis is based on the assumption that a robot remains in contact with its *environment* at all times. Thus, for example, a robot arm moving in free space is in contact with an environment that exhibits *null* dynamic resistance, while, a robot arm whose endpoint has been firmly bolted down onto the floor is in contact with an environment that offers *infinite* dynamic resistance. Pushing a block on a friction table, moving in a viscous medium or against a spring, and working the piston of a pneumatic cylinder are all examples where the environment offers *finite* dynamic resistance. This perpetual mechanical coupling between a robot and its environment will be used as a fundamental invariant in our task analysis.

5.8.1 Object Task Primitives

In general, the actions performed by a component during the acquisition or parts mating phases can be classified into six modes depending on the component's interactions with *its* environment. These modes are free motion, guarded motion, fine motion, free force application, guarded force application and fine force application [Venkataraman 1988]. The meanings of these action modes are elucidated below:

- **Free Motion** occurs when a body moves about in an unconstrained fashion and there is neither any obstacle in its path, nor surfaces against which it establishes contact. Robotic free motions will not pass through structural, kinematic or algorithmic singularities.

- **Guarded Motion** occurs when a body has to *guard* itself from establishing *contact* while moving about. For example, the body avoids obstacles by guarding itself from establishing mechanical contact with any obstacle present in its motion path. Note that throughout this chapter, the word *contact* could have a meaning more general than its immediate mechanical connotation. Kinematic and algorithmic singularities, and joint limits may be modeled as contact locations.

- **Fine Motion** occurs when a body has the ability to move about while remaining in stable contact with its environment. Sliding motion of an arm while tracing a friction surface is an example of fine motion under mechanical contact.

- **Free Force Application** occurs when a body applies forces upon its environment in an unconstrained manner. A robot arm bolted to the ground is an example of this free force application. Note that most practical situations involving forces may never fall under the category of free force application, and its introduction at this level is for completeness of the task primitive vocabulary.

- **Guarded Force Application** occurs when a body has to *guard* itself from losing contact while applying forces upon its environment. A robot arm maintaining contact with a moving object is an example of guarded force application.

- **Fine Force Application** occurs when an arm applies forces upon the environment that it is in contact with. A robot arm pushing a block on a friction surface is an example of fine force application.

At the real hand level in Figure 5.5, vocabulary of hand task primitives necessary for the execution of the six object task primitives is developed now.

5.8.2 Hand Task Primitives

Each hand task primitive describes the motions and/or forces imparted upon its immediate environment. During flexible assembly, this implies consideration of interactions between hand and the component that needs to be mated with a sub-assembly. We describe a vocabulary of task primitives using which interactions that occur between the hand and the component during acquisition and parts-mating phases may be described.

- **Preshape**: During the acquisition phase, the fingers of a hand establish contact with the component to be acquired, and stably overcome the latter's gravitational effects. Prior to the establishment of mechanical contact, the hand must *preshape* to a configuration that will enable its fingers to come in contact with the component at prespecified points on the object. The importance of preshaping, as well as the existence of a strong relationship between preshape, and the geometric and physical characteristics of the component, and subsequent tasks, has been discussed in literature. So, preshape is chosen as our first hand task primitive.

- **Grasping** and **Acquisition**: Once the hand is preshaped, its fingers are required to close in towards the object until stable mechanical contacts have been established with the object. We desire that all fingers establish simultaneous contact, since this will result in maximum interaction stability. Thereafter, in a *coordinated* fashion, the fingers will be required to overcome the component's weight, so that states of the component may be modified thereafter. Aforementioned hand actions are sometimes coupled together and referred to as *grasping* [Lyons 1985; Li *et al* 1987; Yoshikawa and Nagai, this volume]. In [Mason and Salisbury 1985; Nguyen 1988], only the actual imparting of forces upon the object, once mechanical contacts have been established, is referred to as grasping.

  In this work, grasping implies closing the fingers in towards the component until mechanical contacts have been established, while the term acquisition is used for actually imparting the finger contact forces required to overcome the component's weight. The main reason for this distinction is the difference in their control requirements (discussed in [Venkataraman 1988]).

- **Manipulation**: Once the component has been acquired, the hand is required to effect changes in its state, subject to interactions between the component and *its* environment. Interactions between the component and its environment may be described using free, guarded and fine motion task primitives and free, guarded and fine force application task primitives. We refer to this whole suite of actions performed by a hand as *manipulation*. More specifically, we denote actions of the hand that result in free, guarded, or fine motions of the component as *motion manipulation*, and those that result in free, guarded and fine force applications of the component as *force application manipulation*. Note that in both cases, the fingers have to be *coordinated* and satisfy object level performance requirements.

During the execution of the grasping, acquisition and manipulation task primitives, the actions performed by the fingers of a dextrous hand are not decoupled from each other. This was indicated in the discussions above with words such as *coordinated*. We will expand on this in the next section.

5.8.3 Finger, Joint and Actuator Task Primitives

We do not discuss the decomposition of virtual to real fingers task primitives in this work. However, what we develop here for real fingers will be directly applicable to virtual fingers also.

Each real finger, when completely decoupled from the others, is a manipulator. For *precision grasps* [Lyons 1985], it has only one point of contact with the object, and interactions between each finger and its environment may be adequately described using the six task primitives described at the object-level. For *lateral* and *power* applications, the same vocabulary can be used, except that information contained in each of the task primitives must include inter-phalanx constraints as well. This implies that the whole spectrum of interactions between each finger and *its* environment may be captured using the finger task primitives, free, guarded, and fine finger motions and free, guarded, and fine finger force applications.

It is the responsibility of every joint, and in turn, every actuator, to actually execute the finger task primitives. In general, a vocabulary of free, guarded and fine joint motions and free, guarded and fine joint force applications would provide a sufficiently rich vocabulary at the joint-level. Likewise, for the actuator, free, guarded and fine actuator motions and free, guarded and fine actuator force applications would do splendidly.

5.9 Task Decomposition

We now describe the decomposition of object task primitives into joint-level task primitives. In [Vijaykumar *et al* 1987], we described a production rule to perform this decomposition, and concentrated on the language aspects. We also presented the notion of a *task invariant* that would turn true when sensory conditions indicates an error in task execution. In this chapter, we discuss details of the actual decomposition process.

5.9.1 Reach Phase

The reach phase of assembly is executed in two parts. During the first, denoted *gross reaching* here, the robot arm is commanded to execute a free motion (or guarded motion if the assembly environment is ridden with obstacles), and the dextrous hand mounted on it is commanded to execute a preshape task primitive *concurrently*. Upon completion of the gross reaching phase, local sensing systems (proximity sensors and local cameras) may be used to *tune* the preshape configuration, or arm position or pose. This is referred to in this work as *fine reaching*, and involves *concurrently* moving the arm guardedly to bring the component within the workspace of the hand, and fingers of the hand to correct, if necessary, the preshape posture.

5.9.2 Acquisition Phase

The acquisition phase involves active usage of only the dextrous hand. The robot arm acts as a compliant *seat* on which the dextrous hand is placed. Acquisition phase involves the execution of a grasp, followed by an acquisition task primitive with the dextrous hand.

5.9.3 Transportation Phase

During the transportation phase, just as in the reaching phase, the arm first transports the component to the proximity of the parts-mating site, during *gross transportation*, and then guardedly aligns the component for the parts-mating operation, during *fine transportation*. This involves a free arm motion task primitive, followed by a guarded arm motion task primitive. During this the hand continually executes an acquisition task primitive to retain the component stably within itself.

5.9.4 Parts-Mating Phase

Part-mating involves, first, a guarded move by the arm to bring the component in contact with the sub-assembly. We denote this with the term *pre-mating*. Several sequences, each involving the concurrent execution of a manipulation task primitive by the hand, and a guarded force application by the arm follow. Nominally, the number of such sequences is derived from the plan that describes the parts-mating operation.

At any point during execution, if sensory information flags a task execution error, replanning may require the execution of an alternate task primitive. For example, during the first part of the parts-mating phase, if the component makes unexpected contact with its environment due to geometric modeling errors, the acquisition hand task primitive would be modified to fine force application during replanning. We are currently addressing this issue of feedback in a formal manner.

5.9.5 Concurrency Constraints during Decomposition

In the previous sections, as we developed the necessary task primitive vocabularies, we employed terminologies such as *coordinated* finger actions to describe concurrency constraints that must be satisfied during task primitive execution. In addition, we also imposed similar constraints during the decomposition of arm and hand tasks for the four phases of assembly. We now describe two types of concurrency constraints that will be accommodated in the task architecture.

Concurrency constraints that arise during assembly can be categorized into two types. The first involves state relationships, and the second, time relationships. We denote the former using the terminology *Coordination*, and the latter *Synchronocity*, and define the two as follows.

- **Coordination**: Actions performed by two bodies are coordinated when a prescribed relationship in the states of the two bodies is maintained throughout. Consider for example, an acquisition hand task primitive. Fingers of the dextrous hand are required to apply forces upon the component (being acquired) in a coordinated fashion to overcome object level gravitational forces.

- **Synchronocity**: Actions performed by two bodies are synchronous if they begin and terminate executing the action at the same instant. For example, during grasping, when fingers establish contact with the component, maximum stability results only if all fingers establish contact at the same instant. This can be achieved by considering grasping to be made of synchronous finger guarded motions.

5.9.6 Task Decomposition Information

We will now describe the decomposition of ideal hand commands into the respective joint-level actuation commands. Similarities may be drawn for the decomposition of arm commands.

- Preshape consists of free finger motions, which in turn may be executed using free joint force application task primitive.

- Grasping consists of synchronous guarded finger motions, which may be executed using free joint force applications task primitive.

- Acquisition consists of coordinated guarded finger force applications, which may be executed using free joint force applications task primitive.

- Motion manipulation implies that free, guarded or fine object motions are required. This is realized using fine finger force applications, which in turn implies that free joint force applications are required.

- Finally, force application manipulation implies that free, guarded or fine object force applications are required. This may be realized using fine finger force applications, which in turn implies that free joint force applications are required.

5.10 Task Primitive Representation

A task primitive must capture the essence of interactions that occur between a robot and its environment, and provide information pertinent to the control of these interactions. In addition, it must allow the specification of states of the assembly in question throughout the period during which the task is executed. Consider, for example, free motions of a robot arm. Due to the *null* resistance offered by the environment, only robot motions are necessary to completely describe interactions that occur in-between. In addition, a specification of the desired manipulator trajectories during task execution completes the description of the free motion task primitives. When mechanical contacts are present, the *frame* in which task information is provided usually coincides with the location of contact itself. The choice of coordinates (position *vs* force along a direction, etc.) reflect the interactions between robot and its environment, and the desired motion and force trajectories (or constraints) specify state changes (or constraints to be satisfied) during task executions.

Task representations are developed in a two-stage process. The first involves encoding the information necessary for a complete representation of the corresponding task primitive, and the second involves the description of a *task variable* which can be used for *task-dependent* control strategies [Venkataraman 88].

Note that in this work, as well as [Venkataraman 88], the two concurrency constraints will be implemented at the control level. Insofar as task representations as concerned, this implies that hand task primitive representations will simply be a collection of those of individual fingers, and hand task variables will simply be a collection of finger task variables. And, object task variables will be identical to finger task variable, but contain information at a higher level of abstraction. We now briefly describe what a task varible means, and develop mathematical representations for task variables for the object, hand and finger task primitives.

5.10.1 Task Variable

Task variable contains information necessary for the execution of corresponding task primitive. For example, for free motions, an arm or a finger would be required to follow pre-specified motion trajectories. Here, the task variable would simply be motions, because controlling motions (in a task frame) would directly imply control of the task itself.

Many approaches have been proposed for implementing guarded motion. The approach that lends itself most naturally to integration into control is the use of generalized *compliance* (related to the work by Mason in [Mason and Salisbury 1985]), or generalized *admittance* [Hogan 1985]. Guarded force application would use generalized *stiffness* (related to the work by Salisbury in [Mason and Salisbury 1985]) or generalized *impedance*

[Hogan 1985] instead. For these schemes, the task variable would be constructed as a *sum* of actual motions (forces) and a force-error-corrective-motion (motion-error-corrective-force-application) due to compliance (stiffness), and task-dependent control could be achieved by controlling the summed *compliant* motions.

Most of the work in fine motion and fine force application has pertained itself to planning. Very little emphasis has been placed thus far on the execution of these task primitives. During fine motion, the satisfaction of force constraint equations will need to be monitored continually during task executions, in addition to controlling the robot's motions. Fine force applications would have a *dual* requirement where motion constraints will be continually monitored during executions, along with force control. Regardless of the the task primitive, both motion and force information must be stored in the task variable for control, and this what will be assumed in this chapter. Note that that for fine motion (fine force application) the task variable contains motions trajectories (force trajectories) along with force (motion) constraint equations.

5.10.2 Object Task Primitive Representation

Let the task variable at the ideal hand level be denoted by Y. Let further, the motions and forces allowable by the contact in question be represented by $\mathcal{M}$ and $\mathcal{F}$ respectively. For example, for free motions, $\mathcal{M}$ will contain the three translations and three rotations of a rigid body, and $\mathcal{F}$ will consist of zero elements. For free force applications, $\mathcal{F}$ will represent the three forces at, and three torques about the contact point, and $\mathcal{M}$ will consist of zero elements. Similar arguments may be made for other *bilateral* and *unilateral* contacts [Venkataraman 1988], with environments that offer null, finite, or infinite resistances.

Let M and F represent the motions and forces necessary for the execution of the task primitive. For example, a free motion task primitive that takes a robot from position P_1 to P_2 is given by $M = P_1 \longrightarrow P_2$, and zero entries for F. The following two conditions must be satisfied for consistency between the contact representation, and the task primitive representation:

$$\begin{aligned} M &\subseteq \mathcal{M} \\ F &\subseteq \mathcal{F} \end{aligned}$$

With these, task representations may be derived as given below:

$$\begin{aligned}
Y &= [M] && \text{for Free Motion} \\
Y &= [M + C\,(F_t,\, F)] && \text{for Guarded Motion} \\
Y &= \left[M^T,\, \eta^T\,(F,\, P)\right]^T && \text{for Fine Motion} \\
Y &= [F] && \text{for Free Force Application} \\
Y &= [K\,(M_t,\, M) + F] && \text{for Guarded Force Application} \\
Y &= \left[\eta^T\,(M,\, P)\,,\, F^T\right]^T && \text{for Fine Force Application}
\end{aligned}$$

$C(.)$ and $K(.)$ are compliance and stiffness operators that operate upon the actual and threshold force and motion values respectively. The corresponding threshold forces and motions are given by F_t and M_t respectively. For example, during compliant motions, the operator C will *compare* the actual forces F with F_t, and produce a motion that will cause the actual forces to converge towards F_t. Note that in [Mason and Salisbury 1985], the compliance and stiffness operators are linear and operate upon the difference between threshold and actual values. If we apply this model, then the corresponding task variable equations become

$$\begin{array}{rcll} Y & = & [M + C\,\Delta F] & \text{for Guarded Motion} \\ Y & = & [K\,\Delta M + F] & \text{for Guarded Force Application} \end{array}$$

where, ΔM and δF are $(M_t - M)$ and $(F_t - F)$ respectively.

$\eta(.)$ describes inequality constraints for fine motion/fine force application. This function is used to evaluate if inequality constraints are satisfied, and output a signal that is proportional to the *extent* to which there is an error in satisfying the constraint. Note that in this case, the function itself (instead of a trajectory) is the planner specification. P in the parameter list of η describes a list of constants that may be used in the function. η is an important *mapping function.*

5.10.3 Hand Task Primitive Representation

Finger task primitives will be identical to the ones described above for the task primitives at the ideal hand level. For dextrous hand usage, let the motions, forces, and the task variable of the ith finger be represented by M_i, F_i and Y_i respectively. With this:

$$\begin{array}{rcll} Y_i & = & [M_i]^T & \text{for Free Finger Motion} \\ Y_i & = & [M_i + C_i\,(F_{t_i}, F_i)] & \text{for Guarded Finger Motion} \\ Y_i & = & [M_i,\, \eta_i\,(F_i)] & \text{for Fine Finger Motion} \\ Y_i & = & [F_i] & \text{for Free Finger Force Application} \\ Y_i & = & [K_i\,(M_{t_i}, M_i),\, F_i] & \text{for Guarded Finger Force Application} \\ Y_i & = & [\eta_i\,(M_i),\, F_i] & \text{for Fine Finger Force Application} \end{array}$$

Accumulating over all the fingers, we have:

$$\begin{array}{rcll} Y & = & [M] & \text{for Preshape} \\ Y & = & [M + C\,(F_t, F)] & \text{for Grasp} \\ Y & = & [K\,(M_t, M)] & \text{for Acquisition} \\ Y & = & [\eta\,(M),\, F] & \text{for Manipulation} \end{array}$$

where,

$$\begin{array}{rcl} Y & = & Col([Y_1, Y_2, \ldots\ldots, Y_i, \ldots\ldots, Y_{n-1}, Y_n]) \\ M & = & Col([M_1, M_2, \ldots\ldots, M_i, \ldots\ldots, M_{n-1}, M_n]) \\ F & = & Col([F_1, F_2, \ldots\ldots, F_i, \ldots\ldots, F_{n-1}, F_n]) \end{array}$$

$$\eta = Col([\eta_1, \eta_2, \ldots\ldots, \eta_i, \ldots\ldots, \eta_{n-1}, \eta_n])$$

$$C = \begin{bmatrix} C_1 & & & & & & & \\ & C_2 & & & & & & \\ & & \cdot & & & & & \\ & & & \cdot & & & & \\ & & & & C_i & & & \\ & & & & & \cdot & & \\ & & & & & & \cdot & \\ & & & & & & C_{n-1} & \\ & & & & & & & C_n \end{bmatrix}$$

$$K = \begin{bmatrix} K_1 & & & & & & & \\ & K_2 & & & & & & \\ & & \cdot & & & & & \\ & & & \cdot & & & & \\ & & & & K_i & & & \\ & & & & & \cdot & & \\ & & & & & & \cdot & \\ & & & & & & K_{n-1} & \\ & & & & & & & K_n \end{bmatrix}$$

Note that we have not assumed at this point that the off-diagonal terms of the C or K matrices are zero. In actual implementation, these off-diagonal reflect inter-finger dependencies [Venkataraman 1988].

5.11 Conclusion

In this chapter, we have developed a controller architecture for a flexible assembly cell equipped with dextrous manipulation capabilities. The central idea was to use constraints from the description of the task to be executed to select an appropriate abstract model and physical robot hand-arm system with which to execute the task plan. With this idea as target, we described how allocation can be done on the basis of a task criterion for a task plan, and we introduced a set of control models for a dextrous hand that can be allocated and interfaced to real hand-arm systems.

Implementation has been proceeding in both of the main components of this chapter: the execution of task plans with task criterion, and the construction of useful hand models containing task-specific interaction dynamics. The task plan representation work has been carried out using the $\mathcal{RS}$ model [Lyons and Arbib 1989] as both a formal tool [Lyons 1986] and a multiprocessor implementation language [Lyons 1987]. The computational complexities of the architecture we have described are very high, especially the issue of remapping robot models. $\mathcal{RS}$ allows us to structure our plan representation as a network of concurrent cooperating computing agents (small grain processes). An $\mathcal{RS}$ formulation of this architecture is described in [Lyons 1988].

We have chosen the Stanford/JPL hand [Mason and Salisbury 1985] as representative of dextrous hands. All implementations done so far are strongly bottom-up. Simplified preshape and grasp primitives have already been executed. The simplification is that fingers were commanded at the joint level, and inter-finger motion dependencies were neglected. The main reason for these simplifications was the scarcity of computational resource. So far, all implementations have been performed on a single MicroVax CPU executing control programs written in C, and using the VAXELN real-time environment. Simple grasps have been executed without finger-tip force feedback to demonstrate the structural compliance of the hand. Grasp stability was experimentally demonstrated through robustness to force perturbations on the object. In terms of analysis, we have applied the theoretical framework developed in this chapter as well as Chapters 4 and 5 of [Venkataraman 1988] to the Stanford/JPL hand turning a bolt. Chapter 6 of [Venkataraman 1988] gives a detailed discussion of this, as well as all the implementations described above.

We see this chapter as a natural extension of our work on formal ways to represent the notion of *task context* (the use of information related to a task at hand). We argue that our view of some parts of the task planning problem as an allocation/interfacing of set of task-oriented models is a novel and useful way to use task information in dealing with complex robots, and dextrous hands in particular.

6

CONDOR: A Computational Architecture for Robots

Sundar Narasimhan[1]
David M. Siegel[1]
John M. Hollerbach[1]

ABSTRACT This chapter presents an overview of the *CONDOR* architecture, a fully implemented system that controls the Utah/MIT hand. The architecture is especially suited for complex robots that are characterized by a number of joints and consequently demand powerful computer architectures to be controlled and utilized effectively. Using the *CONDOR* as a representative example, this chapter attempts to introduce the hardware and software issues that are relevant to the design of real-time control systems for robot hands.

6.1 Introduction

Computer architectures for robotics and indeed for other real-time systems have not received the attention they should. Partly, this has been due to a perception that present-day computers are fast enough and present-day robots are simple enough that investing time and money into developing yet another system is not warranted. Real-time software is also equated with assembly level programming and low level device driver software. Consequently, issues involved in real-time programming have not been considered to be academically interesting by computer scientists.

The first perception is simply not true for the kinds of robots that this book addresses. Robot hands that usually have nine to twenty-five degrees of freedom, are complex mechanical devices. For example, the Utah/MIT hand has 16 joints, each with 2 actuators. There are 32 tendon tension sensors, and 16 joint position encoders. Typically, a servo rate on the order of 400 Hz is required, which necessitates reading 19,200 sensor values per second and sending out 12,800 actuator values per second. If the operating system introduces anything but the most minimal overhead for servicing these rapid events, the desired update rates will not be met.

---

[1]MIT Artificial Intelligence Laboratory, 545 Technology Sq., Cambridge, MA 02139

Coupled with the need for performance is the need for flexible software development environments in robotics. Most robots are simply too tedious to program – robot hands are even more so. While a number of efforts are under way to automate the task of a robot programmer, researchers need a good solution in the interim, when no such *task-level* programming languages are generally available.

This chapter describes the *CONDOR* system, a programming environment developed for robotics research [Narasimhan 1988]. Our earlier attempt at providing a standardized architecture for robotics was named the Muse [Narasimhan *et al* 1986]. The *CONDOR* system was designed to achieve efficiency in terms of hardware performance and flexibility in terms of software development. While these two goals were sometimes conflicting, we feel that the computational architecture that has resulted satisfies the requirements posed by complex robots built in research environments. In what follows, we present the lessons we have learned from our attempts. Since the *CONDOR* has now been operational for over a year, we feel that our experiences will be helpful to other researchers who have similarly complex robots to control and program.

6.2 The Utah/MIT Hand Project - A brief update

The *CONDOR* and the Muse were both designed initially as architectures to control the Utah/MIT dextrous robotic hand. This hand is a four-fingered hand, developed by the Center for Engineering Design at the University of Utah and the Artificial Intelligence Laboratory at the Massachusetts Institute of Technology, to pursue the achievement of advanced robot dexterity [Jacobsen *et al* 1984]. There have been a series of papers published on different aspects of this project; on design and construction of the hand [Jacobsen *et al* 1984; Jacobsen *et al* 1986], on low-level control issues [Biggers, Jacobsen and Gerpheide 1986], on algorithms for force control [Hollerbach, Narasimhan and Wood 1986], on tactile sensor design and construction [Siegel, Garabieta, Hollerbach 1985; Kriegman *et al* 1985], and on the software architecture used to control the hand [Narasimhan *et al* 1986].

The hand has now been mounted on a four degree of freedom x-y-z table (see Figure 6.1). We have completed re-implementing the position control schemes that had originally been implemented on the MUSE on the new *CONDOR* system. We have also implemented the force controller based on the algorithm described in [Hollerbach, Narasimhan and Wood 1986] on this system. The results from the initial experiments we have performed using the force controller indicate that the computational architecture presented in this chapter is adequate to address problems of this nature. The *CONDOR* has also been used to control other robots, such as the MIT Serial Link Direct Drive Arm and the Whole Arm Manipulation system,

Figure 6.1: The Utah/MIT Hand Arm system.

and plans are under way to use it for two other robots that are under construction.

6.3 Hardware

The *CONDOR*'s hardware architecture consists of a *development* host connected up to a *multi-microprocessor* real-time component through a high-bandwidth standardized *interconnect*. The development host runs a time-sharing operating system (both versions use derivatives of Berkeley 4.2 BSD Unix). The *real-time* component consists of one or more single board microcomputers plugged into a standard bus (the Hand Controller uses six such processors). When the need for more computational power arises, more processor boards can be added to the real-time end. Both components of the system are based on the Motorola 68020 processor to ensure a high degree of hardware and software compatibility. Such an architecture can be characterized as a coarse-grained, shared-memory multi-processor based on a common bus. The block diagram in Figure 6.2 illustrates the main components of the architecture.

There are a number of features in the *CONDOR*'s design. These are summarized below:

1. The *CONDOR* is based on an industry standard interconnect. While the Muse was based on the MultiBus I, the *CONDOR* is based on the VME Bus.

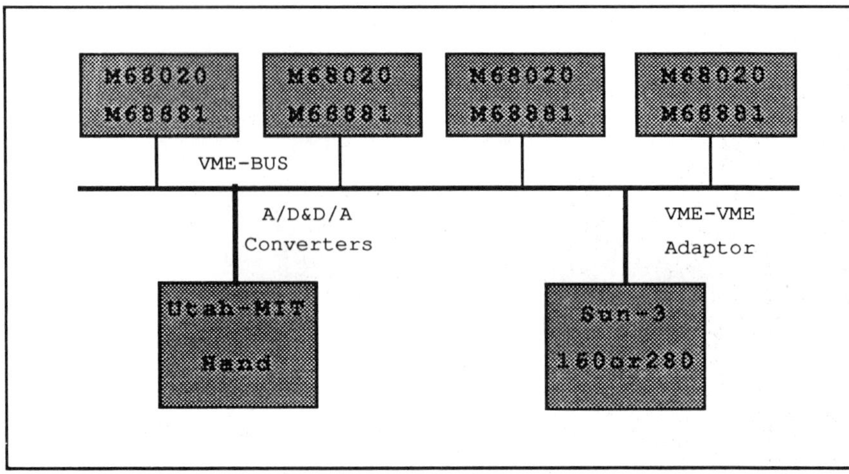

Figure 6.2: *CONDOR* hardware block diagram (adapted from [Narasimhan *et al* 1988], ©IEEE).

2. The *CONDOR* uses a bus-to-bus adaptor to facilitate a real-time environment to communicate with high bandwidth across to a development host.

3. The processor architectures on which the development host and the real-time single board computers are based is the same, obviating the need for special purpose cross compilers and linkers.

4. The *CONDOR* uses a *mailbox* hardware interrupt to facilitate communication between its real-time processors.

5. The use of *dual-ported* memories make it extremely efficient to implement shared memory based communication schemes as well.

In what follows, we present a set of simple guidelines. They are presented in a decreasing order of importance as we view them. Our guidelines may seem more like common sense on hindsight, but they are nevertheless a distillation of our experience with three versions of our hardware and software configurations.

- **Do not build custom hardware unless you really have to.** The reasons for this guideline are obvious. The amount of effort invested in building and maintaining custom hardware is often disproportionate compared with the increased performance they provide. Most robotics research laboratories are also staffed not by hardware designers, but by researchers whose primary focus is robotics. Hardware development is therefore best left to companies that specialize at it.

As applied to robot hands, which are characterized by a number of different joints (and consequently actuators), any custom hardware solution must be replicated many times. While modular designs are indeed possible, the increased number of components not only increases the cost of design and maintenance, it also increases the chances of failure.

We do not wish to imply that special purpose hardware has no place in robotics research. Time and space constraints may necessitate custom VLSI implementations in some cases. The *CONDOR* is open-ended to allow such additions at a later date should they become necessary; however, we feel that we are at an early stage in research with robot hands where control algorithms have not matured sufficiently. Consequently, we feel that custom architectural solutions are not suitable for the kinds of robot hands that are currently being built.

- **Use simple, standardized components.** There are many reasons for choosing standard components to be the basis of any hardware subsystem. Firstly, we wanted a standard computer platform based on which many research robots could be built. Using standard components ensures that the same controller and development software can be reused and only those portions of the software that are robot specific need to be reimplemented. Duplication of the hardware also becomes easier: one of our *CONDOR* systems was built, tested and running from scratch, within a week.

  Both versions of our controller for the Utah/MIT hand were built based on industry standard busses [Narasimhan *et al* 1986; Narasimhan 1988]. Using a standard bus ensured the availability of peripherals for interfacing to our robots, such as analog to digital converters, digital to analog converters, resolvers, and shaft encoders.

- **Use multi-micro-processors but be conservative.** The use of micro-processors in real-time control systems is certainly not something new. There have been a few attempts to use a number of these powerful micro-processors [Chen *et al* 1986; Gauthier *et al* 1987; Kim *et al* 1987; Leung and Shanblatt 1987]. While the Muse was based on the Motorola 68000 processor, the *CONDOR* is based on the Motorola 68020 processor coupled with the fast Motorola 68881 floating point unit. As we mentioned earlier, the computational task involved in controlling a robot such as the Utah/MIT hand is quite burdensome for a single processor. Rather than having a monolithic single processor and program to control robot hands, we advocate using tightly-coupled systems.

Processor Type	Speed	Cost	Comments
Microvax II	1 MIP	mod	interconnect problems
Vax 11/750	1 MIP	high	interconnect problems
Symbolics 3600	1 MIPS	high	lacks real-time support
National 32032	1 MIPS	low	
Motorola 68000	1 MIPS	low	lacks floating point
Motorola 68020	2.5 MIPS	moderate	has fp co-processor
Motorola 68030	4 MIPS	high	unavailable

Table 6.1: Comparisons of processing power available from alternative hardware configurations (July 87). (©IEEE, Narasimhan *et al* 1986).

Our system differs from most of the other systems built under university environments in two important ways:

1. We attempt to avoid duplicating software and hardware components that can be found on conventional, non-real-time, systems.
2. We provide an extremely efficient computation environment.

The power of the individual microprocessor that is to be chosen is of some importance. Listed in Table 6.1 are a few CPU's that we looked at before deciding on our current architecture. This list of processors is representative. One of the important considerations in choosing the hardware is the software support available for it. A commercially available fast processor is not really useful without a good, optimizing high level language compiler and development environment.

From amongst this list, we chose Ironics Corporation's Motorola 68020 based single board computers coupled with the Motorola 68881 floating point unit. There were a couple of architectural options that we ruled out at the outset when we were putting together the second version of our system. We ruled out RISC architectures, partly because most commercially available CPU's still do not deliver fast floating point performance. Such architectures are also highly non-standard and usually require specialized compilers and development environments (although this may have changed by the time this book is published). We also ruled out processors that looked promising but were very much in the alpha or beta stages of their releases, to reduce the time spent on debugging hardware problems.

Tightly coupled systems facilitate easy software development. With dual ported memory, multi-processor programs can be written using the shared memory model. Robot hand control involves *coarse-grain* parallelism and this can be exploited by partitioning the tasks appropriately. When computational requirements increase, additional power can be obtained by simply adding more processor boards to the system and repartitioning the control algorithms.

- **Get floating point support if you can.** Another realization that has become more obvious to us is that hardware support for floating point is useful in typical hand control applications. In the Muse, there was no such support. All arithmetic was performed in scaled integers. Although such implementations did demonstrate the feasibility of such schemes, the maintainability of the hand control software could be rated only as poor. Consequently, in the *CONDOR*, we chose the Motorola 68020 mainly to use its co-processor interface to the fast floating point Motorola 68881 chip.

- **Use a standard bus.** The interconnect used in the first version of our system was the then standard Multibus I. Our second version was based on the VME-Bus. It must be mentioned that these industry-standard interconnects provide a very high bandwidth connection between the individual microprocessors, and with dual-ported memory shared memory communication becomes available for free. Peripheral boards, such as serial and parallel interfaces, A/D and D/A converters, and additional memory boards, are all readily available for such standard busses from a number of vendors.

  It is true that such tightly-coupled schemes do not generalize beyond around ten microprocessors. Such schemes also do not work very well with cached systems, necessitating special solutions. However, we feel that other schemes for communication do not offer the bandwidth needed for the real-time computations that a robot hand typically requires. For the Utah/MIT hand, our initial estimates indicated that the bus had to be able to transfer approximately 320K per second (5 channels per joint, at a rate of 1KHz for 16 joints).

  The first version of our solution used a DMA connection between a VAX 11/750, which was our development host to the control microprocessors. This we found was too slow to do high bandwidth data transfers because of the setup overhead associated with each transfer, and because of the incompatible byte formats on the two machines. Serial and parallel ports were not considered because of their low bandwidth, while we overruled devices like ethernet interconnections because of their complicated software requirements.

- **Choose your development host carefully.** The choice of a development host ought to be dictated primarily by the software development environment that you wish to have. One important feature of the *CONDOR* is that it separates out, rather cleanly, the real-time component, from the development environment. Obviously, if these two environments are compatible to a large extent then much is to be gained, since programs can be written and debugged on the development host and then run on the control microprocessors with small modifications. Capabilities that a good development host ought to provide include high level language compilers for the real-time system, a good debugger, and bit-mapped graphics. We feel that the separation of the development from the real-time environment has enabled us to optimize the former for ease of writing and debugging real-time control programs and the latter for efficiency.

- **Be prepared to get your hands dirty.** It would be nice if our guideline for using standardized, board level products resulted in a scenario wherein end users would need to have only a minimum knowledge of computer hardware. They could then concentrate on what they presumably want to do with the minimum of wasted effort. Unfortunately, however, this is only *almost* the case. The *CONDOR* requires the user to have some knowledge of the multiprocessor for which the programs are being written. Debugging and maintaining the hardware has, on the other hand, become extremely easy.

To conclude this section on hardware, we would like to mention that the particular choice of components often seems to be dictated by intangibles or by variables that are hard to measure, such as availability of peripheral products, software compatibility, price, etc. Our system, whose block diagram is shown in Figure 6.2, has worked surprisingly well, and is now being used to control a wider variety of robots as we indicated earlier.

6.4 Software

An important piece in any computational architecture is the software that is available to run on the hardware. Our previous comments regarding flexibility and efficiency of robot control hardware apply equally as well to the software components. In this section, we will provide an overview of the software environment provided by the *CONDOR* system concentrating on what we feel are the innovative aspects of this part of the system.

Our approach to robot programming has been `library` oriented: at the lowest level are the system libraries that change relatively infrequently. Built on top of this is a set of libraries that are common across different robots; for example, routines to generate trajectories, low level kinematics calculations etc. Finally come the robot specific libraries or user programs that we expect researchers to write for particular robots.

The *CONDOR*'s software development environment is almost entirely written in the C programming language. The design goals of the software system were to:

1. provide a flexible environment in which control programs can be written, debugged, and run,
2. provide efficient and low overhead means of doing often repeated tasks,
3. provide easy to use programmer libraries for dealing with data transfer hardware, and real-time interaction with robotic devices, and
4. provide a graphics based user interface.

The *CONDOR*'s system software is structured around a few relatively simple organizing principles. In a typical program development scenario, the user is expected to write and compile a program on the development Sun-3 host. Since the Sun runs the Unix operating system[2], the programmer has access to all the standard Unix software development tools. Once the program has been compiled, it is linked with the real-time library, and then downloaded onto the slave microprocessors for execution. The run-time environment provides the user with access to a number of program libraries for performing common tasks in a portable manner. For example, libraries are provided for control loop scheduling, file serving, and plotting data, to name just a few.

Thus, the *CONDOR*'s environment is really *two* different environments. First, it is an environment on the Sun known as the *development* environment, and then it is a *run-time* environment for both the Sun and the slave microprocessors. Much effort has been put into making the Sun and the slave microprocessor run-time environments as compatible as possible. In fact, most programs that run on the real-time controllers will run on the Sun simply by relinking them with the appropriate library.

The hardware architecture is tightly-coupled, and is a true MIMD machine. As such, it requires a program to be partitioned into segments that can run on multiple processors. Our approach to managing this multiplicity problem in terms of software engineering has been to largely ignore it. The number of processors in a *CONDOR* system rarely exceeds five or six and we do not expect this architecture to be applied to problems requiring more than a dozen processors. In essence, the coarse-grain parallelism applied to control programs is managed by the programmer directly.

6.4.1 Innovative aspects of the software

Detailed accounts of the various libraries that comprise the *CONDOR*'s real-time controller can be found in [Narasimhan 1988]. In this section, we

---

[2]Most of the system runs on Sun 0S 3.4, which is closely compatible with and is based on Berkeley 4.2 BSD Unix.

would like to summarize the novel aspects of the software that we have found to be extremely useful:

1. The system contains a device mechanism modeled after the device switch interface commonly found in Unix. The *CONDOR* does not support dynamic loading of device drivers but allows configuration of new devices rather easily. By providing a well-documented interface that is easy to use, device drivers can be written and configured into the system by programmers who do not have a detailed understanding of how the *CONDOR*'s system libraries work. Although dynamic loading is not supported, the *CONDOR* does provide a mechanism (similar to `probe` routines found in conventional Unix) for dynamically finding and configuring devices. Besides this, the *CONDOR* also provides mechanisms to handle interrupting devices, handle shared interrupt vectors and emulate most of the standard Unix system calls, such as `open`, `read`, `write`, etc. Such a capability is extremely important to support the variety of interfaces and sensors found on robots today.

2. An extremely efficient message passing scheme upon which all interprocessor communication is modeled. We will explain this scheme later on in this chapter since it forms the basis for much of our software tools.

3. A static servo loop scheduler that is based on a rate monotone algorithm.

4. Higher level facilities built on top of the shared memory architecture using message passing. This includes support for *virtual terminals*, *file operations* and *debugging support*.

5. Software tools and programs that allow easy configuration, downloading and execution of user level programs.

6.4.2 Scheduling

Real-time control programs are different from conventional programs since they require that certain events be scheduled to execute withing strict time constraints. For example, most robot control involves loops that read data from analog to digital converters, perform some control law computation, and then write out computed values to the actuators using digital to analog converters. Low level control loops need to be scheduled so as to run at specified sampling rates.

The *CONDOR* provides two such mechanisms by which servo loops can be scheduled on the real-time microprocessors. Both mechanisms rely on scheduling at the C **procedure** level; i.e. the user has the ability to specify that a particular C procedure has to be executed at a specified rate. Other systems that provide scheduling at a higher level (task or process level) suffer from higher task-switching overhead and hence lower real-time performance.

Simple Servos

The simple mechanism for scheduling servo loops on the *CONDOR* provides support only for two running tasks on a single *CONDOR* micro processor. One of these tasks is the control loop that is scheduled to run at a specified rate, while the other is a background task that runs whenever the control loop is not running. This background task is usually dedicated to running a command loop that interacts with the user, and can perform various functions like enabling or disabling the real-time task. The background task can also reschedule the real-time loop to run at different rates. This mechanism is provided for those tasks that need to run on a processor with very little overhead.

The MOS

For users that need to schedule more than one real-time loop to run on a single *CONDOR* microprocessor, a more complex interface called the MOS is provided. This second set of library routines allows a processor to run various control loops at different rates, in a highly efficient manner. Since a typical hand control program will have several servo loops running at rates in excess of 400 Hz, it is important for each scheduler invocation to be fast. To achieve this, scheduling flexibility has been limited to minimize the execution overhead that it requires. In fact, it is a gross overstatement to call this an operating system. It is, in fact, just an efficient utility for programming a system timer and for starting procedures based on precomputed rate information.

To minimize execution overhead, the MOS is table driven. An *event table* is automatically generated by the system when the **mos_start** command is issued. This table lists the elapsed time between invocations of the scheduled servo loops. For example, the event table for two servo loops, one running every ten seconds and the other running every five seconds, has two entries. The first entry indicates that both loops are to start, and five seconds elapse until the next event. The second entry indicates that the five second loop should start, and another five seconds are to elapse before the next event. After this, the cycle repeats, and the first entry of the event table is reused.

With the system outlined so far, it is possible for more than one loop to be runnable at the same time. The system must have an orderly method for selecting the actual loop that will be run from the set of runnable loops. A *process table* is maintained for this purpose. All the tasks in the system are arranged, in order of decreasing servo rate, in this table. When the event table indicates a loop is ready to run, it is marked runnable in the process table. The system then searches down the process table, and starts the fastest rate loop that is marked runnable.

The time to the next event stored in the event table is loaded into a timer on the processor. When the time has elapsed, the running task is interrupted, and the scheduler is reinvoked. The next tasks in the event table that are scheduled to start are marked runnable in the process table. If a loop with a speed slower than the interrupted loop is made runnable, the interrupted loop will be resumed. If a higher speed servo loop is made runnable the slower loop that was interrupted will be temporarily suspended, until higher speed loops that are runnable complete.

An implication of assigning a priority to a process based on its rate is that a loop can only be interrupted by a higher speed loop, and hence, no coprocessing can take place. This is not considered to be a problem. The rate specified for a servo loop is a request that the loop be run that number of times a second. The exact time that a loop is invoked is not important, as long as it is runs within its specified time slice. In other words, a loop scheduled to run every second is only a guarantee that the loop will run *sometime* within a second. A finer precision in selecting the time at which a procedure will run is not needed within our control programming scenario.

Eliminating coprocessing results in a convenient simplification to the system; only one stack need be maintained for all the servo processes running on a processor. Stack pointers are not changed when a new process is invoked, or a suspended process is resumed.

When a loop terminates, the scheduler is also invoked. The terminating loop is marked idle in the process table, and a new loop is selected to run. If no servo loops in the process table are runnable, the background job is activated.

We have found that the mechanisms described above satisfy most of our requirements when it comes to scheduling real-time servo loops on the *CONDOR*.

6.4.3 Message Passing

Most of the software facilities we have described thus far provide support for running a program on a single processor. The *CONDOR*'s message passing system addresses the issue of multiple processors and communication between them. It provides a simple and low-overhead manner in which communication of data can occur between multiple processors, and between processes on the Sun. Since robot control is almost always compute bound, a system for communication between such tasks has to be extremely time-efficient. Consequently our implementation of the message passing system has been tuned for efficiency.

As mentioned by [Gauthier *et al* 1987], interprocess communication for robotics, has been tackled in a variety of ways. Architectures based on RS-232 serial lines rely on slow and primitive forms of communication. More recently, schemes based on parallel ports and LAN's have begun to appear. Shared memory based schemes are popular in tightly coupled systems such as those described by [Chen *et al* 1986], while combinations of both shared memory and message passing have been used in others.

The present system is, to a large extent, a redesign of the system described in [Narasimhan *et al* 1986]. Although the functionality provided by the earlier system was far greater than that provided by the present version, the new scheme is much more efficient.

Messages

From our description of the *CONDOR*'s hardware architecture it should be easy to observe that since both the Sun and the slave microprocessors are bus masters on a common VME bus, each machine has access to the every other's dual-ported memory. Interprocessor communication could be implemented purely through the bus directly using shared memory. Unfortunately, while this unrestricted access is highly efficient, it is hard to control. To overcome the problems of unrestricted memory access, we designed and implemented a mailbox-based message passing system. Mailbox interrupts can be thought of as a software extension to the processor's hardware level interrupts. Another way of thinking about them conceptually is to regard mailbox numbers as port numbers that map to specific remote procedure calls.

A mailbox interrupt has a vector number and a handler routine. When a particular mailbox vector arrives, the handler associated with that vector is invoked. The handler is passed the processor number that initiated the mailbox interrupt and a one integer data value. This integer data value is the message's data[3].

The earlier Version I message passing system was substantially more complex. Messages could be of arbitrary size, and they were addressed to virtual devices that corresponded to the mailbox handler routines in Version II. These handler routines were assigned to processors by a preprocessor that took as input an assignment file, that configured the routines available on each processor. The preprocessor then generated a routing table that had to be linked in with each program. The routing table mapped a virtual device number to a processor that could handle the function. The *CONDOR* redesign was done because the complexity and the overhead of the earlier system prevented most control programs from using message passing.

Another important capability that the Version I implementation lacked was the ability to reply to messages. A program could also not determine if a particular message succeeded or failed. Implementation of messages with

[3]Integers are currently any 32 bit data quantity.

replies could be done in the Version I system through an ad-hoc process that was very inflexible. The *CONDOR* provides support for messages with or without replies. The present implementation uses a reverse send from the recipient processor to the sending processor to acknowledge the receipt of a message. The implementation has been written carefully to be reentrant so that nested sends will indeed work correctly.

The following examples illustrate the operation of the message passing system. In the example, a message will be sent to a handler to read the value of a memory location. The location to be read is passed to the handler as the data portion of the message. The reply from the handler is the contents of that memory location. The code fragment for the handler would be:

```
simple_decoder(proc, data)
int proc;
int data;
{
    return(*(unsigned int *)data);
}
```

The handler is associated with a vector number using **mboxvectorset**:

```
mbox_set_vector(12, simple_decoder, ''A test handler'');
```

Now, any message sent to this processor for vector 12 will be handled by the **simpledecoder** handler.

Another processor can invoke the decoder by using the **mboxsend** routine. If the **simpledecoder** routine is available on processor 0 one can execute the following piece of code on any of the processors (including 0 itself) to invoke the service:

```
value = mbox_send_with_reply(0, 12, address);
```

This will cause the handler that corresponds to the number 12 to be invoked on processor 0 with the second argument being **address**. The call will not return until the other processor has responded with the value found at the given address. This call can be used to provide synchronization. For services that do not require synchronization, and hence do not return a value, the **mboxsend** call can be used.

In summary, the following are the key features of the message passing system:

1. Since message sending happens asynchronously, the execution of a handler resembles an interrupt. Consequently, message handler routines are expected to be short pieces of code that almost never have to modify the interrupt system.

2. The system is extensible in the sense that more complicated protocols can be built on top of it. For example, the underlying system does

not support queueing of messages, although one can easily build one for mailboxes that require this.

3. Since the message system is based on shared memory, sending long messages is usually handled by sending a pointer to the beginning of a long piece of data.

4. Where efficiency is important, the message handling system can be used to set up pointers from one processor into another's memory. The processors can read and write this shared memory, without the minimal overhead of message passing.

There are two hardware requirements imposed by our design of the message passing system on our system architecture. In fact, we did have to overrule at least one microprocessor system because it did not provide the basic capabilities described below. In our design, message sending and the invoking of message handlers is implemented using a *mailbox interrupt* which is a hardware interrupt that is invoked by writing into a particular memory location in a processor's memory. This hardware support is critical for the implementation's efficiency. To protect the integrity of certain critical data structures, the *test-and-set* instruction is used. It is important that this instruction be supported truly indivisible by the hardware across the bus.

6.4.4 Support for Message Passing on the Sun

The Sun development host supports the same primitives for message passing available on the control microprocessors. Any number of processes on the Sun may communicate with each of the slave processors using the message passing protocol.

A message that arrives on the Sun must be mapped to a particular Unix process that can handle it. While each microprocessor is thought of as a single message-receiving processor, the Sun supports the notion of *virtual processors*. Each Sun process that receives interrupts is assigned a unique hardware interrupt vector. Each process on the Sun which participates in message passing must register itself with the Sun kernel, indicating which hardware interrupt vector corresponds to its messages. When a slave processor interrupts the Sun it does so using this vector. The Unix kernel traps on this interrupt vector and signals all processes that have expressed an interest in receiving the interrupt.

Our present implementation of this mechanism is written in the fashion of a normal Unix device driver, requiring no modifications to the Unix kernel. This facilitates the addition of this mechanism to almost all derivatives of 4.2 BSD Unix with very little work.

From the Sun, the *CONDOR* system maps the entire VME 24D32 space into the user address space of the control process (using the **mmap** system

Type of Operation	Msecs/Message	Variation (Msecs)
Ironics to Sun	34	5
Ironics to Sun(R)	38	10
Sun to Ironics	3.9	0
Sun to Ironics(R)	4	0
Ironics to Ironics	0.2	0
Ironics to Ironics(R)	0.25	0

Table 6.2: Performance of the Message Passing System[4] (©IEEE, Narasimhan *et al* 1988).

call). Memory references to any of the control processor's memory, or to the additional one megabyte memory board allocated in the VME backplane for Sun use, become simple array references. The `PROC_RAM` macro returns a pointer to the beginning of memory for the particular processor. For example, to write a value to location 100 in processor 3's memory one would use the following code:

```
int *processor3_ram = (int *)(PROC_RAM(3));
processor3_ram[100] = value;
```

The code above would work, in fact, on any processor in the system.

In Table 6.2 we provide a summary of the performance of the message passing system as benchmarked by a variety of routines. As can be seen from this table, the performance of the message passing system is extremely fast between two control processors. The slower speed for messages sent from the control processors to the Sun host is caused by overhead present in the Unix timesharing operating system.

It should also be noted that the message rates are not as high as servo rates. Consequently, messages are used only as signals to start and stop processes or control the flow of computation and are not used as timing pulses. Servo communication is done using shared memory structures that are initialized by messages.

6.4.5 Message Passing and its Implication for Control

Using the facilities provided by the message passing system it is possible to treat a hierarchical controller as an object-oriented system that responds to control messages. For example, a low-level joint PID controller could be described as an object that responds to two different kinds of messages: messages that alter internal parameters such as gains and position set points, and messages that control the execution of the servo loops. Using this scheme, any processor, including the Sun host, can control the execution of any servo loop in the system.

6.4.6 Higher Level Protocols

The *CONDOR* system requires the concurrent operation of several control microprocessors to perform its tasks. Programming such a complex MIMD machine would certainly be a nightmare were it not for the numerous services that were built on top of the base system using the message passing facilities. These services are flexible interaction with a number of slave microprocessors at a time, provide fileserver capabilities on the development host, and symbolic debugging.

Debugging

One of the most important utilities in the *CONDOR* is the symbolic debugger. To implement a debugger in a flexible manner it was decided to emulate the Unix ptrace system call on the slave microprocessors. Ptrace is a system called used by Unix debuggers to examine registers, set breakpoints, and control program execution of a slave process. Our emulated ptrace is linked into all control programs as part of the standard library.

The emulated ptrace handler communicates with the Sun-3 development host using the message passing system. For example, if the Sun-3 host wishes to examine a control program's registers, it sends a message down to the processor's ptrace handler. The ptrace handler would reply with the desired information. The debugging program thus runs on the host Sun computer where it has access to the downloaded program's symbol table information. Only the low level ptrace routine runs on the microprocessors. With the ptrace handler, almost all Unix based debuggers can be adapted, with little modification, to debug programs on the controller microprocessors. Currently we use the Gnu Debugger. It provides fully symbolic, C source code level debugging tools. The debugger not only helps debug higher level application programs, but has also been used to find low level system bugs.

The debugger can be used in three basic modes. The first allows the user to start executing a program directly under the debugger's control. This mode would be used when a bug is actively being tracked down, and setting initial break points might be necessary. The second mode allows the debugger to be attached to a processor after execution has begun. For example, if a program were in an infinite loop, the debugger could be attached to the running program to determine what went wrong. Finally, if a program receives a fatal exception, the debugger can be attached after the error occurred to help analyze the problem.

Once the debugger has attached to a process on the slave control processor, it's complete capabilities can be used to debug the program running on the remote host, just as if it were running on the development host.

File Serving

Control programs often need access to data stored in files, and need to store collected data in files for further analysis. To allow these types of access, the *CONDOR* provides a fileserver protocol, built using the low level message passing system. From a programmer's point of view, all the standard Unix file operations are available, both buffered and unbuffered. Typically, each file operation results in a message being sent to the Sun, where a server process performs the equivalent Unix file operation. The result is then sent back to the microprocessor as a reply.

To make the fileserver efficient, shared memory is extensively used to avoid copying data. When a microprocessor performs a read or a write, the fileserver process running on the Sun reads or writes the data directly to or from the microprocessor's memory. No intermediate data copy is required. To perform a read, for example, the microprocessor would send a message to the Sun giving the address to read the data into, and the number of bytes to read. The data is directly transferred into the processors buffer.

The fileserver is designed to operate in a stateless manner. All data necessary is stored on the microprocessor. The Sun server does cache some information and if necessary, it can make a request to the microprocessor. Thus the Sun fileserver process can be terminated and restarted, without affecting the state of programs running on various slave microprocessors.

Virtual Terminals

Many microprocessors are used in a *CONDOR* system. The Sun-3 host computer provides a window-system based interface to access these computers. One window for terminal input and output to each computer is provided. A virtual terminal protocol, built using the message passing system, routes data between the Sun and the microprocessors. This avoids the bank of terminals that would otherwise be connected to the real-time controller.

The virtual terminal protocol works in both directions. The Sun can send a message to a microprocessor that contains an input character. The character is added to the terminal input queue by the virtual terminal message handler. Each character is sent in one message, since transfers in this direction are limited by the rate at which a person can type.

Terminal output from a microprocessor to a Sun virtual terminal window is sent in blocks. This is done purely for efficiency purposes. The output message contains a pointer into the terminal driver's memory, and the number of characters to output.

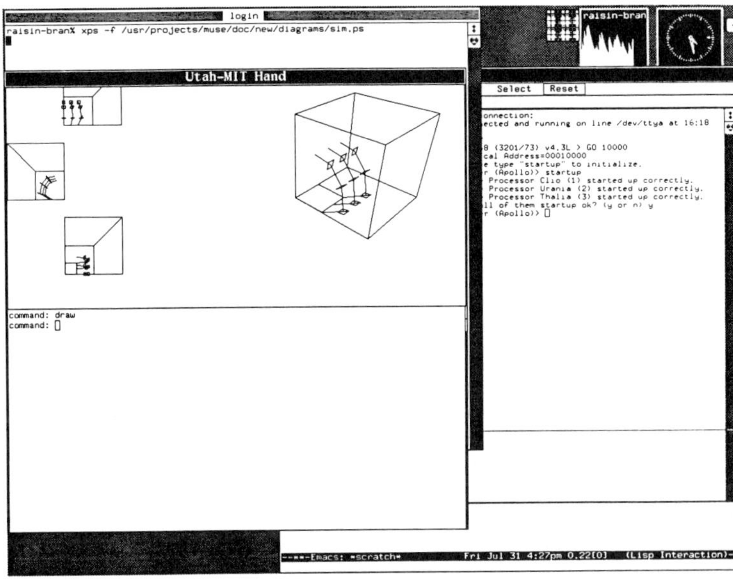

Figure 6.3: ***CONDOR*** **running under the X Window System, showing the screen of a typical session.**

6.4.7 THE *CONDOR* USER INTERFACE

The fileserver, debugger and virtual terminals combined together form the *CONDOR* user interface. This program is an X-window system based application (see Figure 6.3) that programmers utilize to interact with the slave microprocessors. The user interface provides one virtual terminal for each slave processor, it runs a fileserver, and it can start debuggers.

6.5 Conclusion

In this chapter, we have presented a multi-microprocessor system architecture that has been designed explicitly for research in robotics and particularly for robot hands. The hardware and software systems have been designed to provide flexibility and ease of maintenance as well as performance. Our goals were achieved by decoupling the real-time component of our system from the development environment and making sure that each system was highly optimized for its task. We believe that the *CONDOR* will form the basis for a new type of standardized robot controller in research laboratories. Such a common architecture will then enable a sharing of work and duplication of results that has not been possible until now.

7

Control Architecture for the Belgrade/USC Hand

George A. Bekey[1]
Rajko Tomovic[2]
Ilija Zeljkovic[1]

ABSTRACT This chapter describes design and control features of a five-fingered anthropomorphic end-effector designed primarily for grasping tasks. Advantages and limitations of the design are discussed, and special emphasis is placed on its suitability for autonomous, non-numerical or reflex control of grasp. Following a discussion of its mechanical design, we present the controller and sensor features incorporated into the current finger model. A knowledge-based control of hand preshape (prior to grasping) is then outlined, and the hand's suitability as a testbed for the study of human and robot hand motion control is discussed. The final section of this chapter describes future directions.

7.1 Introduction

The Belgrade/USC dextrous hand is an anthropomorphic end effector for robot manipulators. Its design is based on the Belgrade prosthetic hand [Tomovic and Boni 1962], the 1950's. The original Belgrade hand had five fingers controlled by a single motor. When any finger pad contacted an object to be grasped, all the fingers close until the pressure on all the finger pads was approximately equal, thus achieving a high degree of automatic shape adaptability and extreme control simplicity. The hand presented here was designed on similar principles. Its goal was to obtain a hand well-suited to grasping tasks and capable of autonomous shape adaptation, while applying simple control architecture based on the principles of non-numerical or reflex control [Tomovic 1983; Bekey and Tomovic 1986] to drive the hand. Hence, emphasis was placed upon built-in synergies between motions of finger segments and local autonomy during grasping, rather than maximizing

[1]Computer Science Department, University of Southern California, Los Angeles, CA 90089-0782

[2]Electrical Engineering Department, University of Belgrade, Belgrade, Yugoslavia

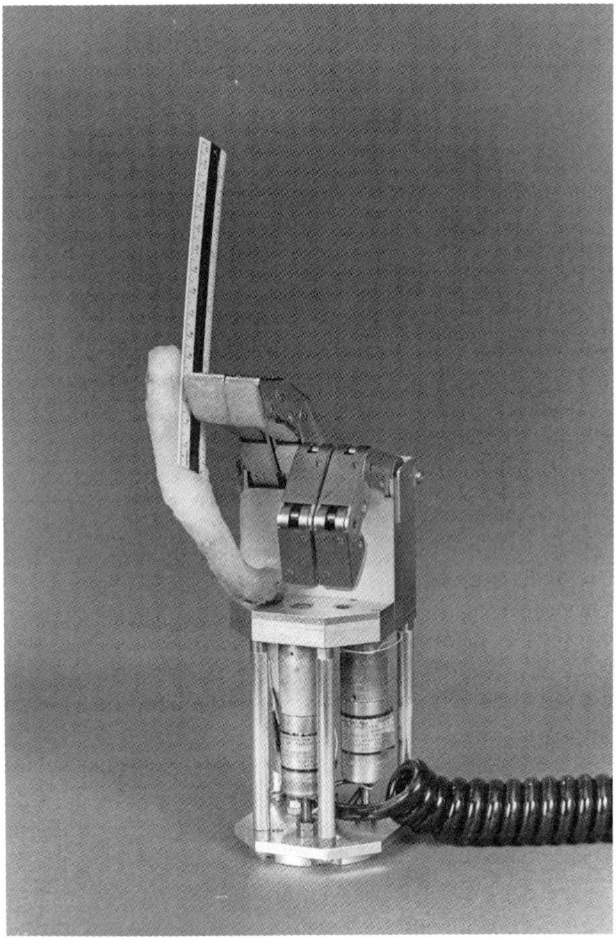

Figure 7.1: Belgrade/USC Hand (Model I)

flexibility and dexterity.

The high-level grasp controller is knowledge-based, selecting a preshape for the hand on the basis of visual information. Finger motions that result in a stable grasp of objects of arbitrary shape (within allowable constraints on size and weight) are determined using local sensor-based control. The hand is designed for mounting on the wrist of a PUMA 560 robot. It is capable of holding a payload of 5 lbs.

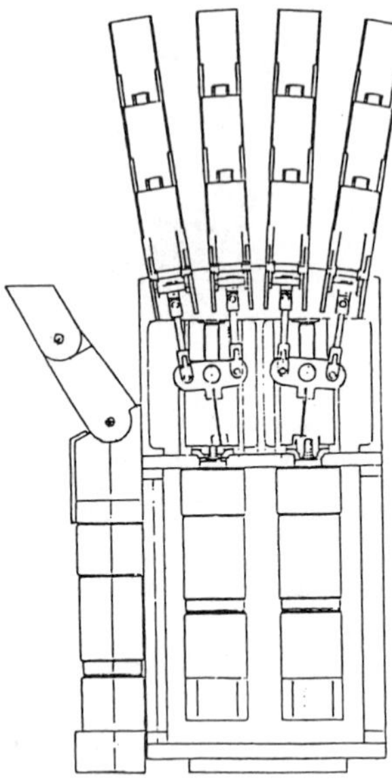

Figure 7.2: Hand mechanical structure illustrating finger spreading and thumb articulation.

7.2 Physical design

Model I of the Belgrade/USC hand was equipped with a rigid thumb and four articulated fingers (see Figure 7.1). Model II of the hand (Figure 7.2) is approximately human-sized, and its five fingers are constructed using aluminum alloy material. Its thumb has two joints and rotates about an axis parallel to the wrist. It can rotate 120° from its fully extended position into opposition to fingers No. 2, 3, or 4. The other four fingers are identical in size and consist of three joints each. All finger joint rotations are supported by miniature ball bearings, and are driven by DC servomotors located within the wrist structure. Actuation, through the reduction gears are directly transmitted to fingers using mechanical linkages, thereby alleviating the need for cables or tendons.

In addition, two other design features make this hand radically different from the Utah/MIT hand [Jacobsen *et al* 1986] or the Stanford/JPL hand [Salisbury 1982]. First, motions of each finger segments are not individually controllable; they are connected by means of linkages in such a way as to display motions similar to those of human fingers during grasping; we term this relationship "internal synergy". The coupling between the finger segments is shown in the schematic diagram of Figure 7.3. It is evident that rotation of the knuckle joint will produce rotation of the two additional finger joints. Figure 7.4 illustrates the synergistic motion of the finger segments by indicating corresponding points on the trajectory of

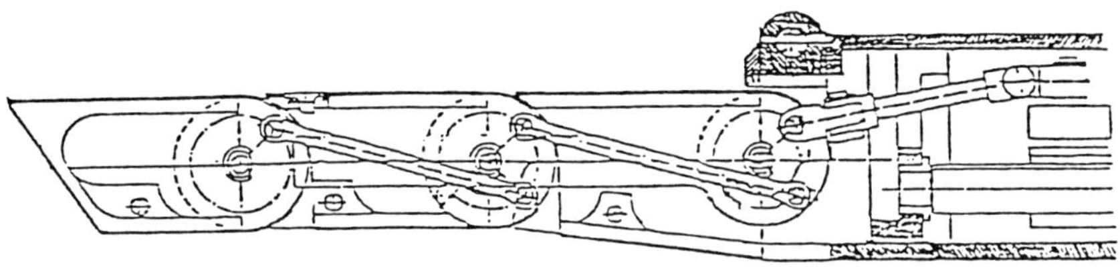

Figure 7.3: Schematic diagram of finger-side view.

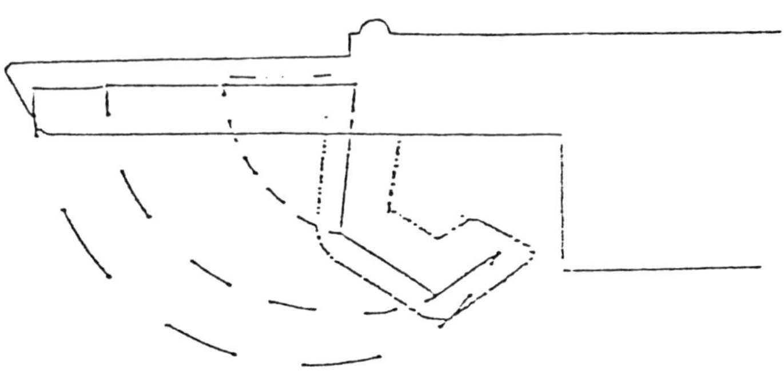

Figure 7.4: Finger motion during flexing (©IEEE, Rao *et al* 1988).

each joint during flexion.

The second unique feature of this hand is its design for autonomous shape adaptation. Four motors mounted in the wrist structure provide the external degrees of freedom. Two motors are used to position and flex the thumb respectively, while the others flex two fingers each. The finger drive is applied through a rocker arm (illustrated in Figure 7.5) designed in such a way that if the motion of one finger of the driven pair is inhibited, the second finger continues to move, thus achieving some shape adaptation without external control. The adaptation feature is illustrated in Figure 7.6, where the solid line shows the actual finger position when a ball or cylinder has been grasped. The adaptation feature provides the equivalent of additional degrees of freedom to the fingers, without active control.

A consequence of the architecture described above is that this hand is well suited to grasping tasks, but does not have the dexterity for applications that require a larger number of externally controllable degrees of freedom. The grasping task begins with spreading of the fingers, as illustrated in

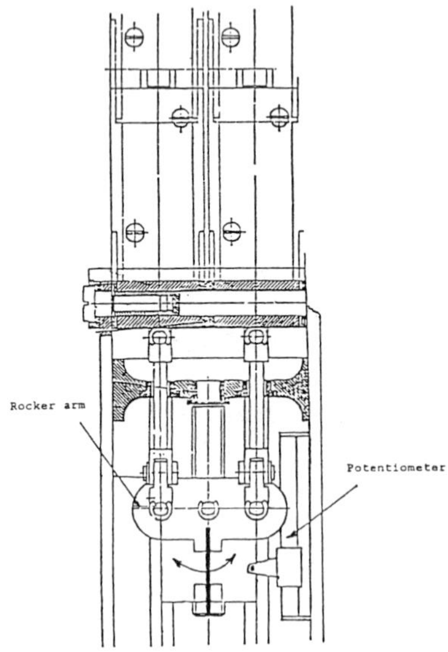

Figure 7.5: Rocker arm for achieving shape adaptation.

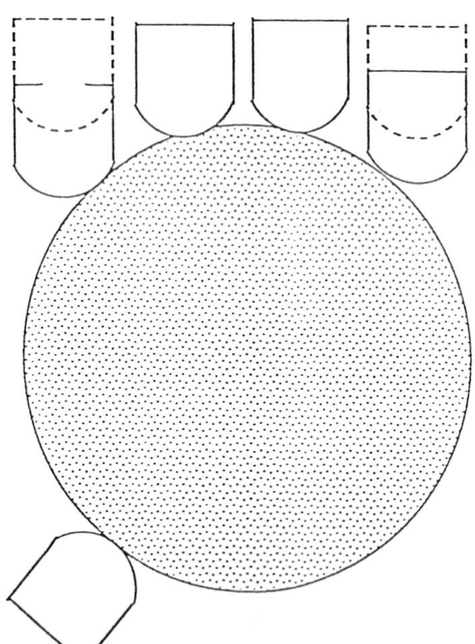

Figure 7.6: Shape adaptation ability.

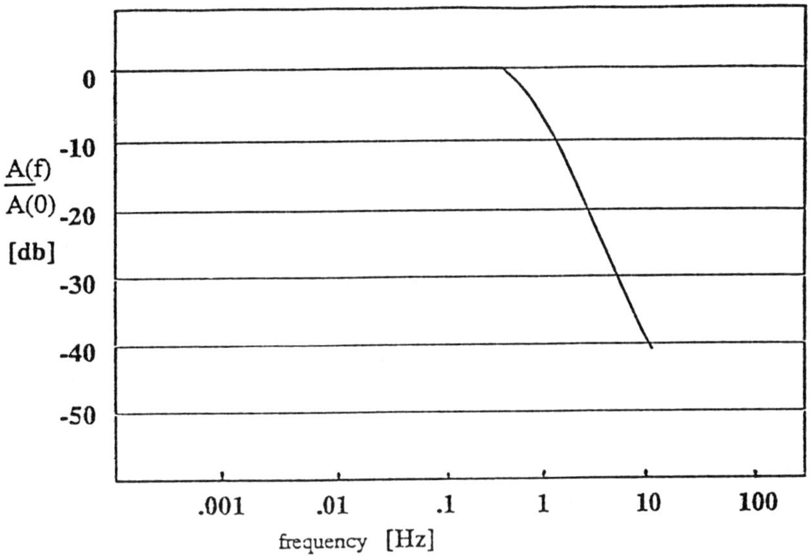

Figure 7.7: Frequency response of a single finger.

Figure 7.2.

The frequency response of the fingers was studied by applying square wave inputs of gradually increasing frequency to the drive servomotors. Due to the nonlinearities in the structure (especially deadzones), sinusoidal inputs were not useful since the fingers did not move at all with excitation signals below some threshold. The peak amplitude of the resulting periodic finger motion was recorded. The ratio of the peak finger motion to the amplitude of the forcing square wave was normalized and plotted in Figure 7.7. The amplitude ratio decreases rapidly after about 0.6 Hz. Complete finger closure is accomplished in approximately 2 seconds. Clearly, this hand is significantly slower than either the human hand or the Utah/MIT hand. Human fingers are capable of tapping at rates approaching 10 Hz. The nonlinearity of the mechanical structure is also evident in this figure, since the rate of decay in the amplitude ratio is between the values characterizing first and second-order systems.

7.3 Sensors

Basic preshape decisions are made using an external vision system [Rao *et al* 1988]. The hand itself is equipped with two sets of sensors:

1. Position sensors, to indicate the rotation of the finger base with respect to the palm plane, and

2. Touch-pressure sensors in the fingertips, to detect contact with an object and the force being exerted

7.3.1 Position sensing

The hand has been equipped with position sensing for both the thumb and other fingers. It is designed for alternate mounting of the thumb, depending on whether it is intended to be a left or right hand. Thumb rotation is indicated by a potentiometer which is installed in the alternate thumb position and coupled to the active thumb base by a gear.

Since the hand is designed for "reflex" grasping, the primary feedback loop is intended to close around those specific signals which determine the hand-object relations: contact event and contact force. (In the future a slippage loop will be included as well). In light of this objective, true servo control which includes stable, high precision position and velocity control is not required. Also a low finger speed facilitates position control design. These considerations have led us, so far, to avoid using shaft encoders. Instead, we use small conductive plastic potentiometers (made by Bourns), which have very high resolution (of the order of 320 ohms/mm) and extremely high life expectancy (of the order of one million cycles). They are also extremely compact (35x10x3mm) and and can be mounted directly within the finger structure, as shown in Figure 7.5. They offer two additional advantages:

(a) Absolute position of the fingers is always available, without initial calibration. This feature makes rapid control possible after unexpected shutdowns.

(b) There is no need for switches to mark the initial position, as would be the case if shaft encoders were used for position sensing.

Potentiometers are side-mounted on the mechanism which drives the rocker arm for each finger pair (Figure 7.5), enabling the position sensors to provide information about composite motion of finger pairs, rather than on the motion of individual fingers. In view of the shape adaptation feature described above, individual finger position feedback is not required.

7.3.2 Force sensing

Finger tips of the hand are presently equipped with Interlink Electronics Force Sensing Resistor (FSR) elements [Interlink] which are capable of sensing forces in the range of 20 grams to 5 Kg. These sensors can also report contact events, when they sense a minimum normal force.

The FSR is available in thin sheets, consisting of a polymer sheet with a layer of sensing film applied by screen printing. The thin flexible sheets are suitable for mounting on various surfaces and can be fabricated in special shapes. The sensing film consists of both electrically conducting and nonconducting particles suspended in a polymer matrix. The particle size ranges from fractions of a micron to several microns. Various type of metallic type conductors and semi-conducting ingredients are combined in order to reduce temperature dependence. "The inactive ingredients are included to provide control of mechanical properties of the film, increase surface durability, promote adhesion to the substrata and enhance electrical standoff properties" [Interlink]. The resulting film changes its resistance in a relatively uniform way as a function of compression.

Figure 7.8 shows the force-resistance curve obtained from a number of measurements for a particular sample of the material under conditions specified in the figure. The initial rapid drop in resistance is used for contact event reporting. Repeatability of the force-resistance curve is within 1%, so that hysteresis is clearly minimal.

7.4 Preshape control

Shape of the hand prior to grasping is controlled by a knowledge-based system. A data structure received from the vision system contains information on the location, orientation and geometry of the target object. The geometric information contained within this data structure is combined with task information to produce the most desirable grasp mode configuration using a high-level control program [Shrader and Andeen 1986; Rao *et al* 1988; Interlink]. Chosen configurations are transferred to the hand control computer (an IBM PC/AT) using a serial port. The overall control configuration is shown in Figure 7.9.

It should be noted that the control of hand shape is based on shape adaptation and sensor feedback. This approach to control implies that neither the kinematic nor the dynamic equations of the fingers need to be solved.

7.5 Finger motion control

Four DC motors are used as actuators to produce the finger motion of the hand. Speed is controlled by changing the voltage level. A 4-channel D/A converter was built for this purpose and connected to an IBM PC/AT via its parallel port. Finger position and force sensing signals are acquired by the 16 A/D converters on an Analog and Digital I/O Board installed in the PC (Figure 7.10).

7.5.1 Finger actuators

The flexion of the thumb and each finger pair are actuated by separate DC torque motors. Presently all three motors have the following characteristics: 36 volts, 6 watts, 8810 rpm no load speed, 43 mNm stall torque, gear reduction ratio 30:1. Thumb rotation is produced by a fourth motor with

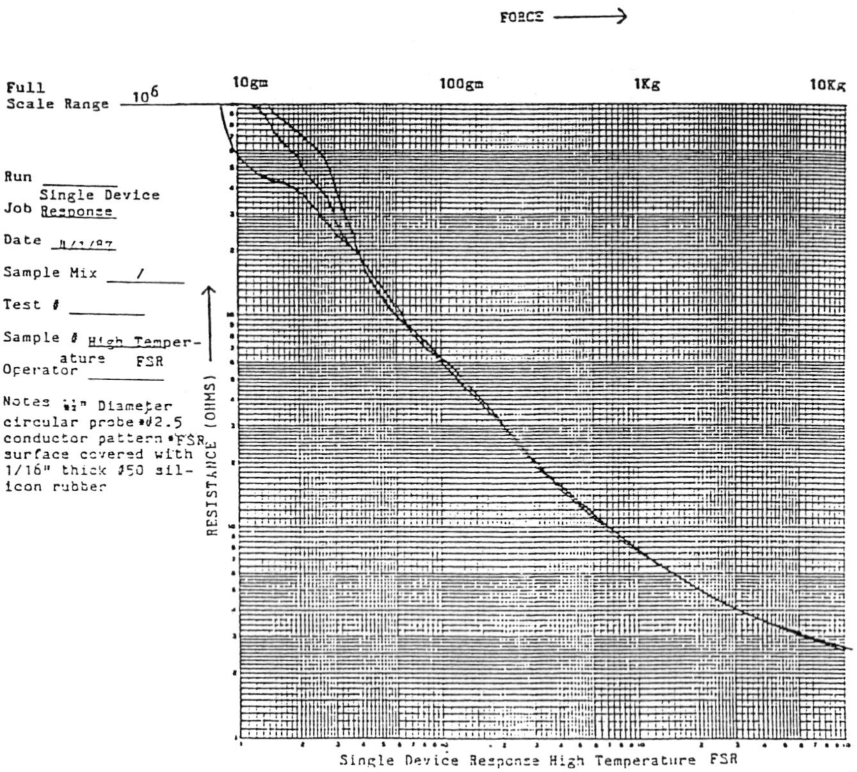

Figure 7.8: Force-resistance characteristic of FSR material. (Reprinted with permission from Interlink Electronics.)

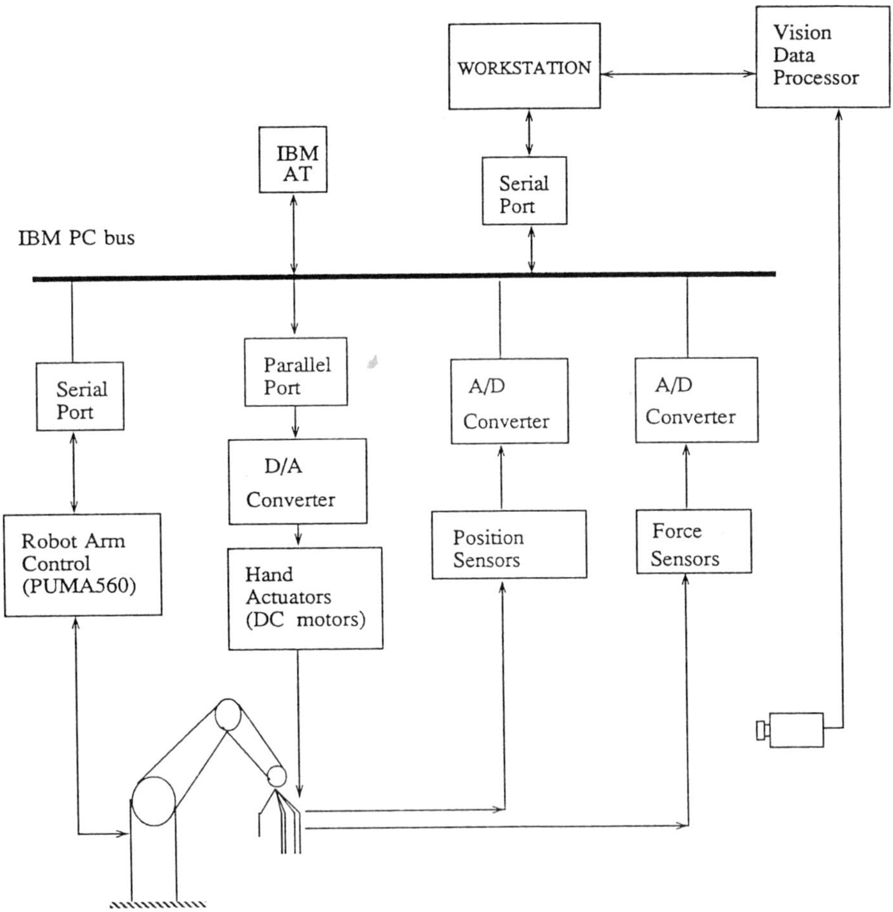

Figure 7.9: Overall control configuration.

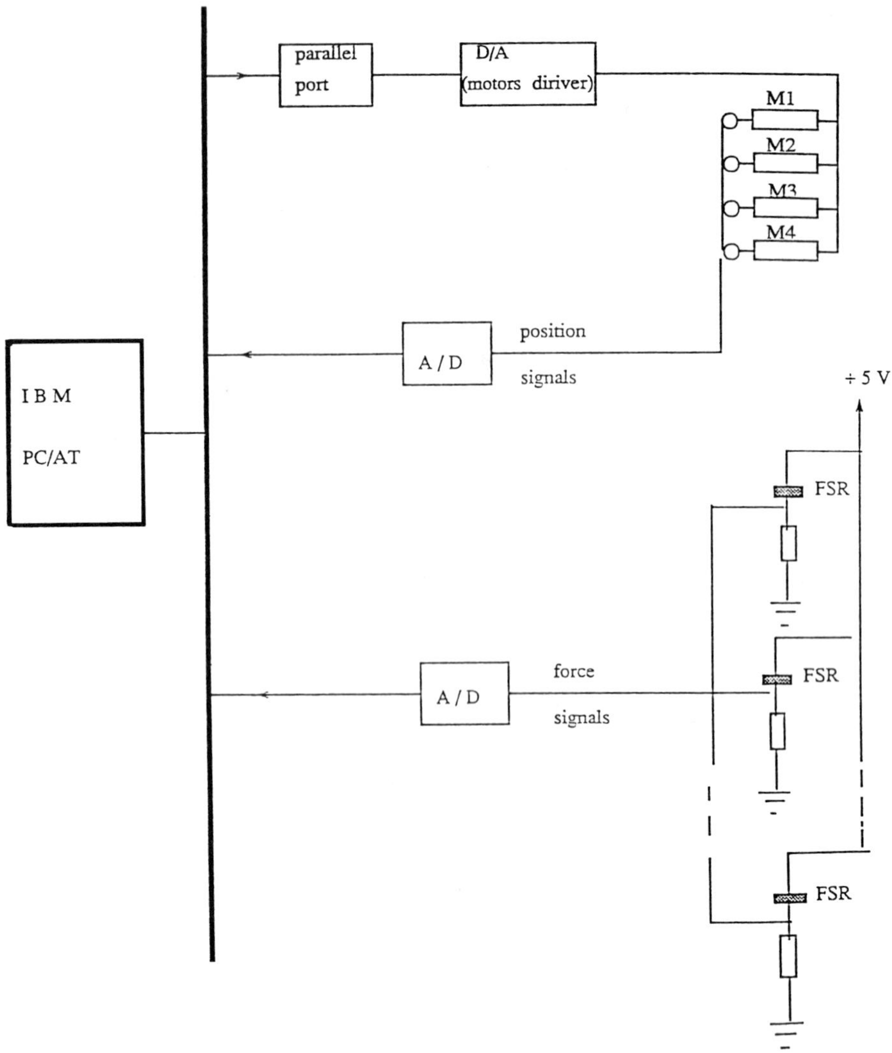

Figure 7.10: Hand controller.

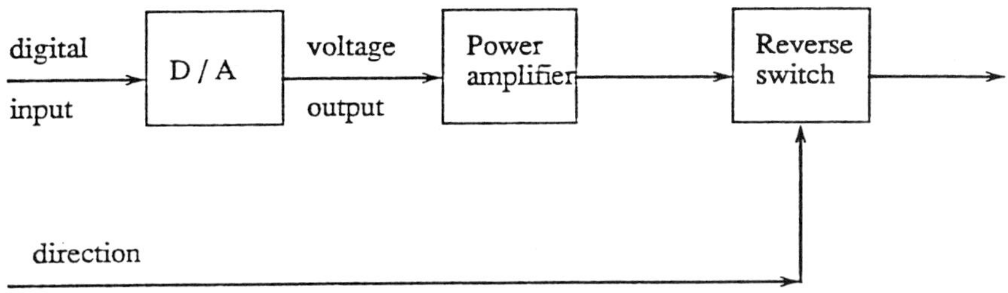

Figure 7.11: Motor driver circuit.

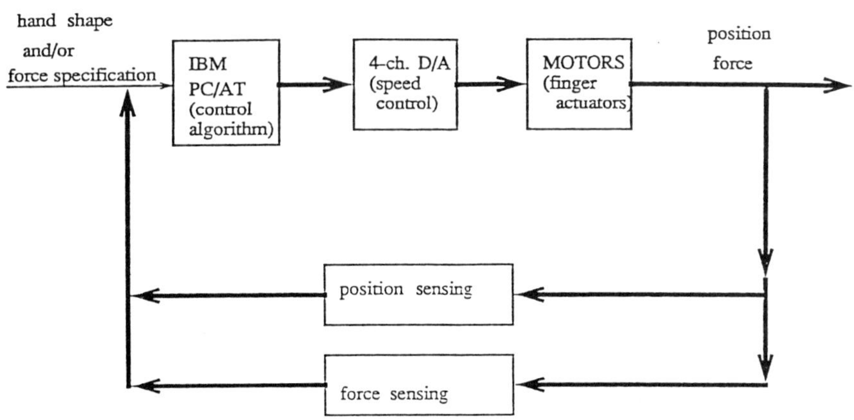

Figure 7.12: Finger motion control.

the same characteristics, except for a gear reduction ratio of 102:1.

7.5.2 Motion-speed control

Each motor is controlled by the circuit shown on Figure 7.11. The complete motor control system consists of four channels identical to this figure, contained on a printed circuit board. The digital inputs to the i-th (i=1,4) D/A converter are TTL compatible. Motor speed is controlled by specifying the output voltage with an appropriate bit combination. The direction of motion of the i-th motor is determined by the input to a reversing switch.

7.5.3 Closed loop control

Position and force sensors described above provide the necessary feedback signals to the controller (Figure 7.12). Slippage sensing will eventually provide the third feedback signal.

Force feedback enables hand positioning in order to reach a desired force balance. Force sensors are located on all the finger tips and at several positions on the palm. Eventually, almost the whole hand surface will be covered with force sensors. This area will be divided into a number of sections which could enable prescribing a force intensity pattern. Force distribution, in a static sense, varies across the hand area with the hand orientation, object weight and shape. Dynamic forces, or more general task dependent forces, are more complex and further research will be required to determine the optimum distribution of force sensors across the hand surface.

Force signals are fed into the controller through A/D converters. The sampling period is less than 5 ms.

The desired force pattern is specified by a set of threshold values, one for each sensor which takes part in the particular grasp execution. Hence, the grasping process proceeds as follows: We issue a command to move the fingers to a desired (target) position. The target position is an estimated position at which the prescribed force thresholds will be exceeded for each sensor by an increment sufficiently small to insure that the hand is not damaged. During the finger motion the force sensors' output is monitored. When prescribed force for a particular sensor is reached, motion of the corresponding finger is stopped. An increase in force is possible by a small amount of additional motion. The "estimated target position" is required to prevent possible damage to hand or object. While the hand is not completely equipped with force sensors, an unconditional force command could produce an uncontrolled contact event between the object and an insensitive hand section.

All programs for closed loop control of the hand and communication with the PUMA controller and Sun or TI control workstation are written in C.

7.6 Future developments

Among the continuing developments are the following:

1. **Slip sensing.** Incorporation of slippage feedback into the control system. Our grasp strategy requires a reflex force adaptation during the initial grasping phase as well as during any later change in force balance which may result in an unstable grasp. "Unstable grasps" will be detected by means of slip sensors. According to the location where the slip event occurs, an adjustment rule will triggered. Within the available force and hand geometric space constraints, feedback control will adjust the hand shape in order to eliminate the slippage. Slippage sensing will be implemented either by thermistors [Shrader and Andeen 1986] or by miniature mechanical elements.

Thermistor-based slip sensors offer the advantages of low cost and small size. However, their response depends on contact quality and pressure as well as on the slip, and the response may be noisy.

We have succeeded in building a small rotating drum sensor (11x4x3 mm) which is small enough to be embedded within a fingertip. Standard shaft encoder techniques are used to obtain a signal which is relatively noise free. Resolution appears to be about 1 mm of slip.

2. **Intelligence.** We also propose increasing the intelligence in the hand shape control software, to allow the system to learn from unsuccessful grasp attempts. Grasp learning using neural networks to relate grasp posture to object geometry has already been implemented for simple cases.

7.7 Conclusion

The Belgrade/USC multifingered robot hand is a multipurpose end effector, suited for a variety of applications where control simplicity, shape adaptation and compactness are desired. It is unique in that the control architecture is designed on biological rather than engineering principles.

Acknowledgements: The development of the Belgrade/USC hand has been supported by the National Science Foundation, the National Institute of Health, the Institute for for Manufacturing and Automation Research, and the University of Novi Sad. The authors gratefully acknowledge the support of these agencies and institutions. In addition, we wish to express our appreciation to numerous colleagues for their ideas, suggestions and inspiration, including Prof. L Rakic from the University of Belgrade, Mr. Dragan Seslija from the University of Novi Sad, and Professors Michael Arbib, Sukhan Lee and Thea Iberall of the University of Southern California. Without the help of these and many other persons, the hand would not exist today.

Part III

Lessons Learned from Dextrous Robot Hands

Subramanian T. Venkataraman

Great strides have been made over the last decade in the design and development of dextrous hands. Significant progress has occurred in the usage of these hands as well. Part three of this book is targeted towards taking stock of where the state-of-the-art in dextrous robot hand research currently is.

Dextrous hand research has been, until recently, a largely experimental field. Some of the work in human hand studies, such as [Schlesinger 1919; Napier 1956], have had the flavor of top-down design. But, by and large, robotics researchers have had a strong leaning towards the development and usage of end effector hardware. Functional requirements from the problem domain were often extracted manually from the specific set of implementation tasks, and hand-coded into the design and development of the end effector. In other cases *fuzzy* specifications such as anthropomorphism were expanded manually by engineers into design data. Examples include [Tomovic and Boni 1962] for prosthetics and [Skinner 1975] for industrial assembly. The Okada Hand [Okada 1982a], designed and developed in the late 1970's, is by far the most complex hand upon which manipulation implementations have been performed. The design was loosely based on the characteristics of a human hand. It was servo-controlled, and the actual manipulation task was hand-coded into the system at the servo-level.

The seminal thesis of Salisbury [1982] and the NSF workshop at MIT AI Laboratory [Hollerbach 1982] may be considered the beginning of the *neo*-dextrous hand decade of the 1980's. Its foundations were laid on: (i) the development of mathematical tools to analyze and instill into the hand design phase requirements posed by an application domain, (ii) the development of hands which will functionally emulate the human hand itself, and (iii) the development of computational architectures for effective usage of a dextrous hand. The Stanford/JPL hand design and development (based on kinematic mobility analysis and static finger-object contact force analysis) [Salisbury

1982], the Utah/MIT Hand (designed to emulate the human hand) [Jacobsen *et al* 1984], and the CONDOR system (a computational architecture that supports implementations on the Utah/MIT hand) [Narasimhan, Siegel and Hollerbach, this volume]; force [Okada 1982a; Brock and Chiu 1985], and tactile sensors [Siegel, Drucker and Garabieta 1987; McCammon and Jacobsen, this volume; Fearing, this volume] for dextrous grasping and manipulation; grasp and manipulation planning algorithms [Lyons 1985; Grupen and Henderson 1988; Nguyen 1988; Li and Sastry, this volume; Yoshikawa and Nagai, this volume], and control implementations [Salisbury 1982, Venkataraman and Djaferis 1987; Loucks *et al* 1987], have all been products of this new generation of dextrous hand researchers. Teleoperation of dextrous hands using data gloves and other input devices [Pao and Speeter 1989]; and the application of dextrous hands for knowledge-based grasping [Stansfield 1989] are also currently being investigated.

In this section, we take stock of the state-of-the-art in robotic dextrous manipulation research. Specifically, we present some of the recent advances in the areas of grasp/manipulation planning and control, and tactile sensing and interpretation. This section is by no means a comprehensive handbook of existing technologies. It, however, definitely highlights some key contributions in the field. Li and Sastry devote Chapter 8 towards the development of mathematical expressions for the transformation of finger contact forces onto a grasped object (and *vice versa*). They then analyze grasp and manipulation stability, and define *quality* of grasps through a set of *ellipsoids*, whose shape reflect corresponding task requirements. They discuss how grasps may be planned through the specification of task ellipsoids and *structured grasp quality measures*, presenting a control law for coordinated control of multifingered dextrous hands. The chapter is concluded with results of dextrous manipulation simulations.

Yoshikawa and Nagai focus upon the analysis of grasping and manipulation in Chapter 9. Tha authors interpret forces applied on an object by the fingers of a dextrous hand in terms of a *manipulation* component, and a *grasping* component. Manipulation forces cause the object to undergo the desired state changes, while the grasping forces map onto the former's *null* space. Physically, the grasping forces *squeeze* the object and lend stability to the interactions that occur between the hand and the object. The authors present an algorithm for the decomposition of fingertip forces into the corresponding grasping and manipulation components and analyze some of the relationships between geometry of the object and stability of grasps.

In Chapter 10, Fearing addresses the interpretation of object shapes using tactile sensor data. The tactile sensor designed and developed by the author [Fearing, Rise and Binford 1986] was mounted on the fingertips of the Stanford/JPL hand [Salisbury 1982], to perform the required experiments. The dextrous hand system utilizes the NYMPH processing system [Chen *et al* 1986], for the acquisition and interpretation of tactile sensor data, and for performing fingertip force control. The author uses generalized cones to represent objects in the world. Object models are developed by fitting the sparse tactile data (from possibly multiple samples) to the closest generalized cone.

A tactile sensor system designed for the Utah/MIT hand [Jacobsen *et al* 1984] is described in Chapter 11. Its authors McCammon and Jacobsen defend the usage of tactile sensing for dextrous manipulation, and describe a sensor processing system design that allows rapid and reliable acquisition of tactile data. The sensor has been designed to overcome problems that result due to aliasing from local object symmetries. Various schemes for interfacing the sensor to a processor is discussed. The multiplexing scheme proposed in this chapter enables rapid acquisition of sensor data.

Finally, in the last chapter, Okada presents a novel tactile sensor design based on suspension shells. The sensor would be co-axially mounted on the phalanges of each finger, with radial springs placed in-between. Forces and torques exerted on the fingers, as well as their location may be computed from strain measurements on the springs. The chapter also includes a unique design to mount these sensors on the fingertips of a dextrous hand. The stiffness and vibration characteristics of these sensors for various spring arrangements are analyzed, and a prototype of the sensor is presented.

The range of topics covered in these chapters (as well as other literature) would lead one to believe, as was also mentioned during the panel discussion (Part IV of this book), that dextrous hand researchers have graduated from this first phase of design and development in the new era with flying colors. This perception is true. However, prior to jumping headlong into the second phase, we must take stock of what has been learned and carefully correlate it with the initial set of philosophical questions and assumptions. The second generation of hands must live up to a higher set of standards. They must be versatile, yet easy to control. They must be electrically robust, yet have a rich set of sensors mounted upon their finger phalanges. Hand control algorithms must integrate model-based techniques with reflexes, yet must be formally analyzable to verify their stability and robustness. The second generation must also bridge the gap between motor behaviorists, experimental psychologists, physiologists, applied mathematicians, engineers, and the often too neglected end-users. We must strategize based on a motivation to understand mathematical and engineering intricacies of dexterity, while constantly drawing upon the insights gathered from human hand studies and consolidating our definitions of problem domains.

8

Issues in Dextrous Robot Hands

Zexiang Li[1]
Shankar Sastry[1]

ABSTRACT In this chapter, we study grasp planning and coordinated manipulation by a multifingered robot hand under various contact constraints. First, we formulate the hand kinematics and establish force/velocity transformation relations for a hand manipulation system. Then, we propose two quality measures for evaluating a grasp. These measures can be used to plan for *good* grasps. Finally, we derive a set of control laws for coordinated manipulation by a robot hand under either fixed points of contact or rolling contact. These control schemes all give decoupled error equations between the position loop and the internal grasp force loop.

8.1 Introduction

A new avenue of progress in the area of robotics is the use of a multifingered robot hand for fine motion manipulation. The versatility of robot hands accrues from the fact that fine motion manipulation can be accomplished through relatively fast and small motions of the fingers and from the fact that they can be used on a wide variety of different objects (obviating the need for a large stockpile of custom end effectors). Several articulated hands, such as the Stanford/JPL hand [Salisbury 1982] and the Utah/MIT hand [Jacobsen *et al* 1984], have recently been developed to explore problems relating to grasping and manipulation of objects. It is of interest to note that the coordinated action of multiple robots in a single manufacturing cell may be treated in the same framework as a multifingered hand.

Grasping and manipulation of objects by a multifingered robot hand is more complicated than the manipulation of an object rigidly attached to the end of a six-axis robotic arm for two reasons: the kinematic relations between the finger joint motion and the object motion are complicated, and the hand has to firmly grasp the object during its motion.

The majority of the literature in multifingered hands has dealt with

[1]Department of Electrical Engineering and Computer Science, University of California, Berkeley, CA. 94720

kinematic design of hands and the automatic generation of stable grasping configurations as also with the use of task requirement as a criterion for choosing grasps (see for example the references [Cutkosky 1985; Kobayashi 1985a; Kerr 1985; Barber, Volz and Desai 1986; Nguyen 1988; Nakamura, Nagai and Yoshikawa 1987; Trinkle 1987; Li, Canny and Sastry 1989]. Some of these references [Cutkosky 1985; Barber, Volz and Desai 1986; Nguyen 1988; Ji 1987; Chiu 1988; Li *et al* 1989] have suggested the use of a task specification as a criterion for choosing a grasp, albeit in a some what preliminary form. A few control schemes for the coordination of a multifingered robot hand or a multiple robot system have been proposed [Zheng and Luh 1985; Hayati 1986; Tarn, Bejczy and Yun 1986; Arimoto, Miyazaki and Kawamura 1987; Nakamura, Nagai and Yoshikawa 1987]. The most developed scheme is the master-slave methodology [Zheng and Luh 1985; Tarn *et al* 1986; Arimoto, Miyazaki and Kawamura 1987] for a two-manipulator system. The schemes developed so far all suffer from the drawback that they either assume rigid attachment of the fingertips to the object or are open loop. The schemes do not account for an appropriate contact model between the fingertips and the object.

In this chapter, we study *grasp planning and the determination of coordinated control laws with point contact models.* We develop dual notions of grasp stability and grasp manipulability and propose a simple procedure for task modeling. Using the task we then define the structured (or task-oriented) grasp quality measures, which are subsequently used for devising a grasp planning algorithm. We give control laws for the coordinated control of a multifingered robot hand manipulating an object, which takes into account both the dynamics of the object and the fingers and assumes a point contact model.

A brief outline of the chapter is as follows: In Section 8.2, we formulate hand kinematics and develop velocity/force transformation relations for a hand manipulation system. In Section 8.3, we propose two task oriented quality measures for evaluating a grasp. These quality measures will then be used to derive a grasp planning algorithm. In Section 8.4, we develop control schemes for coordinated manipulation by a multifingered robot hand. In Section 8.5, we conclude this chapter with several important remarks.

8.2 Kinematics of Multifingered Robot Hands

In this section, we discuss concepts concerning rigid body motion and kinematics of a robot hand. For further treatment of these subjects see [Kobayashi 1985; Kerr and Roth 1986; Montana 1986; Li, Hsu and Sastry 1987].

Notation

Let C_i and C_j be two coordinate frames of $\mathcal{R}^3$, where i, j are arbitrary subscripts. Then, $r_{i,j}$ and $R_{i,j}$ denote the position and orientation of C_i

relative to C_j. Furthermore, $g_{i,j} \triangleq (r_{i,j}, R_{i,j}) \in SE(3)$ denotes the configuration of C_i relative to C_j, where the Euclidean group, $SE(3)$, denotes the configuration space of an object.

Definition 8.2.1 *The velocity of C_i relative to C_j is defined by*

$$\begin{bmatrix} v_{i,j} \\ w_{i,j} \end{bmatrix} = \begin{bmatrix} R_{i,j}^t \dot{r}_{i,j} \\ \mathcal{S}^{-1}(R_{i,j}^t \dot{R}_{i,j}) \end{bmatrix} \tag{8.1}$$

where the map $\mathcal{S}$

$$\mathcal{S} : \mathcal{R}^3 \longrightarrow so(3) : \begin{bmatrix} w_1 \\ w_2 \\ w_3 \end{bmatrix} \longmapsto \begin{bmatrix} 0 & -w_3 & w_2 \\ w_3 & 0 & -w_1 \\ -w_2 & w_1 & 0 \end{bmatrix}$$

identifies $\mathcal{R}^3$ with $so(3)$, the space of 3 by 3 skew-symmetric matrices.

Proposition 8.2.1 *Consider three coordinate frames C_1, C_2 and C_3. The following relation exists between their relative velocities:*

$$\begin{bmatrix} v_{3,1} \\ w_{3,1} \end{bmatrix} = Ad_{g_{3,2}^{-1}} \begin{bmatrix} v_{2,1} \\ w_{2,1} \end{bmatrix} + \begin{bmatrix} v_{3,2} \\ w_{3,2} \end{bmatrix} \tag{8.2}$$

where $Ad_{g_{3,2}^{-1}}$, the Adjoint map of SE(3), is a similarity transformation given by

$$Ad_{g_{3,2}^{-1}} = \begin{bmatrix} R_{3,2}^t & -R_{3,2}^t \mathcal{S}(r_{3,2}) \\ 0 & R_{3,2}^t \end{bmatrix}.$$

Proof. The proof is rather straightforward, see for example [Li *et al* 1987]. □

Corollary 8.2.1 *Consider three coordinate frames C_1, C_2 and C_3. Suppose that C_3 is fixed relative to C_2. Then, the velocity of C_3 relative to C_1 is related to that of C_2 by a constant transformation, given by*

$$\begin{bmatrix} v_{3,1} \\ w_{3,1} \end{bmatrix} = \begin{bmatrix} R_{3,2}^t & -R_{3,2}^t \mathcal{S}(r_{3,2}) \\ 0 & R_{3,2}^t \end{bmatrix} \begin{bmatrix} v_{2,1} \\ w_{2,1} \end{bmatrix} \tag{8.3}$$

In this chapter, we will assume that the object in consideration is smooth and convex. This assumption also applies to the fingers of a robot hand.

Definition 8.2.2 *The boundary of a smooth rigid object is an embedded 2-dimensional manifold $S \subset \mathcal{R}^3$, which can be expressed as the union of finitely many open sets $\{S\}_i$, such that S_i is the image of a diffeomorphism*

$$\varphi : U \subset \mathcal{R}^2 \longrightarrow S_i \subset \mathcal{R}^3.$$

The pair (φ, U) is called a coordinate system of S. The coordinates of a point $s \in S_i$ are $\mathbf{u} = (u, v) = \varphi^{-1}(s)$.

We assume that an object in consideration has an orthogonal coordinate system, in the sense that

$$\varphi_u(\mathbf{u}) \cdot \varphi_v(\mathbf{u}) = 0, \ \ \forall \mathbf{u} \in U,$$

where $\varphi_u(\mathbf{u})$ denotes the partial derivative of φ with respect to u. When (φ, U) is orthogonal, we define the Gauss frame at a point $\mathbf{u} \in U$ as the coordinate frame with origin at $\varphi(\mathbf{u})$ and coordinate axes

$$\mathbf{x}(\mathbf{u}) = \varphi_u(\mathbf{u})/||\varphi_u(\mathbf{u})||, \ \mathbf{y}(\mathbf{u}) = \varphi_v(\mathbf{u})/||\varphi_v(\mathbf{u})|| \text{ and } \mathbf{z}(\mathbf{u}) = \mathbf{x}(\mathbf{u}) \times \mathbf{y}(\mathbf{u}).$$

Definition 8.2.3 *Consider a manifold S with an orthogonal coordinate system (φ, U). At a point $s \in S_i$, the curvature form K is defined as the 2×2 matrix*

$$K = [\mathbf{x}(\mathbf{u}), \mathbf{y}(\mathbf{u})]^t [\mathbf{z}_u(\mathbf{u})/||\varphi_u(\mathbf{u})||, \mathbf{z}_v(\mathbf{u})/||\varphi_v(\mathbf{u})||],$$

the torsion form T as the 1×2 matrix

$$T = \mathbf{y}(\mathbf{u})^t [\mathbf{x}_u(\mathbf{u}) \backslash ||\varphi_u(\mathbf{u})||, \mathbf{x}_v(\mathbf{u}) \backslash ||\varphi_v(\mathbf{u})||],$$

and the metric as the 2×2 diagonal matrix

$$M = diag(||\varphi_u(\mathbf{u})||, ||\varphi_v(\mathbf{u})||).$$

Example 8.2.1 Consider the sphere of radius R, with the following coordinate system

$$U = \{(u, v) \in \mathcal{R}^2 | -\pi/2 < u < \pi/2, -\pi < v < \pi\}$$

and the map

$$\varphi : U \longrightarrow \mathcal{R}^3 : (u, v) \longmapsto (R \cos u \cos v, -R \cos u \sin v, R \sin u).$$

The coordinates u and v are known as the latitude and longitude, respectively. The Gauss frame is defined by

$$\mathbf{x}(\mathbf{u}) = \begin{bmatrix} -\sin u \cos v \\ \sin u \sin v \\ \cos u \end{bmatrix}, \mathbf{y}(\mathbf{u}) = \begin{bmatrix} -\sin v \\ -\cos v \\ 0 \end{bmatrix}, \mathbf{z}(\mathbf{u}) = \begin{bmatrix} \cos u \cos v \\ -\cos u \sin v \\ \sin u \end{bmatrix}.$$

The curvature form, torsion form and metric are, respectively,

$$K = \begin{bmatrix} 1/R & 0 \\ 0 & 1/R \end{bmatrix}, T = [0, \ -\tan u/R], \text{ and } M = \begin{bmatrix} R & 0 \\ 0 & R \cos u \end{bmatrix}.$$

♠

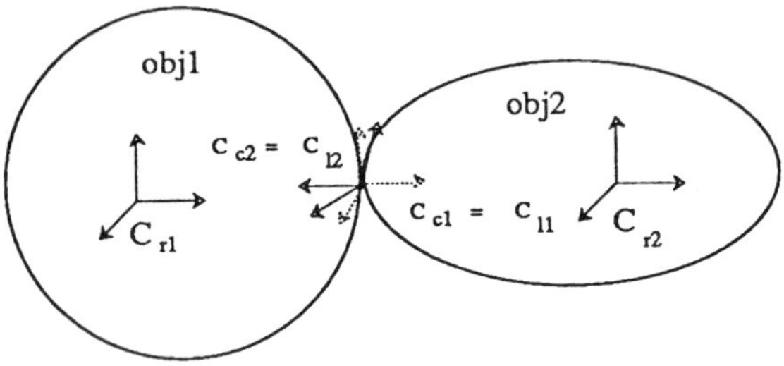

Figure 8.1: Coordinate frames for two objects in contact.

We now consider two objects, called obj1 and obj2, that move while maintaining contact with each other (Figure 8.1). By the earlier convexity assumption the objects will make contact over isolated points. We wish to describe here motion of the contact points in response to a relative motion of the objects.

Let C_{r1} and C_{r2} be the coordinate frames fixed relative to obj1 and obj2, respectively. Let S_1 and S_2 be the embedded 2-D manifolds representing the boundary of obj1 and obj2. S_1 and S_2 can be expressed as the union of open sets, $S_1 = \bigcup S_1^j$, and $S_2 = \bigcup S_2^j$, where S_1^j has orthogonal coordinate system (φ_1^j, U_1^j) and S_2^j has orthogonal coordinate system (φ_2^j, U_2^j). Let $c_1(t) \in S_1$, and $c_2(t) \in S_2$ be the positions at time t of the point of contact relative to C_{r1} and C_{r2}, respectively. We will restrict our attention to an interval I so that $c_i(t)$ belong to a single coordinate system of $S_i, i = 1, 2$.

The coordinate system $(\varphi_i^j, U_i^j), i = 1, 2$, induces a Gauss frame at all points of S_i, which will be denoted by $C_{ci}, i = 1, 2$. We also define a continuous family of coordinate frames, two for each $t \in I$, as follows. Let the local frames at time t, $C_{l1}(t)$ and $C_{l2}(t)$ be the coordinate frames fixed relative to C_{r1} and C_{r2}, respectively, that coincide at time t with the Gauss frames at $c_1(t)$ and $c_2(t)$.

The five parameters that describe the 5 degrees of freedom for the motion of the points of contact are: the coordinates of the point of contact relative to the coordinate system (φ_1^j, U_1^j) and (φ_2^j, U_2^j), given by, respectively,

$$\mathbf{u}_1(t) = (\varphi_1^j)^{-1}(c_1(t)) \in U_1^j, \text{and } \mathbf{u}_2(t) = (\varphi_2^j)^{-1}(c_2(t)) \in U_2^j.$$

and the angle of contact, $\phi(t)$, defined as the angle between the x-axes of C_{c1} and C_{c2}. We chose the sign of ϕ so that a rotation of C_{c1} through angle $-\phi$ around its z-axis aligns the x-axes.

We describe the motion of obj1 relative to obj2 at time t, using the local coordinate frames $C_{l1}(t)$ and $C_{l2}(t)$. Let (v_x, v_y, v_z) and (w_x, w_y, w_z) be the translational and rotational velocities, respectively, of $C_{l1}(t)$ relative to $C_{l2}(t)$. These provide the 6 degrees of freedom for the relative motion between the objects.

The symbols K_1, T_1 and M_1 represent, respectively, the curvature form, torsion form and metric at time t at the point $c_1(t)$ relative to the coordinate system (φ_1^j, U_1^j). We can analogously define K_2, T_2 and M_2. We also let

$$R_\phi = \begin{bmatrix} \cos\phi & -\sin\phi & 0 \\ -\sin\phi & -\cos\phi & 0 \\ 0 & 0 & -1 \end{bmatrix} \triangleq \begin{bmatrix} \hat{R}_\phi & \begin{matrix} 0 \\ 0 \end{matrix} \\ 0\,0 & -1 \end{bmatrix}, \quad \tilde{K}_2 = \hat{R}_\phi K_2 \hat{R}_\phi.$$

Note that R_ϕ is the orientation matrix of C_{c1} relative to C_{c2}. Hence, $\tilde{K}_2$ is the curvature of obj2 seen from obj1. By the convexity assumption, the relative curvature form $K_1 + \tilde{K}_2$ is invertible. The following equations that relate motion of the points of contact to the relative velocity of the objects are due to Montana [1986].

Theorem 8.2.1 *The point of contact and the angle of contact evolve according to*

$$\begin{cases} \dot{\mathbf{u}}_1 = M_1^{-1}(K_1 + \tilde{K}_2)^{-1}\left(\begin{bmatrix} -w_y \\ w_x \end{bmatrix} - \tilde{K}_2 \begin{bmatrix} v_x \\ v_y \end{bmatrix}\right), \\ \dot{\mathbf{u}}_2 = M_2^{-1}\hat{R}_\phi(K_1 + \tilde{K}_2)^{-1}\left(\begin{bmatrix} -w_y \\ w_x \end{bmatrix} + K_1 \begin{bmatrix} v_x \\ v_y \end{bmatrix}\right), \\ \dot{\phi} = w_z + T_1 M_1 \dot{\mathbf{u}}_1 + T_2 M_2 \dot{\mathbf{u}}_2, \\ 0 = v_z. \end{cases} \tag{8.4}$$

Montana [1986] calls the first three equations of (8.4) the kinematic equations of contact, and the last equation the constraint equation.

We define three special modes of contact in terms of the relative velocity components (v_x, v_y, v_z) and (w_x, w_y, w_z) by

(1) Fixed point of contact:

$$\begin{bmatrix} v_x \\ v_z \end{bmatrix} = 0, \text{ and } \begin{bmatrix} w_x \\ w_y \end{bmatrix} = 0; \tag{8.5}$$

(2) Rolling contact:

$$\begin{bmatrix} v_x \\ v_y \end{bmatrix} = 0, \quad \text{and } w_z = 0; \tag{8.6}$$

(3) Sliding contact:

$$\begin{bmatrix} w_x \\ w_y \\ w_z \end{bmatrix} = 0. \tag{8.7}$$

We have from (8.4) that

Corollary 8.2.2 *The kinematic equations of contact correspond to each of the contact modes are*

$$\begin{cases} \dot{\mathbf{u}}_1 = 0, \\ \dot{\mathbf{u}}_2 = 0, \\ \dot{\phi} = w_z, \end{cases} \tag{8.8}$$

for fixed point of contact,

$$\begin{cases} \dot{\mathbf{u}}_1 = M_1^{-1}(K_1 + \tilde{K}_2)^{-1} \begin{bmatrix} -w_y \\ w_x \end{bmatrix}, \\ \dot{\mathbf{u}}_2 = M_2^{-1}\hat{R}_\phi(K_1 + \tilde{K}_2)^{-1} \begin{bmatrix} -w_y \\ w_x \end{bmatrix}, \\ \dot{\phi} = T_1 M_1 \dot{\mathbf{u}}_1 + T_2 M_2 \dot{\mathbf{u}}_2. \end{cases} \tag{8.9}$$

for rolling contact, and

$$\begin{cases} \dot{\mathbf{u}}_1 = -M_1^{-1}(K_1 + \tilde{K}_2)^{-1}\tilde{K}_2 \begin{bmatrix} v_x \\ v_y \end{bmatrix}, \\ \dot{\mathbf{u}}_2 = M_2^{-1}\hat{R}_\phi(K_1 + \tilde{K}_2)^{-1}K_1 \begin{bmatrix} v_x \\ v_y \end{bmatrix}, \\ \dot{\phi} = T_1 M_1 \dot{\mathbf{u}}_1 + T_2 M_2 \dot{\mathbf{u}}_2. \end{cases} \tag{8.10}$$

for sliding contact.

When a robot hand manipulates an object with its fingers contacting the object by one of the above contact modes, it defines, respectively, coordinated manipulation, rolling motion and sliding motion [Li and Sastry 1988].

We shall now derive the kinematics of a multifingered robot hand system and formulate the the velocity/force transformation relations so that control schemes for coordinated manipulation and for rolling motion can be studied.

Consider the hand manipulation system shown in Figure 8.2, which consists of an object and a k-fingered robot hand.

We denote by $m_i, i = 1, ...k$, the number of joints of finger i, and by $\theta_i, \tau_i \in \mathcal{R}^{m_i}$, the vector of joint angles and the vector of joint torque, respectively, of finger i. We define a set of coordinate frames as follows. The reference frame of the system is C_p, which is fixed to the hand palm. The body coordinate frame is C_b, which is fixed to the mass center of the object. At time t the local frame of the object at the point of contact with finger i is $C_{bi}(t)$, which by our earlier convention is fixed relative to C_b. The

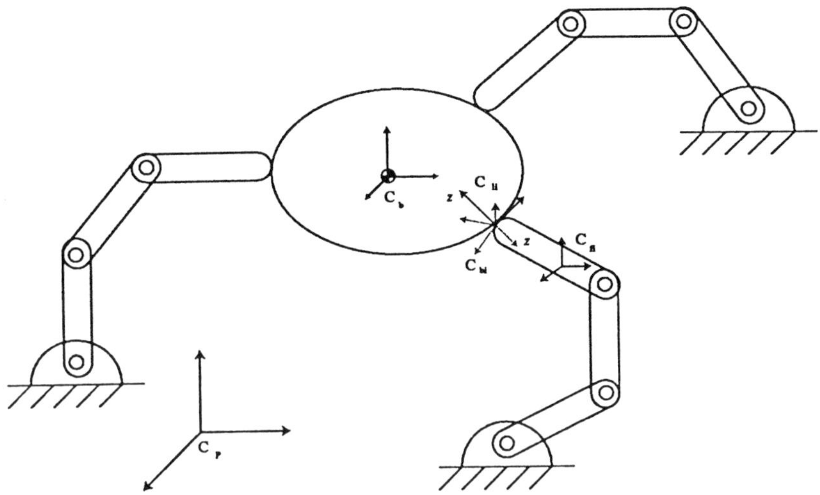

Figure 8.2: A hand manipulation system.

reference frame of finger i is C_{fi}, which is fixed to the last link of finger i, and the local frame of finger i at time t is $C_{li}(t)$, which is fixed relative to C_{fi} and has origin at the point of contact with the object.

Let $c_{oi}(t) \in S_o$ and $c_{fi}(t) \in S_i$ be the positions at time t of the point of contact of the object with finger i relative to C_b and C_{fi}, respectively. Here, S_o denotes the boundary of the object and S_i the boundary of the last link of finger i. We assume that the contact point relative to finger i occurs over the last link only. We will restrict our attentions to a time interval I so that $c_{oi}(t)$ belongs to a single coordinate system (φ_o^j, U_o^j) of S_o, and $c_{fi}(t)$ belongs to a single coordinate system (φ_i^j, U_i^j) of S_i, $i = 1, ...k$. The coordinates of $c_{oi}(t)$ relative to (φ_o^j, U_o^j) will be denoted by $\mathbf{u}_{oi}(t) \in \mathcal{R}^2$, and the coordinates of $c_{fi}(t)$ relative to (φ_i^j, U_i^j) by $\mathbf{u}_{fi}(t) \in \mathcal{R}^2$. Let $\phi_i(t)$ be the angle of contact of the object at time t with respect to finger i.

Let (v_x^i, v_y^i, v_z^i) be the components of translational velocity of $C_{bi}(t)$ relative to $C_{li}(t)$, and (w_x^i, w_y^i, w_z^i) the components of rotational velocity (to apply the results of the previous section, let the object be obj1, and finger i be obj2). Since the local frames $C_{bi}(t)$ and $C_{li}(t)$ share a common origin (i.e., $r_{bi,li} = 0$), we obtain from Proposition 8.2.1 the following velocity constraint relation:

$$\begin{bmatrix} v_{bi,p} \\ \omega_{bi,p} \end{bmatrix} = \begin{bmatrix} R_{\phi_i} & 0 \\ 0 & R_{\phi_i} \end{bmatrix} \begin{bmatrix} v_{li,p} \\ \omega_{li,p} \end{bmatrix} + \begin{bmatrix} v_x^i \\ v_y^i \\ v_z^i \\ w_x^i \\ w_y^i \\ w_z^i \end{bmatrix} \tag{8.11}$$

where

$$R_{\phi_i} = \begin{bmatrix} \cos\phi_i & \sin\phi_i & 0 \\ -\sin\phi_i & -\cos\phi_i & 0 \\ 0 & 0 & -1 \end{bmatrix}$$

is the orientation matrix of C_{bi} relative to C_{li}. Note that by the constraint equation of (8.4), we have in (8.11) that $v_z^i = 0$.

On the other hand, by Corollary 8.2.1, the velocity of C_{bi} is related to the velocity of C_b by a constant transformation

$$\begin{aligned} \begin{bmatrix} v_{bi,p} \\ w_{bi,p} \end{bmatrix} &= \begin{bmatrix} R_{bi,b}^t & -R_{bi,b}^t S(r_{bi,b}) \\ 0 & R_{bi,b}^t \end{bmatrix} \begin{bmatrix} v_{b,p} \\ w_{b,p} \end{bmatrix} \\ &= Ad_{g_{bi,b}^{-1}} \begin{bmatrix} v_{b,p} \\ w_{b,p} \end{bmatrix}, \end{aligned} \tag{8.12}$$

and similarly for finger i one has

$$\begin{aligned} \begin{bmatrix} v_{li,p} \\ w_{li,p} \end{bmatrix} &= \begin{bmatrix} R_{li,fi}^t & -R_{li,fi}^t S(r_{li,fi}) \\ 0 & R_{li,fi}^t \end{bmatrix} \begin{bmatrix} v_{fi,p} \\ w_{fi,p} \end{bmatrix} \\ &= Ad_{g_{li,fi}^{-1}} \begin{bmatrix} v_{fi,p} \\ w_{fi,p} \end{bmatrix}. \end{aligned} \tag{8.13}$$

Moreover, the velocity of the finger reference frame, C_{fi}, is related to the velocity of the finger joints, $\dot\theta_i$, by the finger Jacobian,

$$\begin{bmatrix} v_{fi,p} \\ w_{fi,p} \end{bmatrix} = J_i(\theta_i)\dot\theta_i. \tag{8.14}$$

Substituting (8.12), (8.13) and (8.14) into (8.11) yields the constraint equation relating velocity of the object to the joint velocity of the fingers.

$$Ad_{g_{bi,b}^{-1}} \begin{bmatrix} v_{b,p} \\ w_{b,p} \end{bmatrix} = J_{fi}\dot\theta_i + \left[v_x^i, v_y^i, 0, w_x^i, w_y^i, w_z^i\right]^T \tag{8.15}$$

where

$$J_{fi} \triangleq \begin{bmatrix} R_{\phi_i} & 0 \\ 0 & R_{\phi_i} \end{bmatrix} \cdot Ad_{g_{li,fi}^{-1}} \cdot J_i(\theta_i).$$

For a point contact with friction, a finger can exert linear forces upon the object about the point of contact. Thus, only contact constraints in the directions of the translational velocities can be re-enforced. When finger i contacts the object with a fixed point of contact, substituting (8.5) into (8.15) yields the following specialized constraint equation

$$B_i^t Ad_{g_{bi,b}^{-1}} \begin{bmatrix} v_{b,p} \\ w_{b,p} \end{bmatrix} = B_i^t J_{fi} \dot{\theta}_i, \tag{8.16}$$

where

$$B_i^t = \begin{bmatrix} 1 & 0 & 0 & 0 & 0 & 0 \\ 0 & 1 & 0 & 0 & 0 & 0 \\ 0 & 0 & 1 & 0 & 0 & 0 \end{bmatrix}.$$

Note that if finger i is a soft finger as in [Salisbury 1982], which enables the finger to exert an additional torque upon the object about the contact normal, or if finger i is rigidly attached to the object as in [Zheng and Luh 1985], which enables the finger to exert all six components of forces and torques upon the object, B_i^t can be modified accordingly as in [Kerr 1985; Li *et al* 1987] still holds.

When finger i contacts the object with rolling constraint, the corresponding constraint equation remains the same form as in (8.16), except that the contact coordinates $\mathbf{u}_{oi}(t)$ and $\mathbf{u}_{fi}(t)$ are no longer stationary and evolve according to (8.9).

When finger i slides across the object, the corresponding constraint equation becomes

$$B_i^t Ad_{g_{bi,b}^{-1}} \begin{bmatrix} v_{b,p} \\ w_{b,p} \end{bmatrix} = B_i^t J_{fi} \dot{\theta}_i + \begin{bmatrix} v_x^i \\ v_y^i \\ 0 \end{bmatrix}, \tag{8.17}$$

and the contact coordinates evolve according to (8.10).

We now examine briefly contact constraints in terms of contact wrenches. For a contact model, let n_i denote the total number of independent contact wrenches that finger i can apply to the object. For a point contact with friction, $n_i = 3$ (i.e., a force in the normal direction and two components of frictional forces in the tangent directions) but for a soft finger $n_i = 4$ (i.e., in addition to the three contact wrenches of a frictional point contact, a torque about the contact normal). The resulting body wrench from applied contact wrenches of finger i can be expressed as

$$\begin{bmatrix} f_b \\ m_b \end{bmatrix} = Ad^t_{g_{bi,b}^{-1}} B_i x_i \tag{8.18}$$

where $f_b \in \mathcal{R}^3$ is a linear force and $m_b \in \mathcal{R}^3$ is a torque about the origin of C_b, and $x_i \in \mathcal{R}^{n_i}$ is the magnitude vector of applied contact wrenches along the basis directions of B_i. For a frictional point contact, x_i is constrained to lie in the frictional cone K_i specified by

$$K_i = \left\{ x_i \in \mathcal{R}^{n_i}, x_{i,3} \leq 0, x_{i,1}^2 + x_{i,2}^2 \leq \mu^2 x_{i,3}^2 \right\}$$

where μ is the coefficient of static Coulomb friction. Note that when finger i slides across the object, contact wrenches are restricted to the boundary ∂K_i of the friction cone, given by

$$\partial K_i = \left\{x_i \in \mathcal{R}^{n_i}, x_{i,3} \leq 0, x_{i,1}^2 + x_{i,2}^2 = \mu^2 x_{i,3}^2\right\}.$$

By the *Principle of Virtual Work*, the joint torque required for maintaining static equilibrium in the presence of contact wrench $x_i \in \mathcal{R}^{n_i}$, is given by

$$\tau_i = J_{fi}^t B_i x_i. \tag{8.19}$$

Finally, for the hand manipulation system, we define by $m = \sum_{i=1}^k m_i$, the total number of joints; $n = \sum_{i=1}^k n_i$, the total number of constraints; $\theta = (\theta_1^t, ...\theta_k^t)^t$, $\tau = (\tau_1^t, ...\tau_k^t)^t \in \mathcal{R}^m$, respectively, the hand joint variable and the hand joint torque vectors; $B = \text{diag}(B_1, ...B_k)$ the basis matrix; $x = (x_1^t, ...x_k^t)^t \in \mathcal{R}^n$ the magnitude vector of contact wrenches along the directions of B, and $K = K_1 \oplus ... \oplus K_k$ the force cone. Then, the contact constraint equation (8.16) for both fixed points of contact, and rolling contacts can be concatenated for $i = 1, ...k$ to give,

$$G^t \begin{bmatrix} v_{b,p} \\ w_{b,p} \end{bmatrix} = J_h \dot{\theta}, \tag{8.20}$$

where

$$G = [Ad^t_{g^{-1}_{bi,b}}, ...Ad^t_{g^{-1}_{bi,b}}]B \text{ and } J_h = B^t \text{diag}\{J_{f1}, ...J_{fk}\}$$

is called the grip Jacobian and the hand Jacobian, respectively.

It is important to observe that *(1) for fixed points of contact G is constant, but this is not necessary true for J_h, unless $w_z^i = 0$. (2) For rolling contact both G and J_h depend on the contact coordinates, which are not stationary.*

The equation that relates the resulting body wrench to the applied contact wrenches is

$$\begin{bmatrix} f_b \\ m_b \end{bmatrix} = Gx \tag{8.21}$$

and the equation that relates contact wrenches to the required joint torques for maintaining static equilibrium is

$$\tau = J_h^t x. \tag{8.22}$$

The relations have been summarized in the following table, while Figure 8.3 illustrates transformation of forces and motion in a hand manipulation system. Note that the vector $\lambda \in \mathcal{R}^n$ is called the contact velocity.

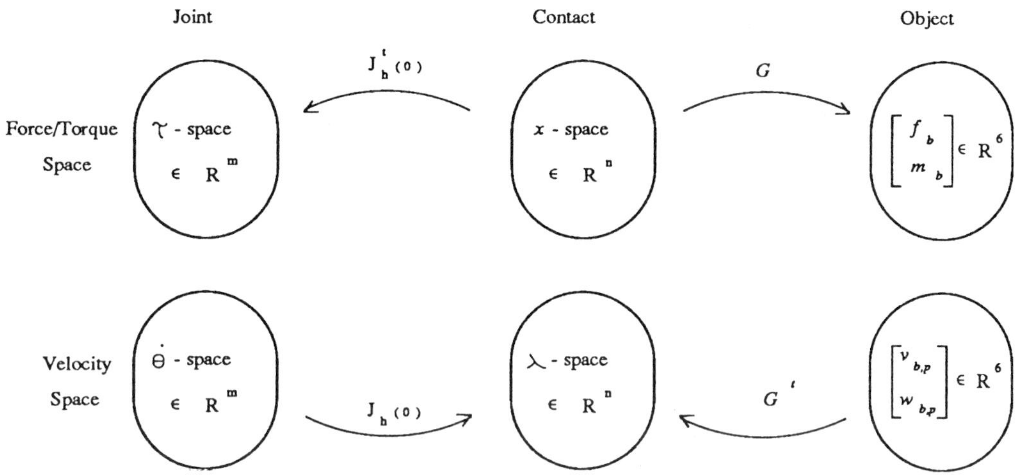

Figure 8.3: Force/velocity transformation for a hand manipulation system.

	Force Torque Relations	Velocity Relations
Body to Fingertip	$\begin{bmatrix} f_b \\ m_b \end{bmatrix} = Gx$	$\lambda = G^t \begin{bmatrix} v_{b,p} \\ w_{b,p} \end{bmatrix}$
Fingertip to Joints	$\tau = J_h^t(\theta)x$	$J_h(\theta)\dot{\theta} = \lambda$

Remark 8.2.1 *(1) The null space of the grip Jacobian G, denoted as $\eta(G)$, is called the space of internal grasping forces [Kerr and Roth 1986]. Any applied finger forces in $\eta(G)$ do not contribute to the motion of the object. However, during the course of manipulation a set of nonzero internal grasping forces is needed to assure that the grasp is maintained. Both Kerr [1985] and Nakamura, Nagai and Yoshikawa [1987] have presented detailed discussions on the optimal selection of internal grasping forces.*

The following dual definitions are now intuitive.

Definition 8.2.4 *(Stability and Manipulability of a Grasp) Define a grasp by a multifingered hand by $\Omega \triangleq (G, K, J_h)$ (see Figure 8.3). Then, for*

$K = R^n$ we have:

1. *The grasp Ω is said to be stable if, for every body wrench $(f_b^t, m_b^t)^t$, there exists a choice of joint torque τ to balance it.*
2. *The grasp Ω is said to be manipulable if, for every body motion $(v_{b,p}^t, w_{b,p})^t$, there exists a choice of joint velocity $\dot{\theta}$ to accommodate this motion without breaking contact.*

Remark 8.2.2 *(1) A stable grasp has been called a force-closure grasp in [Salisbury 1982]. It is important to note that stability is not to be understood in the sense of Lyapunov since we are not discussing stability of a differential equation.*
(2) A manipulable grasp is called a grasp with full mobility in [Salisbury 1982].

Grasp stability and manipulability are now easily characterized for a given position of the fingers by

Proposition 8.2.2 *(1) A grasp is stable if and only if G is onto, i.e. the range space of G is the entire $\mathcal{R}^6$. (2) A grasp is manipulable if and only if $R(J_h) \supset R(G^t)$, where $R(\cdot)$ denotes the range space.*

Proof. See [Li *et al* 1987]. □

Note that condition (1) suffers from the drawback that the force domain is left completely unconstrained. As we have seen earlier that the forces are constrained to lie in a convex cone K, taking into account the unidirectionality of the contact forces and finite frictional forces, in which case the image of $K \cap R(J_h)$ under G should cover all of $\mathcal{R}^6$. Thus, we have

Corollary 8.2.3 *A grasp under unisense and finite frictional forces is both stable and manipulable if and only if*

$$G(K \cap R(J_h)) = \mathcal{R}^6, \text{ and } R(J_h) \supset R(G^t). \tag{8.23}$$

8.3 Grasp Planning

Typical tasks associated with multifingered robot hands include scribing, inserting a peg into a hole, and assembly operations. Common to these tasks are the fact that the robot hand must manipulate an object from one configuration to another, while exerting a set of desired contact forces on the environment. Successful execution of such tasks amounts to having the robot hand perform the following sequence of operations: (1) select a "good" grasp, and (2) coordinate the fingers to manipulate an object along a prescribe trajectory. As we can see, the first operation is essential to the execution of the task. For example, if either the pencil is not grasped at proper positions or the fingers are not at proper postures, it would be extremely difficult to perform a scribing task. In this section, we study the automatic generation of *good* grasps for a given task, and in the next section we study object manipulation based on a *good* grasp chosen in this section. The precise notion of a *good* grasp will be clear as we proceed.

Existing approaches to characterize a grasp can be classified as the *fuzzy* approach, which describes a grasp in linguistic variables such as a "power grasp", a "precision grasp", or a "lateral grasp", and etc., for more details see [Cutkosky 1985; Cutkosky and Howe, this volume; Iberall and MacKenzie, this volume], and the "nonfuzzy" approach, which describes a grasp in terms of some given criteria such as stability and objective functions [Barber *et al* 1986; Li *et al* 1989]. We will study grasp planning based on the later approach only.

In the literature, various stability criteria used to characterize a grasp have been proposed and studied extensively [Hanafusa and Asada 1977a; Salisbury 1982; Nguyen 1988]. But, in many cases such a criterion is too rough as it may generate a large number of stable grasps to a given object. For example, there exist infinitely many choices of stable grasps for a scribing task, and while some are satisfactory some others are not. To solve this problem, additional criteria have been introduced by [Barber *et al* 1986; Li *et al* 1989]. These include the minimum singular value of G, the determinant of GG^t [Li *et al* 1989], and some objective functions defined in [Barber *et al* 1986].

After investigating human grasps, Cutkosky [1985] suggests using the task requirement as *the* criterion for evaluating a grasp. Several other researchers have also studied on the incorporation of task requirement in the selection of a grasp. In [Cutkosky 1985; Nguyen 1988], a task is modeled by a desired compliance matrix and the final grasp is then required to have the desired compliance property. In [Hanafusa and Asada 1977a], a task is modeled by a desired inertia matrix about some operating point, and the final grasp is required to have the desired inertia property. In [Li *et al* 1989], a task is modeled by an ellipsoid, called a task ellipsoid, in the wrench space of the object, and the final grasp is required to maximize the task ellipsoid with unit control effort. While all the above three approaches were concerned with the selection of a task oriented optimal grasp, the nature of the tasks addressed are different. In [Cutkosky 1985; Nguyen 1988], the tasks are quasi static and the system potential energy is assumed to be dominant over the kinetic energy. In [Hanafusa and Asada 1977a] the tasks are purely dynamic and the inertia property rather than the compliance property is the main concern. On the other hand, the approach by [Li *et al* 1989] does apply to both types of tasks, but it considers only the behavior of the grip Jacobian G in the optimization process. Thus, it does not lead to a *truly optimal grasp*.

In this section, we study grasp planning by considering the aggregated behavior of $\Omega = (G, K, J_h)$ in both the wrench space and the twist space. First, we need to obtain a model of a task. We propose to model tasks

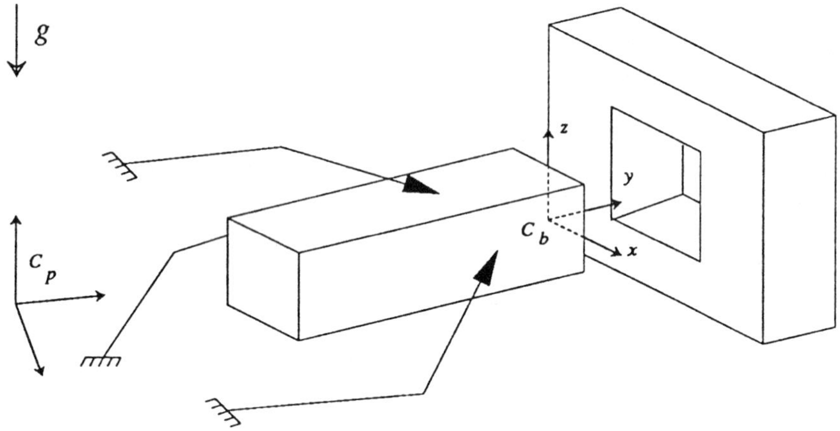

Figure 8.4: Peg-in-hole task (adapted from [Li and Sastry 1987], ©IEEE).

by ellipsoids in both the wrench space and the twist space. Along the line of [Li *et al* 1989] the methodology of modeling a task is to associate each task an ellipsoid (A_α) in the wrench space and another ellipsoid (B_β) in the twist space. The shape of the ellipsoid A_α (or B_β, respectively) reflects the relative force requirement (or the motion requirement) of the task. For example, if the relative force requirement in a certain direction, such as the normal direction of the grinding application in [Hanafusa and Asada 1977a], is large then the task ellipsoid (A_α) has a longer axis in that direction than the rest. To demonstrate the precise implications of this methodology we study task modeling for the following two examples.

Example 8.3.1 *Consider the peg insertion task depicted in Figure 8.4 where the robot grasps the workpiece and inserts it into the hole.*

In order to execute the task, a nominal trajectory is planned before grasping. After grasping the hand follows the planned trajectory until some misalignment of the peg causes the object to deviate from the nominal trajectory and collide with the environment.

With the body coordinate frame chosen as shown, the likehood of collision forces in each force direction of decreasing order would be $-f_y$, $\pm\tau_z$, $\pm\tau_x$, $\pm f_z$, $\pm f_x$, $\pm\tau_y$ *and* $+f_y$. *If we denote by* $(r_i)_{i=1}^6$ *the ratio of maximum expected collision forces in each direction, we obtain a set* A_α, *parametrized by* $\alpha \in [0, \infty)$, *in the wrench space of the object by*

$$A_\alpha = \left\{ (f_x, ...\tau_z) \in \mathcal{R}^6, \frac{(f_y + c_1)^2}{r_1^2} + \frac{\tau_z^2}{r_2^2} + \frac{\tau_x^2}{r_3^2} + \frac{(f_z - c_2)^2}{r_4^2} + \frac{r_x^2}{r_5^2} + \frac{\tau_y^2}{r_6^2} \leq \alpha^2 \right\} \qquad (8.24)$$

where the constant c_1 *reflects the offset of maximum expected collision force between* $+f_y$ *and* $-f_y$ *directions, and* c_2 *reflects the gravitational force on the object. The set* A_α *is an ellipsoid in the wrench space centered at* $(0, c_1, c_2, 0, 0, 0)$*, with the principal axes given by the generalized force directions, and axes lengths by the corresponding ratios* r_i*. The size of the ellipsoid is scaled by the parameter* α*.*

By appropriately assigning a set of values to the constants $(r_i,\ i = 1, ...6)$ *and* $(c_i,\ i = 1, 2)$ *we can decide on the shape of the ellipsoid so that it reflects the task requirement in the wrench space. In particular, the peg insertion task requires that* $(r_i \geq r_j)$ *whenever* $i \geq j$ *and* c_1 *to be large when collision forces in* $+f_y$ *direction are very unlikely.*

On the other hand, since the peg insertion task requires precise positioning the grasp should provide good manipulation capability (or dexterity) in certain directions. First, relatively large motion is needed in the v_y *direction, and then follows* w_y, v_z *and* v_x *directions. If we model by* $(\delta_i)_{i=1}^6$ *the ratio of relative maximum motion requirement among the six generalized velocity directions we obtain an ellipsoid* B_β *in the twist space, parametrized by* $\beta \in [0, \infty)$*, define by*

$$B_\beta = \left\{ (v_x, ..., \omega_z) \in \mathcal{R}^6, \frac{v_x^2}{\delta_1^2} + \frac{v_y^2}{\delta_2^2} + \frac{v_z^2}{\delta_3^2} + \frac{\omega_x^2}{\delta_4^2} + \frac{\omega_y^2}{\delta_5^2} + \frac{\omega_z^2}{\delta_6^2} \leq \beta^2 \right\} \qquad (8.25)$$

The shape of B_β *reflects the task requirement in the twist space. In this case* δ_2 *is the largest of all, and then follows* δ_5, δ_3 *and* δ_1*. Precise values of these constants can be obtained from experiments or experience through error-and-trial procedures.* ♠

Example 8.3.2 *Consider the scribing task shown in Figure 8.5. Human experience tells us that, in order to execute the task efficiently, the grasp should provide, (1) good dexterity at the lead and (2) sufficient normal forces. With the body coordinate frame shown, the task requirement can be translated into requirements on the two task ellipsoids by (a) the task ellipsoid* B_β *in the twist space should have longer axes in* ω_y *and* ω_z *directions than in the other directions, and (b) the task ellipsoid* A_α *in the wrench space should have longer axis in* f_x *direction than in the other directions. Applying this reasoning we obtain in (8.26) and (8.27) two task ellipsoids* A_α *and* B_β *that describe the relative force and velocity ratios of the task.*

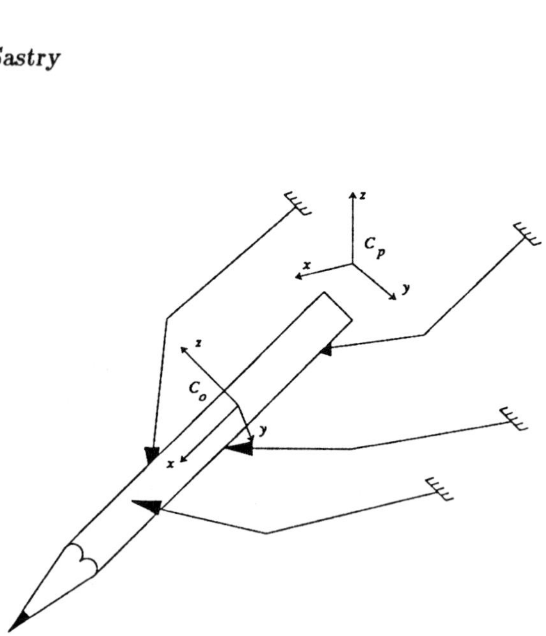

Figure 8.5: A scribing task (adapted from [Li and Sastry 1987], ©IEEE).

$$A_\alpha = \left\{ (f_x, ..., \tau_z) \in \mathcal{R}^6, \frac{(f_x + c)^2}{r_1^2} + \frac{f_y^2}{r_2^2} + \frac{f_z^2}{r_3^2} + \frac{\tau_x^2}{r_4^2} + \frac{\tau_y^2}{r_5^2} + \frac{\tau_z^2}{r_6^2} \le \alpha^2 \right\} \tag{8.26}$$

Wrench Space Task Ellipsoid

$$B_\beta = \left\{ (v_x, ..., \omega_z) \in \mathcal{R}^6, \frac{v_x^2}{\delta_1^2} + \frac{v_y^2}{\delta_2^2} + \frac{v_z^2}{\delta_3^2} + \frac{\omega_x^2}{\delta_4^2} + \frac{\omega_y^2}{\delta_5^2} + \frac{\omega_z^2}{\delta_6^2} \le \beta^2 \right\} \tag{8.27}$$

Twist Space Task Ellipsoid

♠

To conclude these examples, we emphasize that to each task we can associate two task ellipsoids, one in the wrench space that represents the relative force requirement and the other in the twist space that represents the relative motion requirement of the task. The constants (r_i, δ_i, c_i) that determine shapes of these ellipsoids can be obtained from experiments or from experience with similar tasks. Hence, we need only to store in a library a set data for the ellipsoids for a set of interesting tasks.

There are also other approaches to develop task ellipsoids. For example, if stiffness control is used for the hand, then the maximum expected positional uncertainities in each of the task directions may be used to scale the axis of B_β. Also, during parts mating, jamming can be avoided if certain constraints on the ratios of the contact forces are satisfied [Whitney 1983]. Using these constraints on the force ratios to scale the ellipsoid A_α is another approach.

Generalizing from these examples we will assume that a task is modeled by two generalized ellipsoids A_α (in the wrench space) and B_β (in the twist space) of the form

$$A_\alpha = \{\alpha Ax + c, \text{ such that } x, c \in \mathcal{R}^6, |x| \leq 1, \text{ and } A \in \mathcal{R}^{6\times 6}\} \tag{8.28}$$

and

$$B_\beta = \{\beta Bx + d, \text{ such that } h, d \in \mathcal{R}^6, |x| \leq 1, \text{ and } B \in \mathcal{R}^{6\times 6}\} \tag{8.29}$$

where the structure matrices A, B are given by

$$\begin{aligned} A &= [U_1, ...U_6] \operatorname{diag}\{\sigma_1, ...\sigma_6\} \begin{bmatrix} U_1 \\ \cdot \\ \cdot \\ \cdot \\ U_6 \end{bmatrix} \\ &= U\Sigma U^t, U_i \in \mathcal{R}^6, \sigma_i \geq 0, i = 1, ...6. \end{aligned} \tag{8.30}$$

and

$$\begin{aligned} B &= [U_1, ...U_6] \;\; \operatorname{diag}\{\delta_1, ...\delta_6\} \begin{bmatrix} U_1 \\ \cdot \\ \cdot \\ \cdot \\ U_6 \end{bmatrix} \\ &= U\Delta U^t, U_i \in \mathcal{R}^6, i = 1, ...6. \end{aligned} \tag{8.31}$$

Here, U_i, $(i = 1, ...6,)$ is the task direction expressed in the body coordinates and σ_i (or δ_i) is the relative importance ratio in the wrench space (and the twist space, respectively) at direction U_i. The constant $c, d \in \mathcal{R}^6$ is called center of the task ellipsoid A_α and B_β, respectively. In the following development, we may assume without loss of generality that $c = d = 0$, namely the task ellipsoids are all centered at the origin. When the structure matrix, say A, is nonsingular an alternative expression of A_α may be obtained as (where $\langle \cdot, \cdot \rangle$ denotes the inner product in $\mathcal{R}^6$).

$$A_\alpha = \{y \in \mathcal{R}^6, \text{ such that} \langle y - c, \beta^{-2}(AA^t)^{-1}(y - c)\rangle \leq 1\} \tag{8.32}$$

When stiffness control is used for the hand, σ_i is related to δ_i by $\sigma_i = k_i \cdot \delta_i$, where k_i is the desired stiffness in direction U_i. On the other hand, when hybrid position/force control is used for the hand, $\sigma_i = 0$ if direction U_i is position controlled and $\delta_i = 0$ if direction U_i is force controlled. Consequently, we see that our approach to task modeling applies to very general tasks.

8.3.1 The Structured Quality Measures for Grasp Planning

We have shown that a grasp $\Omega = (G, K, J_h)$ contains information about the locations of the fingertips on the object (G and K) and the postures of the fingers (J_h). Also, we have modeled a task by two ellipsoids A_α and B_β. We now integrate these concepts to develop two quality measures for evaluating a grasp, one in the twist space and the other in the wrench space.

Definition 8.3.1 *(The Structured Twist Space Quality Measure μ_t): Following the previous notation we let $O_1^m \subset \mathcal{R}^m$ denote the unit ball in $\mathcal{R}^m$, the finger joint velocity space, and define the structured twist space quality measure $\mu_t(\Omega)$ of Ω by*

$$\mu_t(\Omega) = \sup_{\beta \in \mathcal{R}_+} \{\beta, \text{ such that } J_h(O_1^m) \supset G^t(B_\beta)\} \tag{8.33}$$

Remark 8.3.1 *The geometric meaning of $\mu_t(\Omega)$ is as follows (see Figure 8.6): the unit ball O_1^m in the finger joint velocity space is mapped into the fingertip velocity space by J_h. On the other hand, a task ellipsoid B_β in the twist space is mapped back into the fingertip velocity space by G^t. $\mu_t(\Omega)$ is then the largest β such that $G^t(B_\beta)$ is* contained *in $J_h(O_1^m)$. In other words, $\mu_t(\Omega)$ is the ratio of the* structured output (i.e., the task ellipsoid) over the input (i.e., the finger joint velocity). *We also see from the figure that $\mu_t(\Omega)$ is at its maximum if the inner ellipsoid has the same shape and orientation as the outer ellipsoid.*

Definition 8.3.2 *(The Structured Wrench Space Quality Measure μ_w): Let $O_1^n \subset \mathcal{R}^n$ be the unit ball in the finger wrench space and $\sigma_{\max}(J_h)$ the maximum singular value of J_h. We define the structured wrench space quality measure $\mu_w(\Omega)$ of Ω by*

$$\mu_w(\Omega) = \sup_{\alpha \in \mathcal{R}_+} \{\alpha, \text{such that } G(O_1^n \cap K) \supset A_\alpha\} \cdot \sigma_{\max}^{-1}(J_h) \tag{8.34}$$

Remark 8.3.2 *The geometric meaning of $\mu_w(\Omega)$ can be interpreted as follows (see Figure 8.7): The first term is the largest α such that A_α can be embedded in $G(O_1^n)$ (the output), and the second term is the largest input torque required to generate the finger wrench O_1^n (the input). The structured quality measure is given by the gain factor (output/input).*

These quality measures defined in (8.33) and (8.34) provide useful characterization of a grasp. Clearly, we can say that a grasp Ω is a *good grasp* with respect to a *given task*, modeled by A_α and B_β, if it has higher structured quality measures μ_t and μ_w than other candidate grasps. But due

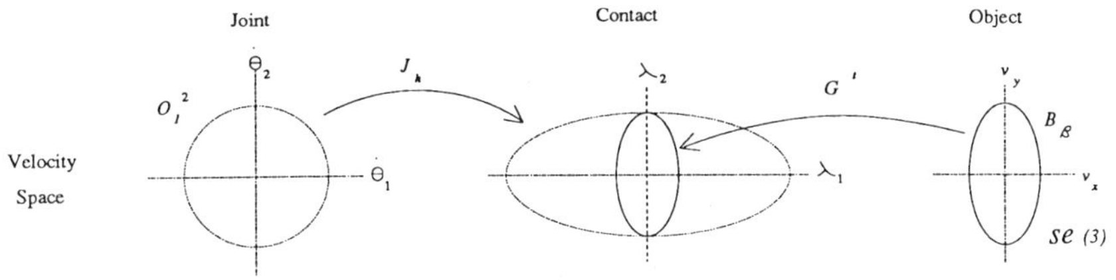

Figure 8.6: Geometrical interpretation of $\mu_t(\Omega)$.

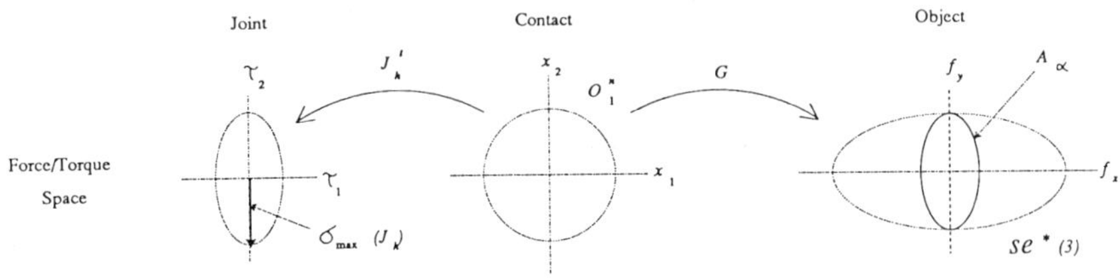

Figure 8.7: Geometrical interpretation of $\mu_w(\Omega)$.

to unidirectionality and finite frictional forces as reflected by K, which is usually a proper subset of $\mathcal{R}^n$, it is not easy to evaluate these measures. To simplify this problem we will assume that $K = \mathcal{R}^n$ and explore the properties of (8.33) and (8.34) further.

Proposition 8.3.1 *Under the assumption that $K = \mathcal{R}^n$ the structured quality measures (8.33) and (8.34) are given by*

$$\mu_t(\Omega) = \sigma_{\max}^{-1/2}\left\{B^t G (J_h J_h^t)^{-1} G^t B\right\} \tag{8.35}$$

and

$$\mu_w(\Omega) = \sigma_{\max}^{-1/2}\left\{A^t (G G^t)^{-1} A\right\} \cdot \sigma_{\max}^{-1}(J_h) \tag{8.36}$$

Proof. Using the following expression

$$J_h(O_1^m) = \left\{y \in \mathcal{R}^n, \langle y,\ (J_h J_h^t)^{-1} y\rangle \le 1\right\}$$

and

$$G^t(B_\beta) = \left\{\beta G^t B x, x \in \mathcal{R}^6, |x| \le 1\right\}$$

in (8.33) and notice that $G^t(B_\beta) \subset J_h(O_1^m)$ if and only if

$$\langle \beta G^t B x,\ (J_h J_h^t)^{-1} \beta G^t B x\rangle \le 1 \tag{8.37}$$

In particular, the above equation must hold for

$$\beta^2 \sup_{|x|=1} \langle G^t B x,\ (J_h J_h^t)^{-1} G^t B x\rangle = \beta^2 \sup_{|x|=1} \langle x,\ (G^t B)^t (J_h J_h^t)^{-1} G^t B x\rangle \le 1$$

which is equivalent to

$$\beta \le \sigma_{\max}^{-1/2}\left\{B^t G (J_h J_h^t)^{-1} G^t B\right\}$$

By (8.33), we have (8.35). The proof of (8.36) follows immediately. □

The quality measures given here can be easily computed using singular value decomposition. If we also want to consider the manipulability of a grasp in a certain direction, say U_i, we may simply apply U_i to (8.37) and obtain that

$$\beta_i = \langle G^t B U_i,\ (J_h J_h^t)^{-1} G^t B U_i\rangle^{-1/2} = \delta_i^{-1} \langle G^t U_i,\ (J_h J_h^t)^{-1} G^t U_i\rangle^{-1/2}.$$

Here, β_i measures the effectiveness of the grasp in imparting motion at direction U_i, and a similar relation holds for the stability measure.

Note that (8.35) and (8.36) exhibit an interesting *dual relation* in the following sense: let the task ellipsoids be the unit balls, if we hold G constant but vary J_h then $\mu_t(\Omega)$ is directly propotional to $\sigma_{\min}(J_h)$ and $\mu_w(\Omega)$ is inversely propotional to $\sigma_{\max}(J_h)$. On the other hand, if we hold J_h constant but vary G then $\mu_w(\Omega)$ is directly propotional to $\sigma_{\min}(G)$ and $\mu_t(\Omega)$ is inversely propotional to $\sigma_{\max}(G)$. This observation implies that to a certain point it is in general not possible to increase the two quality measures simultaneously by varying G and J_h. Namely, increasing one quality measure will sacrifice the other. For instance, if we select a "power grasp" in the scribing task, i.e., a grasp with high quality measure in the wrench space, then the grasp will be very inefficient in imparting motion at the pencil lead. Conversely, if we choose a "precision grasp", that is, *a grasp with high quality measure in the twist space*, the grasp will be very poor in rejecting disturbance forces. The objective of grasp planning, therefore, is to search for a grasp which maximizes some performance measure (PM) defined by

$$\text{PM} = [\mu_t(\Omega)]^\gamma \cdot [\mu_w(\Omega)]^{1-\gamma}, \quad \gamma \in [0,1] \tag{8.38}$$

Here, γ is called the selection parameter. $\gamma > 0.5$ indicates that the task is motion oriented and $\gamma < 0.5$ indicates that the task is force oriented. A grasp that maximizes PM with γ close to 1 will be a "precision grasp" and a grasp that maximizes PM with γ close to 0 will be a "power grasp". More generally, a grasp that maximizes PM with γ close to 0.5 will be both stable and manipulable. Depending on the nature of the task we can set the parameter γ so that the final grasp satisfies the task requirement.

In principle, we can use the definition of PM, geometry of the object and structures of the hand to formulate a standard optimization problem (see [Li *et al* 1989] for details) and by solving this optimization problem we obtain the final grasp. However, simultaneously placing all k fingers optimally usually involves considerable computation. We simplify the problem by considering plans for sequentially placing fingers around the periphery of the object. The suboptimal algorithm for grasp planning is: *place fingers one after another on the object periphery using the constraint that the objective function PM be maximized.* Details of this algorithm has been reported in [Li *et al* 1987].

Example 8.3.3 *Consider again the planar Peg-in-Hole task, shown in Figure 8.8. For simplicity we assume that the body orientation coincides with the orientation of C_p. Hence, the task direction matrix U is given by $U = I \in \mathcal{R}^{3\times3}$. To execute the task, stiffness control is used for the hand [Whitney 1983] and we assume that the desired stiffness matrix is given by*

$$K = diag(k_x, k_y k_\theta) = diag(5, 7, 100). \tag{8.39}$$

From Whitney [1983] and Example 8.3.1 we model the task by

$$U = I \in \mathcal{R}^{3\times3}, \Sigma = diag(4, 52), \Delta = diag(0.8, 0.7, 0.02) = K^{-1} \cdot \Sigma$$

The objective is to search for grasps that maximize quality measures (8.35) and (8.36).

Let the contact be modeled as a point contact with friction, the block width and the finger spacing be 2. To simplify the problem further we make additional assumptions: G is fixed as in the figure and the object is constrained

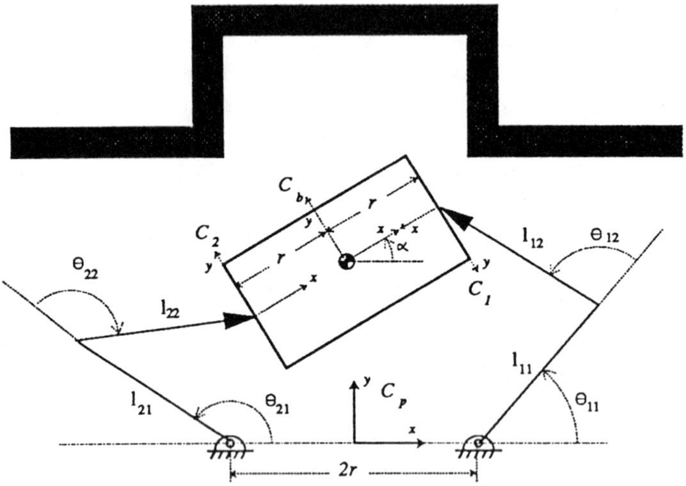

Figure 8.8: Planar peg-in-hole task.

to move vertically. This leaves the system with a single degree of freedom. Let θ_{11} be the generalized coordinate of the system and we study how θ_{11} affecting the structured grasp quality measures.

As shown in the figure, the grasp matrix is given by

$$G = \begin{bmatrix} -1 & 0 & 1 & 0 \\ 0 & -1 & 0 & 1 \\ 0 & -1 & 0 & -1 \end{bmatrix}$$

and the hand Jacobian J_h is

$$J_h = diag(J_1, J_2)$$

where

$$J_1 = \begin{bmatrix} \cos\alpha & -\sin\alpha \\ \sin\alpha & \cos\alpha \end{bmatrix} \begin{bmatrix} -\sin\theta_{11} - \sin(\theta_{11}+\theta_{12}) & -\sin(\theta_{11}+\theta_{12}) \\ \cos\theta_{11} + \cos(\theta_{11}+\theta_{12}) & \cos(\theta_{11}+\theta_{12}) \end{bmatrix}$$

and

$$J_2 = \begin{bmatrix} \cos\alpha & \sin\alpha \\ -\sin\alpha & \cos\alpha \end{bmatrix} \begin{bmatrix} \sin\theta_{21} - \sin(\theta_{21}-\theta_{22}) & \sin(\theta_{21}-\theta_{22}) \\ \cos\theta_{21} + \cos(\theta_{21}-\theta_{22}) & -\cos(\theta_{21}+\theta_{22}) \end{bmatrix}$$

Where α is the orientation angle of the object. The previous assumptions impose the following constraints: $\alpha = 0$, $\theta_{12} = \pi - 2\theta_{11}$, $\theta_{21} = \pi - \theta_{11}$, and $\theta_{11} - \theta_{22} = \pi - (\theta_{11} + \theta_{12})$. Figure 8.9 shows plots of the quality measures and the performance measure ($\gamma = 0.5$) as functions of θ_{11}. The structured

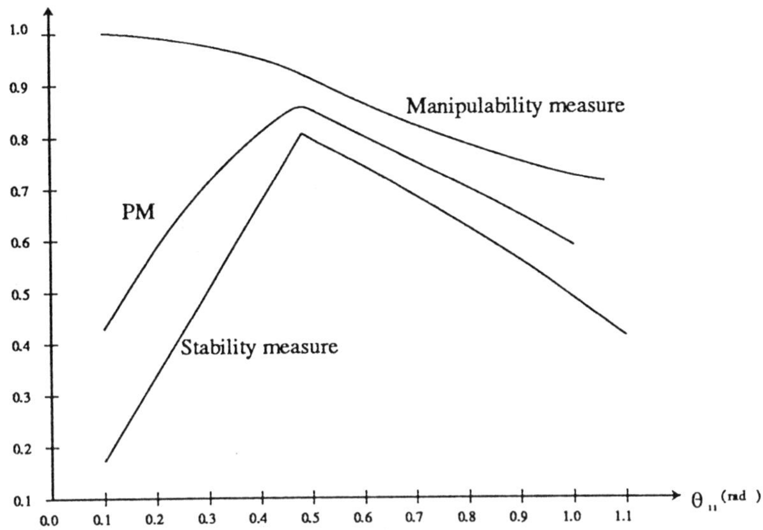

Figure 8.9: Performance measure, grasp quality measures versus θ_{11}

measure $\mu_t(\Omega)$ *and PM attain their maximum at* $\theta_{11} = 0.475$ *radian* (27° *degree). Since G is held constant the task structures have no effect on* μ_w*, which is still inversely propotional to* $\sigma_{\max}(J_h)$*. Clearly, the optimal posture (or grasp) is* $\theta_{11} = 27^o$. ♠

8.4 Control Algorithms for Dextrous Manipulation

In this section, we develop control algorithms for dextrous manipulation. The objectives for each of the manipulation modes are:

1. *Coordinated Manipulation:* Control the fingers, *with fixed points of contact*, so that the object can be manipulated along a prescribed trajectory in $SE(3)$ while exerting possibly a set of desired contact forces on the environment.

2. *Rolling Motion*: Control the fingers, *with rolling contacts*, so that the object can be manipulated along a prescribed trajectory in $SE(3)$, while exerting a set of desired contact forces on the environment.

We will first propose the control scheme for coordinated manipulation, and then show that, after minor modifications, it also applies to rolling motion.

8.4.1 A Control Algorithm for Coordinated Manipulation

We assume that the desired trajectory of the object is

$$g_{b,p}^d(t) = (r_{b,p}^d(t), A_{b,p}^d(t)) \in SE(3), t \in [t_0, t_f]. \tag{8.40}$$

Since $SO(3)$ is a three dimensional manifold, we may choose pitch-roll-yaw variables, or the exponential coordinates to parameterize it. Let $\phi_{b,p} = (\phi_1, \phi_2, \phi_3)^t$ be one of these locally nonsingular parameterization. Then, we can express a trajectory of the object in terms of the parameterization variables

$$g_{b,p}(t) = (r_{b,p}(t), A_{b,p}(\phi_{b,p}(t))) \in SE(3),$$

and the body velocity in terms of the derivatives of the parameterization variables

$$\begin{bmatrix} v_{b,p} \\ w_{b,p} \end{bmatrix} = U(r_{b,p}, \phi_{b,p}) \begin{bmatrix} \dot{r}_{b,p} \\ \dot{\phi}_{b,p} \end{bmatrix} \tag{8.41}$$

Clearly, $U(r_{b,p}, \phi_{b,p})$ is a nonsingular 6×6 matrix. Differentiating (8.41) with respect to time t, yields the acceleration relation

$$\begin{bmatrix} \dot{v}_{b,p} \\ \dot{w}_{b,p} \end{bmatrix} = U(r_{b,p}, \phi_{b,p}) \begin{bmatrix} \ddot{r}_{b,p} \\ \ddot{\phi}_{b,p} \end{bmatrix} + \dot{U}(r_{b,p}, \phi_{b,p}) \begin{bmatrix} \dot{r}_{b,p} \\ \dot{\phi}_{b,p} \end{bmatrix}. \tag{8.42}$$

Note that to realize a desired trajectory of the object it is important that contact constraints be retained. On the other hand, validity of the contact constraints depends on if a proper set of internal grasp forces can be maintained. Thus, the second goal of the control algorithm is to regulate the internal grasp force to the following desired value

$$x_o^d(t) \in \eta(G), t \in [t_0, t_f], \tag{8.43}$$

where $\eta(G)$ is the null space of G. For fixed points of contact G is time independent, and a set of constant internal grasp forces suffices to guarantee contact constraints. Nevertheless, one may choose $x_o^d(t)$ to realize any other criteria.

We now proceed to formulate the control algorithms with (8.40) and (8.43) as our objectives.

From Section 8.2, the finger joint velocity $\dot{\theta}$ and the object velocity $(v_{b,p}^t, w_{b,p}^t)^t$ are related by the following constraint equation

$$J_h(\theta)\dot{\theta} = G^t \begin{bmatrix} v_{b,p} \\ w_{b,p} \end{bmatrix} \tag{8.44}$$

(8.44) is valid regardless of rolling or fixed points of contact. Differentiating (8.44) with respect to time t, yields the following acceleration constraint equation

$$J_h(\theta)\ddot{\theta} + \dot{J}_h(\theta)\dot{\theta} = G^t \begin{bmatrix} \dot{v}_{b,p} \\ \dot{w}_{b,p} \end{bmatrix} + \dot{G}^t \begin{bmatrix} v_{b,p} \\ w_{b,p} \end{bmatrix}. \tag{8.45}$$

Note that for fixed points of contact G is constant and the second term to the right hand side of (8.45) vanishes.

We will need the following assumption about the grasp.

A1. The grasp is both stable and manipulable.

Generation of object/finger trajectories so that assumption (A1) can be satisfied is discussed in [Li and Sastry 1988].

By Assumption (A1), we have that $R(J_h) \supset R(G^t)$ and we may express the joint acceleration $\ddot{\theta}$ in terms of the object acceleration by

$$\ddot{\theta} = J_h^+ G^t \begin{bmatrix} \dot{v}_{b,p} \\ \dot{w}_{b,p} \end{bmatrix} + J_h^+ \dot{G}^t \begin{bmatrix} v_{b,p} \\ w_{b,p} \end{bmatrix} - J_h^+ \dot{J}_h \dot{\theta} + \ddot{\theta}_o. \tag{8.46}$$

Here $J_h^+ = J_h^t (J_h J_h^t)^{-1}$ is the generalized inverse of J_h, and $\ddot{\theta}_0 \in \eta(J_h)$ is the internal motion of redundant joints not affecting the object motion.

Remark 8.4.1 *(1) Using (8.46) will lead to a control algorithm in the task space. But if we express the object acceleration in terms of $\ddot{\theta}$ by*

$$\begin{bmatrix} \dot{v}_{b,p} \\ \dot{w}_{b,p} \end{bmatrix} = (GG^t)^{-1} G \left(J_h \ddot{\theta} + \dot{J}_h \dot{\theta} - \dot{G}^t \begin{bmatrix} v_{b,p} \\ w_{b,p} \end{bmatrix} \right)$$

a control algorithm in the joint space of the fingers can be developed.
(2) When J_h is square, its generalized inverse J_h^+ is just the usual inverse, and $\ddot{\theta}_o$ disappears from (8.46). This also implies that the joint motion is determined uniquely by the motion of the object.

The dynamics of the object are given by the Newton-Euler equations

$$\begin{bmatrix} \hat{m} & 0 \\ 0 & \mathcal{I} \end{bmatrix} \begin{bmatrix} \dot{v}_{b,p} \\ \dot{w}_{b,p} \end{bmatrix} + \begin{bmatrix} w_{b,p} \times \hat{m} v_{b,p} \\ w_{b,p} \times \mathcal{I} w_{b,p} \end{bmatrix} = \begin{bmatrix} f_b \\ m_b \end{bmatrix}, \tag{8.47}$$

where $\hat{m} \in \mathcal{R}^{3\times 3}$ is the diagonal matrix with the object mass in the diagonal, $\mathcal{I} \in \mathcal{R}^{3\times 3}$ is the object inertia matrix with respect to the body coordinates, and $[f_b^t, m_b^t]^t$ is the applied body wrench in the body coordinates which is also related to the applied finger wrench $x \in \mathcal{R}^n$ through

$$Gx = \begin{bmatrix} f_b \\ m_b \end{bmatrix}. \tag{8.48}$$

Notice that gravity and interaction forces from the environment can always be added to the right hand side of (8.47), and corresponding contact wrenches will be generated to counteract them.

Assumption A1 also implies that G is onto, and we can solve (8.48) as

$$x = G^+ \begin{bmatrix} f_b \\ m_b \end{bmatrix} + x_o, \tag{8.49}$$

where $G^+ = G^t (GG^t)^{-1}$ is the left inverse of G, and $x_o \in \eta(G)$ is the internal grasping force. The second control goal is to steer the internal grasping force x_o to the desired value $x_o^d(t) \in \eta(G)$.

Combining (8.47) and (8.49) yields

$$x = G^{+}\left\{\begin{bmatrix} \hat{m} & 0 \\ 0 & \mathcal{I} \end{bmatrix}\begin{bmatrix} \dot{v}_{b,p} \\ \dot{w}_{b,p} \end{bmatrix} + \begin{bmatrix} w_{b,p} \times \hat{m} v_{b,p} \\ w_{b,p} \times \mathcal{I} w_{b,p} \end{bmatrix}\right\} + x_o. \tag{8.50}$$

The dynamics of the ith finger manipulator is given by

$$M_i(\theta_i)\ddot{\theta}_i + N_i(\theta_i, \dot{\theta}_i) = \tau_i - J_i^t(\theta_i)B_i x_i. \tag{8.51}$$

Here, as is common in the literature, $M_i(\theta_i) \in \mathcal{R}^{m_i \times m_i}$ is the moment of inertia matrix of the ith finger manipulator, $N_i(\theta_i, \dot{\theta}_i) \in \mathcal{R}^{m_i}$ is the centrifugal, Coriolis and gravitational force terms, τ_i is the vector of joint torque inputs and $B_i x_i \in \mathcal{R}^6$ the vector of applied finger wrenches. Define

$$M(\theta) = \begin{bmatrix} M_1(\theta_1) & \cdot & \cdot & \cdot & 0 \\ \cdot & & \cdot & \cdot & \cdot \\ \cdot & & \cdot & \cdot & \cdot \\ \cdot & & \cdot & \cdot & \cdot \\ 0 & & \cdot & \cdot & M_k(\theta_k) \end{bmatrix},$$

$$N(\theta, \dot{\theta}) = \begin{bmatrix} N_1(\theta_1, \dot{\theta}_1) \\ \cdot \\ \cdot \\ \cdot \\ N_k(\theta_k, \dot{\theta}_k) \end{bmatrix}$$

and

$$\tau = \begin{bmatrix} \tau_1 \\ \cdot \\ \cdot \\ \cdot \\ \tau_k \end{bmatrix}. \tag{8.52}$$

Then, the finger dynamics can be grouped to yield

$$M(\theta)\ddot{\theta} + N(\theta, \dot{\theta}) = \tau - J_h^t(\theta)x. \tag{8.53}$$

The control objectives are to specify a set of joint torque inputs τ so that both the desired body trajectory (8.40) and the desired internal grasping force (8.43) can be realized.

The first main theorem of the chapter is the following control algorithm for coordinated manipulation by robot hands with *non-redundant fingers.*

Theorem 8.4.1 *Assume that (A1) holds and the fingers are non-redundant, i.e., $m_i = n_i$, for $i = 1, ...k$. Denoting the position trajectory tracking error by $e_p \in \mathcal{R}^6$, and the internal grasping force error by $e_f \in \mathcal{R}^6$,*

$$e_p(t) = \begin{bmatrix} r_{b,p}(t) \\ \phi_{b,p}(t) \end{bmatrix} - \begin{bmatrix} r_{b,p}^d(t) \\ \phi_{b,p}^d(t) \end{bmatrix} \quad \text{and } e_f(t) = x_o(t) - x_o^d(t). \tag{8.54}$$

Then, the control law specified by (8.55) realizes, with fixed points of contact, not only the desired trajectory of the object but also the desired internal grasping force.

$$\begin{aligned} \tau &= N(\theta,\dot\theta) + J_h^t G^+ \begin{bmatrix} w_{b,p} \times \hat m v_{b,p} \\ w_{b,p} \times \mathcal{I} w_{b,p} \end{bmatrix} - M(\theta) J_h^{-1} \dot J_h \dot\theta \\ &+ M_h \dot U \begin{bmatrix} \dot r_{b,p} \\ \dot\phi_{b,p} \end{bmatrix} + J_h^t (x_o^d - K_I \int e_f) \\ &+ M_h U \left\{ \begin{bmatrix} \ddot r_{b,p}^d \\ \ddot\phi_{b,p}^d \end{bmatrix} - K_v \dot e_p - K_p e_p \right\}, \end{aligned} \tag{8.55}$$

where

$$M_h = M(\theta) J_h^{-1} G^t + J_h^t G^+ \begin{bmatrix} \hat m & 0 \\ 0 & \mathcal{I} \end{bmatrix} \tag{8.56}$$

and K_I is a matrix such that the null space of G is K_I-invariant.

Note that generlized inertia matrix of the hand system is:

$$G J_h^{-t} M_h = G J_h^{-t} M(\theta) J_h^{-1} G^t + \begin{bmatrix} \hat m & 0 \\ 0 & \mathcal{I} \end{bmatrix}$$

Remark 8.4.2 *The first four components in (8.55) are used for cancellation of Coriolis, gravitational and centrifugal forces. These terms behave exactly like the nonlinearity cancellation terms in the computed torque control for a single manipulator; the term $J_h^t(x_o^d - K_I \int e_f)$ is the compensation for the internal grasping force loop, and the last term is the compensation for the position loop. We will see in the proof that the dynamics of the internal grasping force loop and that of the position loop are mutually decoupled. Consequently, we can design the force error integral gain K_I independently from the position feedback gains K_v and K_p.*

Proof. The proof is very procedural and straightforward. First, for nonredundant fingers with fixed points of contact, internal motion of the fingers reflected by $\dot\theta_o$ in (8.46) disappear and G is constant. Thus, the acceleration constraint equation (8.46) simplifies to

$$\ddot\theta = J_h^{-1} G^t \begin{bmatrix} \dot v_{b,p} \\ \dot w_{b,p} \end{bmatrix} - J_h^{-1} \dot J_h \dot\theta. \tag{8.57}$$

Substitute (8.57) and (8.50) into (8.53) we have

$$\begin{aligned} & M \left\{ J_h^{-1} G^t \begin{bmatrix} \dot v_{b,p} \\ \dot w_{b,p} \end{bmatrix} - J_h^{-1} \dot J_h \dot\theta \right\} + N = \\ & \tau - J_h^t x_o - J_h^t \left\{ G^+ \begin{bmatrix} \hat m & 0 \\ 0 & \mathcal{I} \end{bmatrix} \begin{bmatrix} \dot v_{b,p} \\ \dot w_{b,p} \end{bmatrix} \right. \\ & \qquad \left. + G^+ \begin{bmatrix} w_{b,p} \times \hat m v_{b,p} \\ w_{b,p} \times \mathcal{I} w_{b,p} \end{bmatrix} \right\}. \end{aligned} \tag{8.58}$$

Linearizing (8.58) with the following control

$$\tau = N(\theta,\dot\theta) + J_h^t G^+ \begin{bmatrix} w_{b,p} \times \hat m v_{b,p} \\ w_{b,p} \times \mathcal{I} w_{b,p} \end{bmatrix} - M(\theta) J_h^{-1} \dot J_h \dot\theta + \tau_1 \tag{8.59}$$

where τ_1 is to be determined, we have that

$$\left\{ M(\theta) J_h^{-1} G^t + J_h^t G^+ \begin{bmatrix} \hat m & 0 \\ 0 & \mathcal{I} \end{bmatrix} \right\} \begin{bmatrix} \dot v_{b,p} \\ \dot w_{b,p} \end{bmatrix} = \tau_1 - J_h^t x_o, \tag{8.60}$$

or

$$M_h \begin{bmatrix} \dot v_{b,p} \\ \dot w_{b,p} \end{bmatrix} = \tau_1 - J_h^t x_o.$$

Substitute (8.42) into the above equation, we have

$$M_h \left\{ U \begin{bmatrix} \ddot r_{b,p} \\ \ddot\phi_{b,p} \end{bmatrix} + \dot U \begin{bmatrix} \dot r_{b,p} \\ \dot\phi_{b,p} \end{bmatrix} \right\} = \tau_1 - J_h^t x_o. \tag{8.61}$$

Further, let the control input τ_1 be

$$\begin{aligned} \tau_1 &= M_h U \left\{ \begin{bmatrix} \ddot r^d_{b,p} \\ \ddot\phi^d_{b,p} \end{bmatrix} - K_v \dot e_p - K_p e_p \right\} \\ &+ M_h \dot U \begin{bmatrix} \dot r_{b,p} \\ \dot\phi_{b,p} \end{bmatrix} + J_h^t \left(x_o^d - K_I \int e_f \right) \end{aligned} \tag{8.62}$$

and apply it to (8.61) to yield:

$$M_h U \{\ddot e_p + K_v \dot e_p + K_p e_p\} = -J_h^t (e_f + K_I \int e_f). \tag{8.63}$$

Multiply (8.63) by $G J_h^{-t}$, we obtain the following equation.

$$G J_h^{-t} M_h U \{\ddot e_p + K_v \dot e_p + K_p e_p\} = -G(e_f + K_I \int e_f) = 0 \tag{8.64}$$

where we have used the facts that $\eta(G)$ is constant and the internal grasping forces lie in the null space of G, i.e.,

$$G(e_f + K_I \int e_f) = 0. \tag{8.65}$$

Since $G J_h^{-t} M_h = G J_h^{-t} M(\theta) J_h^{-1} G^t + \begin{bmatrix} \hat m & 0 \\ 0 & \mathcal{I} \end{bmatrix}$ is positive definite and U is non-singular, (8.64) implies that

$$\ddot e_p + K_v \dot e_p + K_p e_p = 0. \tag{8.66}$$

Thus, we have shown that the position trajectory tracking error e_p can be driven to zero with proper choice of the feedback gain matrices K_v and K_p.

The last step is to show that e_f also goes to zero. If we substitute (8.66) into (8.63) and notice that J_h is nonsingular, we have the following equation.

$$e_f + K_I \int e_f = 0. \tag{8.67}$$

With proper choice of K_I, the above equation implies that the internal grasping force error e_f converges to zero. □

Theorem 8.4.1 provides a control law for coordinated manipulation by robot hands with non-redundant fingers. This implies that, for a point contact with friction, each finger of the hand has exactly three joints. But, in many industrial applications, several robots which often have more than three degrees of freedom are integrated to maneuver a massive load, or to perform a sophisticated task. Under the point contact model assumption the system has redundant degrees of freedom. It is desirable to have a control law that works for a robot hand with redundant degrees of freedom. The control law of Theorem 8.4.1 can be modified for this purpose.

Corollary 8.4.1 *For a robot hand with redundant degrees of freedom, i.e., $m_i \geq n_i, i = 1, ...k$ assume that assumption A1 holds. Then, the control law given by (8.68) will realize both the desired object trajectory and the desired internal grasp force, with fixed points of contact.*

$$\begin{aligned}
\tau &= N(\theta,\dot\theta) + J_h^t G^+ \begin{bmatrix} w_{b,p} \times \hat m v_{b,p} \\ w_{b,p} \times \mathcal{I} w_{b,p} \end{bmatrix} - M J_h^+ \dot J_h \dot\theta \\
&+ M J_h^+ (J_h M^{-1} J_h^t) M_h \dot U \begin{bmatrix} \dot r_{b,p} \\ \dot\phi_{b,p} \end{bmatrix} \\
&+ M J_h^+ (J_h M^{-1} J_h^t)(x_o^d - K_I \int e_f) \\
&+ M J_h^+ (J_h M^{-1} J_h^t) \hat M_h \left\{ \begin{bmatrix} \ddot r_{b,p}^d \\ \ddot\phi_{b,p}^d \end{bmatrix} - K_v \dot e_p - K_p e_p \right\},
\end{aligned} \tag{8.68}$$

where

$$\hat M_h = (J_h M^{-1} J_h^t)^{-1} G^t + G^+ \begin{bmatrix} \hat m & 0 \\ 0 & \mathcal{I} \end{bmatrix}. \tag{8.69}$$

and $J_h^+ = J_h^t (J_h J_h^t)^{-1}$.

8.4.2 A Control Algorithm for Rolling Motion

Controls for dextrous manipulation with rolling constraints have been studied in [Kerr 1985; Cole, Hauser and Sastry 1988]. Here, we modify Theorem 8.4.1 to give a control law for rolling motion. The main difference of rolling motion from coordinated manipulation is that the grip Jacobian G, which depends on the contact coordinates given by (8.9), is time varying. Consequently, the null space of $G(t)$, denoted by $V(t) \subset \mathcal{R}^n$, is also time varying.

In general $e_f(t) = x_o(t) - x_o^d(t) \in V(t)$ does not imply, unless $V(t)$ is time independent, that $\int_0^t e_f(\tau)d\tau \in V(t)$, nor $\dot{e}_f(t) \in V(t)$. Thus, we can not introduce dynamic feedback in the force loop, as we did in Theorem 8.4.1, to create linear force error equation. But,

$$G(t)e_f(t) = 0 \text{ implies that } G(t)\dot{e}_f(t) + \dot{G}(t)e_f(t) = 0. \tag{8.70}$$

Lemma 8.4.1 *Consider the following differential equation*

$$\dot{x}(t) = A(t)x(t). \tag{8.71}$$

Let $\mu(A(t)) = \lambda_{max}(A^*(t) + A(t))/2$ *be the matrix measure of* $A(t)$, *where* λ_{max} *stands for the maximum eigenvalue value. Then,*

$$\|x(t)\| \leq \|x(t_0)\| \exp \int_{t_0}^{t} \mu(A(\tau))d\tau.$$

In other words, if $\mu(A(t)) < 0, \forall t$, and $A(t)$ is sufficiently slow time-varying, then, the system (8.71) is exponentially stable.

Proposition 8.4.1 *Assume that Assumption (A1) holds for a robot hand with non-redundant degrees of freedom. Then, the following control law, along with the kinematic equations of contact given by (8.9), realizes both the desired position trajectory and the desired internal grasp force, for rolling contacts.*

$$\begin{aligned} \tau &= J_h^t G^+ \begin{bmatrix} w_{b,p} \times \hat{m} v_{b,p} \\ w_{b,p} \times \mathcal{I} w_{b,p} \end{bmatrix} - M(\theta)J_h^{-1}\dot{J}_h\dot{\theta} + M_h\dot{U}\begin{bmatrix} \dot{r}_{b,p} \\ \dot{\phi}_{b,p} \end{bmatrix} \\ &+ \underline{M(\theta)J_h^{-1}\dot{G}^t \begin{bmatrix} v_{b,p} \\ w_{b,p} \end{bmatrix}} + J_h^t \left(x_o^d - \underline{\dot{e}_f/\delta - G^+\dot{G}e_f/\delta}\right) \\ &+ M_h U \left\{ \begin{bmatrix} \ddot{r}_{b,p}^d \\ \ddot{\phi}_{b,p}^d \end{bmatrix} - K_v\dot{e}_p - K_p e_p \right\} + N(\theta, \dot{\theta}), \end{aligned} \tag{8.72}$$

where

$$M_h = M(\theta)J_h^{-1}G^t + J_h^t G^+ \begin{bmatrix} \hat{m} & 0 \\ 0 & \mathcal{I} \end{bmatrix} \tag{8.73}$$

and δ *is a sufficiently large number so that the force error equation can be made to be exponentially stable.*

Remark 8.4.3 *We have underlined terms that are different from Theorem 8.4.1.*

Proof. The proof is very similar to that of Theorem 8.4.1 and an outline of it is given here. Under rolling constraints, (8.46) becomes

$$\ddot{\theta} = J_h^{-1}G^t \begin{bmatrix} \dot{v}_{b,p} \\ \dot{w}_{b,p} \end{bmatrix} + J_h^{-1}\dot{G}^t \begin{bmatrix} v_{b,p} \\ w_{b,p} \end{bmatrix} - J_h^{-1}\dot{J}_h\dot{\theta}. \tag{8.74}$$

Substitute (8.74) and (8.50) into the system dynamic equation (8.53) and linearize the resulting equation with the appropriate terms in the control inputs (8.72), we get

$$M_h \begin{bmatrix} \dot{v}_{b,p} \\ \dot{w}_{b,p} \end{bmatrix} = \tau_1 - J_h^t x_o. \tag{8.75}$$

Substituting (8.42) into the above equation and applying the rest of the control inputs yield

$$M_h\{\ddot{e}_p + K_v\dot{e}_p + K_p e_p\} = -J_h^t\left\{e_f + \dot{e}_f/\delta + G^+\dot{G}e_f/\delta\right\}. \tag{8.76}$$

Multiply (8.76) by GJ_h^{-t} and notice that because

$$Ge_f = 0, \text{ and } (G\dot{e}_f + \dot{G}e_f)/\delta = 0.$$

we have

$$G(e_f + \dot{e}_f/\delta + G^+\dot{G}e_f/\delta) = 0$$

which implies that

$$\ddot{e}_p + K_v\dot{e}_p + K_p e_p = 0. \tag{8.77}$$

This shows that the position error goes to zero. On the other hand, substituting (8.77) into (8.76), and using the fact that J_h is of full rank, we conclude that

$$(\delta I + G^+\dot{G})e_f + \dot{e}_f = 0. \tag{8.78}$$

Let $A(t) = -(\delta I + G^+\dot{G})$. It is easy to see that by choosing δ sufficiently large, $\mu(A(t))$ is negative for all $t \in [t_0, t_f]$. Consequently, by Lemma 8.4.1 force error e_f also goes to zero. □

Combining Proposition 8.4.1 with Corollary 8.4.1 produces a control law for rolling motion by a robot hand with redundant degrees of freedom.

8.4.3 Simulation

Consider the two-fingered planar manipulation system shown in Figure 8.8, where the two fingers are assumed to be identical. We model the contact to be a point contact with friction. Let the object width and the finger spacing be 2 units. The grip Jacobian and the hand Jacobian are given in Example 8.3.3.

The grasp will be stable and manipulable for the object along the following trajectory

$$x(t) = c_1\sin(t),\ y(t) = c_2 + c_1\cos(t), \alpha(t) = c_3\sin(t).$$

With the control law given by Theorem 8.4.1, we have simulated the system using a program designed to integrate differential equations with algebraic constraints. Figure 8.10 shows that the initial position error diminishes exponentially as predicted by (8.66).

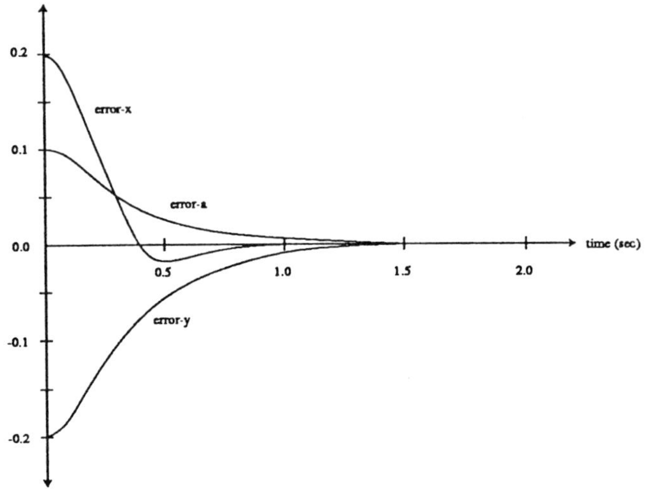

Figure 8.10: Simulated position error

8.5 Conclusion

In this chapter, we have studied grasp planning and coordinated controls for manipulation by a multifingered robot hand. In Section 8.3 of the chapter, we saw that it is unnecessary to classify grasps by "power grasp" or "precision grasp". Contrary to the popular belief, a "power grasp", according to our study, is not strictly in the complement to a "precision grasp". In other words, these exist grasps with are both power grasps and precision grasps. These are called grasps with high structured stability and manipulability measures. In Section 8.4, we have proposed basic control laws for coordinated manipulation by a multifingered robot hand under both fixed points of contact and rolling contact. These control laws give decoupled error equations in the position loop and the internal grasp force loop.

Acknowledgements: This Research was supported in part by National Science Foundation PYI Grant DMC 8451129.

9

Analysis of Multi-fingered Grasping and Manipulation

Tsuneo Yoshikawa[1]
Kiyoshi Nagai[2]

ABSTRACT Grasping and manipulating forces for skillful manipulation of objects by multi-fingered robot hands are discussed in this chapter. Firstly a short discussion of grasping and manipulating forces for two-fingered hands with linear motion is given. Then, for three-fingered hands, a new representation of the internal force among the fingers is given. Based on this representation, the grasping force is defined as an internal force which satisfies the static friction constraint. The concept of grasp mode is also introduced. Manipulating force is then defined as a fingertip force which satisfies the following three conditions: (1) It produces the specified resultant force. (2) It is not in the inverse direction of the grasping force. (3) It does not contain any grasping force component. An algorithm for decomposing a given fingertip force into manipulating and grasping forces is presented.

9.1 Introduction

Various multi-fingered hands for manipulating objects skillfully have been developed [Hanafusa and Asada 1977b; Okada 1979; Salisbury and Craig 1982; Jacobsen *et al* 1985; Kobayashi 1985a] Analytical studies of grasping and manipulation by robot hands have also been done by many researchers. Hanafusa and Asada [1977b] proposed the prehension potential for obtaining a stable finger position of an arbitrarily shaped object. Salisbury and Craig [1982] provide a basic kinematic analysis of articulated hands using the concept of mobility and connectivity. They have also given an expression relating the total resultant force to fingertip forces and having an explicit parametrization of internal force. To determine a suitable grasping posture, Salisbury and Craig [1982] proposed the use of the condition number of a Jacobian matrix, while Yoshikawa [1985] proposed the

[1]Automation Research Laboratory, Kyoto University, Uji, Kyoto 611, Japan
[2]Faculty of Science and Engineering, Ritsumeikan University, Kyoto, Kyoto 603, Japan

use of manipulability measures. Kobayashi [1985b] defined the norm of the handling force for an expected external force, and proposed to use it for obtaining a suitable grasping posture. Hanafusa *et al* [1985] defined the magnitude of the grasping force as the square root of the area of a triangle formed by the internal forces of three fingers. Kerr and Roth [1986] proposed to determine the optimal internal forces as those with the minimum norm under an approximated frictional constraints. Kobayashi [1985a] used the terms of "manipulating force" and "grasping force" with respect to the joint torques of the fingers. In spite of these studies, the authors still think that the forces involved in grasping and manipulation have not been fully understood. This understanding is important for designing hands and for developing grasping and manipulation algorithms.

In this chapter, grasping and manipulating forces are defined explicitly for articulated multi-fingered hands, and the relation between these two forces and fingertip force is made clear. Firstly, the definition of manipulating and grasping forces for two-fingered hands with linear motion is given such that they agree with our intuitive image of these forces. Then, a new representation of the internal force is given for three-fingered hands. Based on this representation, the grasping force and the manipulating force are defined. Finally, an algorithm for decomposing a given fingertip force into manipulating and grasping forces is presented. This chapter is a revised and extended version of [Yoshikawa and Nagai 1987].

9.2 Two-fingered Hand

First we consider a very simple case of two-fingered hands as shown in Figure 9.1 to explain the motivation and the main idea of our study. Assume that the two fingers have one degree of freedom and can move in x direction, and the two fingertips make point contacts with friction with a grasped object. Hence the i-th finger $i = 1,2$ applies force $f_i \epsilon R^1$ (R^n is the set of all n-dimensional real vectors) on the object along the x axis. We first assume that we do not need to consider forces perpendicular to the x-axis. Then the relation between fingertip force $[f_1, f_2]^T$ and the resultant force $t \epsilon R^1$ exerted on the object is given by

$$t = f_1 + f_2 \quad = \quad A \begin{bmatrix} f_1 \\ f_2 \end{bmatrix} \tag{9.1}$$

$$A = [1\ 1] \tag{9.2}$$

Therefore the fingertip force $[f_1, f_2]^T$ which satisfies equation 9.1 for a specified resultant force t is given by the general solution

$$\begin{bmatrix} f_1 \\ f_2 \end{bmatrix} \quad = \quad A^+ t \, + \, [E_2 - A^+ A] \begin{bmatrix} y_1 \\ y_2 \end{bmatrix} \tag{9.3}$$

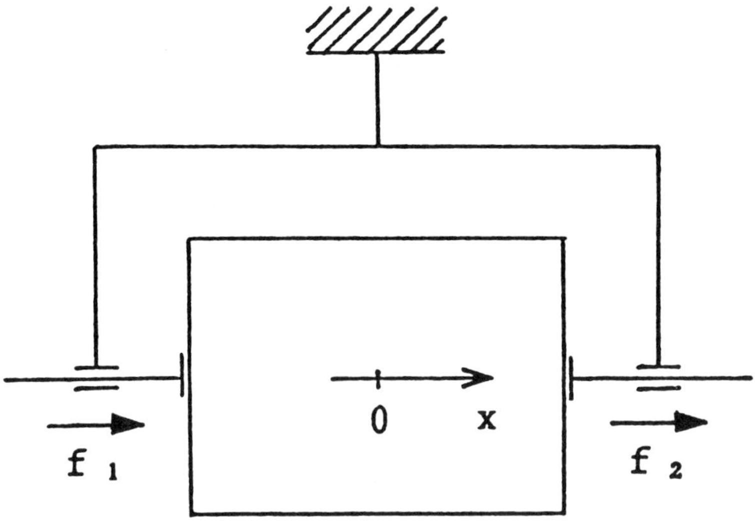

Figure 9.1: Two-fingered hand.

where A^+ is the pseudoinverse of A [Boullion and Odell 1977, pp. 1-11], $[y_1, y_2]^T \epsilon R^2$ is an arbitrary constant vector, and $E_n \epsilon R^{n\times n}(R^{n\times n}$ is the set of all $n\times n$ real matrices) denotes the unity matrix. The first term of equation 9.3 represents the fingertip force with the minimum norm among those which produces the resultant force t. The second term for any $[y_1, y_2]^T$ represents the fingertip force which produces zero resultant force, that is, this fingertip force is an internal force. Substituting equation 9.1 into 9.3 yields

$$\begin{bmatrix} f_1 \\ f_2 \end{bmatrix} = A^+A\begin{bmatrix} f_1 \\ f_2 \end{bmatrix} + [E_2 - A^+A]\begin{bmatrix} f_1 \\ f_2 \end{bmatrix} \tag{9.4}$$

This equation represents a decomposition of any fingertip force into two components. The term $A^+A\,[f_1,\ f_2]^T \triangleq \left[\tilde{f}_{m1}, \tilde{f}_{m2}\right]^T$ might be considered as the component for manipulation, and $[E_2 - A^+A]\,[f_1, f_2]^T \triangleq \left[\tilde{f}_{g1}, \tilde{f}_{g2}\right]^T$, as the component for grasping. Substituting $A^+ = A^T(AA^T)^{-1} = \left[\frac{1}{2}, \frac{1}{2}\right]^T$ from equation 9.2 into equation 9.4, we have

$$\begin{bmatrix} \tilde{f}_{m1} \\ \tilde{f}_{m2} \end{bmatrix} = \begin{bmatrix} (f_1 + f_2)/2 \\ (f_1 + f_2)/2 \end{bmatrix} \tag{9.5}$$

$$\begin{bmatrix} \tilde{f}_{g1} \\ \tilde{f}_{g2} \end{bmatrix} = \begin{bmatrix} (f_1 - f_2)/2 \\ -(f_1 - f_2)/2 \end{bmatrix} \tag{9.6}$$

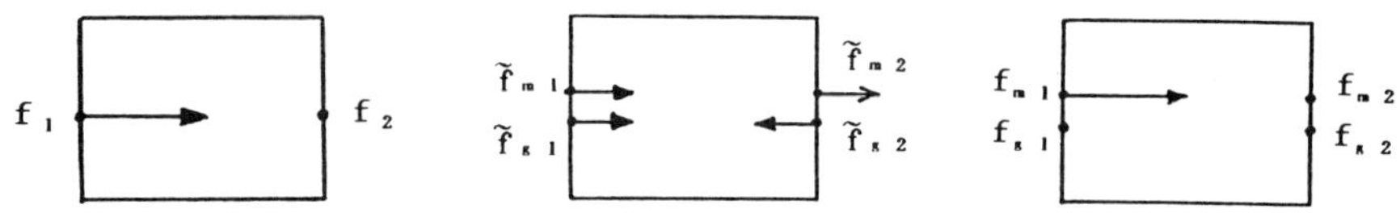

(a) Fingertip force (b) Decomposition by (5) and (6) (c) Decomposition by (8) and (9)

Figure 9.2: Decomposition of fingertip force.

If we consider the case $f_1 = f_0 > 0$ and $f_2 = 0$, equations 9.5 and 9.6 become

$$\begin{bmatrix} \tilde{f}_{m1} \\ \tilde{f}_{m2} \end{bmatrix} = \begin{bmatrix} f_0/2 \\ f_0/2 \end{bmatrix} \tag{9.7}$$

$$\begin{bmatrix} \tilde{f}_{g1} \\ \tilde{f}_{g2} \end{bmatrix} = \begin{bmatrix} f_0/2 \\ -f_0/2 \end{bmatrix} \tag{9.8}$$

Hence the force component for grasping $[\tilde{f}_{g1}, \tilde{f}_{g2}]^T$ is not zero.

Figure 9.2(a) shows this result schematically. It is easily seen that, under this state of grasping, since $f_2 = 0$ the object will easily slip out of the two fingers by an application of any small external force which is not parallel with the x axis, although according to our intuitive image, the grasping force for this case will be zero. Hence equation 9.3 will not be used to define grasping force. Based on this consideration, we define grasping force and manipulating force with the following equations.

$$\begin{bmatrix} f_{m1} \\ f_{m2} \end{bmatrix} = \begin{bmatrix} \tilde{l} \\ -(1-\tilde{l}) \end{bmatrix} |f_1 + f_2| \tag{9.9}$$

$$\begin{bmatrix} f_{g1} \\ f_{g2} \end{bmatrix} = \begin{bmatrix} 1 \\ -1 \end{bmatrix} min(|f_1|, |f_2|) \tag{9.10}$$

where $\tilde{l}$ is an auxiliary parameter which takes value 1 when $f_1 + f_2 \geq 0$ and value 0 when $f_1 + f_2 < 0$. According to these definitions, the fingertip force $[f_1, f_2]^T = [f_0, 0]^T$ is decomposed into

$$\begin{bmatrix} f_{m1} \\ f_{m2} \end{bmatrix} = \begin{bmatrix} f_0 \\ 0 \end{bmatrix} \tag{9.11}$$

$$\begin{bmatrix} f_{g1} \\ f_{g2} \end{bmatrix} = \begin{bmatrix} 0 \\ 0 \end{bmatrix} \tag{9.12}$$

This result shown in Figure 9.2(b) agrees with our intuitive image of the grasping forces. In the following section, these definitions will be extended to the three-fingered hands.

9.2.1 Three-fingered Hand

Fingertip Force and Resultant Force

Basic relations of forces in manipulation of a rigid object by a three-fingered hand will be given under the following assumption.

1. Each fingertip makes a frictional point contact with the object.
2. These contact points on the object do not change their positions due to slip or relative rotation between the object and the fingertips.
3. The three contact points are to located on a straight line.
4. The mechanism of each finger is such that each fingertip can exert a force to the object in any direction.

Assumption 1 implies that each finger can apply a force but not a moment to the object. Assumption 2 implies that the curvature of the fingertip is infinitely large. Otherwise, relative rotation between the object and the fingertip during manipulation will cause some change of the contact position, resulting in a motion with a non-holonomic constraint which is difficult to analyze [Kerr and Roth 1986]. Assumptions 3 and 4 are necessary for arbitrary manipulation of the object.

Figure 9.3 shows a robot hand and an object under consideration. In the figure, $\sum_0$ is the object coordinate frame fixed to the object, C_i is the contact point of the i-th finger (i=1,2,3), $r_i \epsilon R^3$ is the position vector of C_i from the orgin of $\sum_0$, and $f_i \epsilon R^3$ is the fingertip force applied to the object by the i-th finger.

The resultant force $f \epsilon R^3$ and resultant moment $n \epsilon R^3$ due to $f_i, i = 1,2,3$ are given by

$$f = \sum_{i=1}^{3} f_i \tag{9.13}$$

$$n = \sum_{i=1}^{3} (r_i \times f_i) \tag{9.14}$$

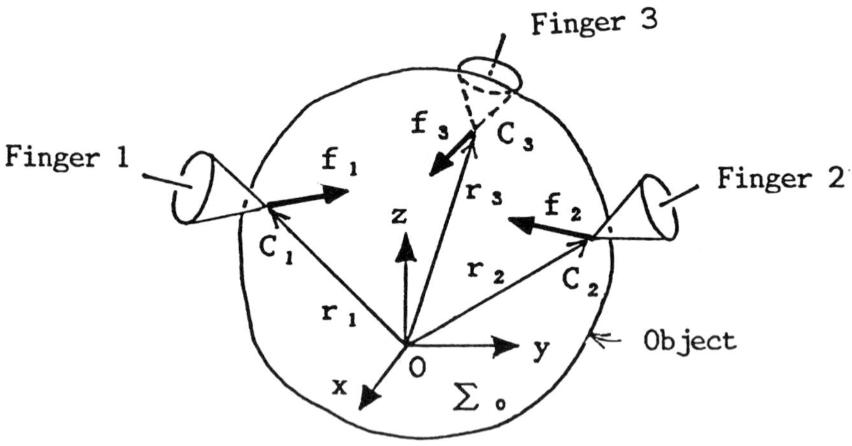

Figure 9.3: Three-fingered hand and object (©IEEE, Yoshikawa and Nagai 1987).

From equations 9.1 and 9.2, the relation between total resultant force $T \triangleq [f^T, n^T]^T$, and the total fingertip force $F \triangleq [f_1^T, f_2^T, f_3^T]^T$ is expressed as

$$T = AF \tag{9.15}$$

Coefficient matrix A is given by

$$A \triangleq \begin{bmatrix} E_3 & E_3 & E_3 \\ R_1 & R_2 & R_3 \end{bmatrix} \epsilon R^{6\times 9} \tag{9.16}$$

where E_n is the n x n unity matrix and

$$R_i \triangleq \begin{bmatrix} 0 & -r_{iz} & r_{iy} \\ r_{iz} & 0 & -r_{ix} \\ -r_{iy} & r_{ix} & 0 \end{bmatrix} \epsilon R^{3\times 3} \tag{9.17}$$

$$r_i \triangleq [r_{ix}, r_{iy}, r_{iz}] \tag{9.18}$$

The manipulation of the object is done through an appropriate choice of the total resultant force T.

Internal Force

The general solution of equation 9.15 is given by

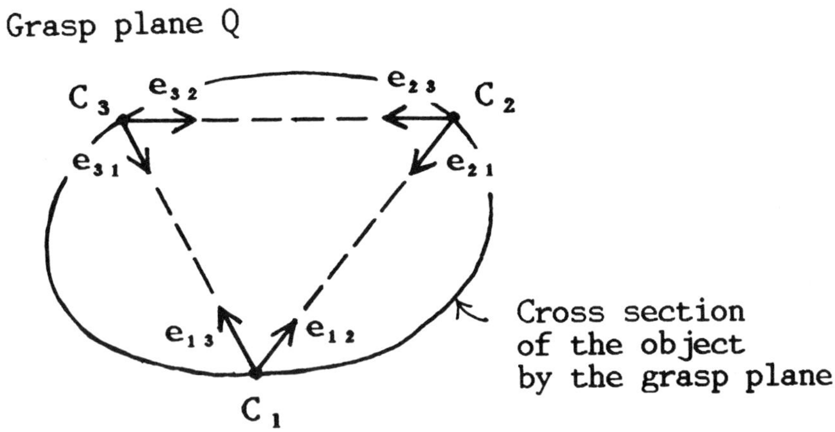

Figure 9.4: Vector e_{ij} (©IEEE, Yoshikawa and Nagai 1987).

$$F = A^+T + (E_9 - A^+A)y \tag{9.19}$$

where $y \epsilon R^9$ is an arbitrary constant vector. The second term $(E_9 - A^+A)y$ represents the internal force. However, this expression of the internal force is not convenient due to the following fact. From assumption 3 it can be easily shown that the rank of A is 6 and

$$rank(E_9 - A^+A) = 3 \tag{9.20}$$

Hence the essential arbitrariness of the internal force is three. Therefore, determining the internal force by the vector y means determining a three-dimensional quantity by operating a nine-dimensional quantity. In order to avoid this inconvenience, a new expression for internal force will be given. Let

$$e_{ij} \triangleq (r_j - r_i)/||r_j - r_i||, i, j = 1, 2, 3, i \neq j \tag{9.21}$$

where $||r||$ denotes the Euclidian norm of vector r. Then e_{ij} is the unit vector directing from C_i to C_j on the grasp plane Q including the three contact points, and $e_{ij} = -e_{ji}$. Figure 9.4 shows these unit vectors on the plane Q. The following proposition holds

Proposition 1: A total fingertip force F is an internal force if and only if there exist a vector $z = [z_{23}, z_{31}, z_{12}]^T \epsilon R^3$ such that

$$F = Gz \tag{9.22}$$

where

$$G \triangleq \begin{bmatrix} 0 & e_{13} & e_{12} \\ e_{23} & 0 & e_{21} \\ e_{32} & e_{31} & 0 \end{bmatrix} \epsilon R^{9\times 3} \tag{9.23}$$

Proof: The set of all internal forces S_1 is given by

$$S_1 \triangleq Range(E_9 - A^+ A) = Null(A) \tag{9.24}$$

where $Range(\cdot)$ and $Null(\cdot)$ are, respectively, the range space and the null space of a matrix. Therefore, it is sufficient to show that

$$S_1 = Range(G) \tag{9.25}$$

From equations 9.16, 9.17, 9.21, and 9.23 we obtain

$$AG = 0 \tag{9.26}$$

Hence

$$Range(G) \subset Null(A) \quad = \quad S_1 \tag{9.27}$$

On the other hand,

$$rank(G) = 3 \tag{9.28}$$

From equations 9.20, 9.27 and 9.28 we finally obtain equation 9.25, completing the proof.

The internal forces $\hat{F} \triangleq [\hat{f}_1^T, \hat{f}_2^T, \hat{f}_3^T]^T$ have the geometric characteristics shown in Figure 9.5. The load lines of the three fingertip forces $\hat{f}_i$ intersect at a point P on the plane Q and the three forces form a closed triangle (force triangle) due to the balance of force and moment. The three degrees of freedom of the internal force can be interpreted as a sum of the two degrees of freedom in the position of the point P on the plane Q, and one degree of freedom in the scale of the force triangle. Therefore, there is one-to-one relation between choosing an internal force and determining both the position of P and the scale of the force triangle. The point P will be called the focus of the internal force hereafter.

9.2.2 Grasping Force

By experience we know that, when we handle an object by our fingers, not only the manipulating force but also the grasping force is necessary in order not to drop the object. In this section, the grasping force is defined and the concept of grasp mode is introduced.

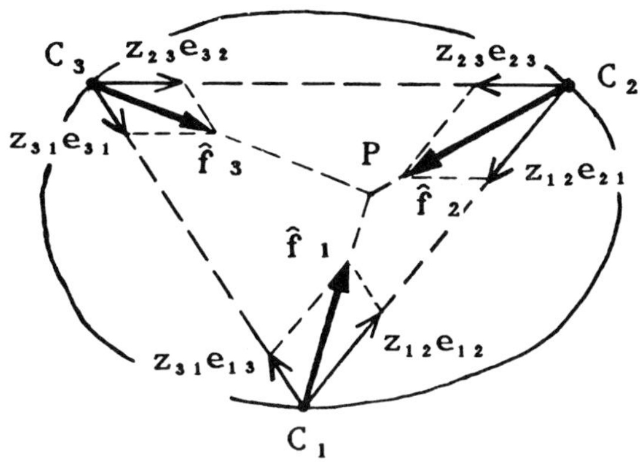

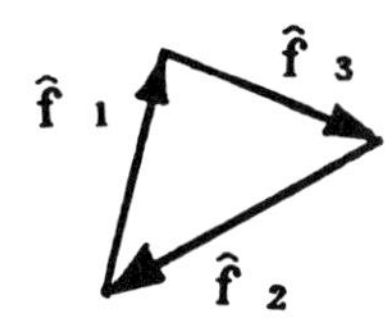

Figure 9.5: Internal force and force triangle (9.5(a) ©IEEE, Yoshikawa and Nagai 1987).

Definition 1: A fingertip force F is called a grasping force if the following two conditions are satisfied.

$$AF = 0 \tag{9.29}$$

$$\frac{f_i^T a_i}{||f_i||} > \frac{1}{\sqrt{1+\mu_i^2}}, i = 1, 2, 3 \tag{9.30}$$

where a_i is the inward unit normal vector of the object surface and μ_i is the static friction coefficient at the contact point C_i.

Equation 9.29 implies that F is an internal force. Equation 9.30 means that each fingertip force satisfies the friction constraint. Due to this constraint, the location of point P on the plane Q in Figure 9.5 for a grasping force F is also restricted. This restriction will now be analyzed. Let the position vector of focus P be r_ρ, and

$$e_{\rho i} \triangleq \frac{r_\rho - r_i}{||r_\rho - r_i||}, i = 1, 2, 3 \tag{9.31}$$

The vector $e_{\rho i}$ is the unit vector directing from contact point C_i to focus P. In the case of Figure 9.5, the fingertip force f_i and the $e_{\rho i}$ are in the same direction. However, this does not necessarily hold for all cases. For example, in the case of Figure 9.6, f_2 and $e_{\rho 2}$ and f_3 and $e_{\rho 3}$ are in the opposite

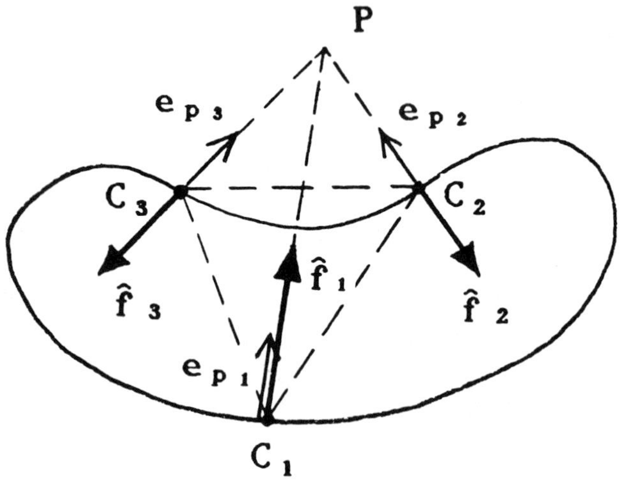

Figure 9.6: Relation between $\hat{f}_i$ and $e_{\rho i}$ (©IEEE, Yoshikawa and Nagai 1987).

direction. Taking this point into consideration, the constraint equation 9.30 can be shown to be equivalent to the following two relations:

$$f_i = sgn(e_{\rho i}^T a_i)||f_i||e_{\rho i} \tag{9.32}$$

$$|e_{\rho i}^T a_i| > \frac{1}{\sqrt{1+\mu_i^2}}, i = 1,2,3 \tag{9.33}$$

In order to express the state of the internal force described by equation 9.22, let

$$\alpha \triangleq [\alpha_1, \alpha_2, \alpha_3] \tag{9.34}$$

$$\alpha_i \triangleq sgn(z_{(i+1)(i+2)}), i = 1,2,3 \tag{9.35}$$

where the subscript i is interpreted as (i-3) when $i \geq 4$ for notational convenience, and sgn(a) denotes the sign of a,i.e.,

$$sgn(a) = \begin{cases} +1, & if \quad a > 0 \\ \pm 1, & if \quad a = 0 \\ -1, & if \quad a < 0 \end{cases} \tag{9.36}$$

A fingertip force which satisfies equation 9.22 will be called the internal force of mode α. For example, the force in Figure 9.5 is of mode [+1,+1,+1]. When F is a grasping force, since this mode is useful for classifying it, a will be called grasp mode. Figure 9.7 shows the four essentially different grasp modes. Let β be

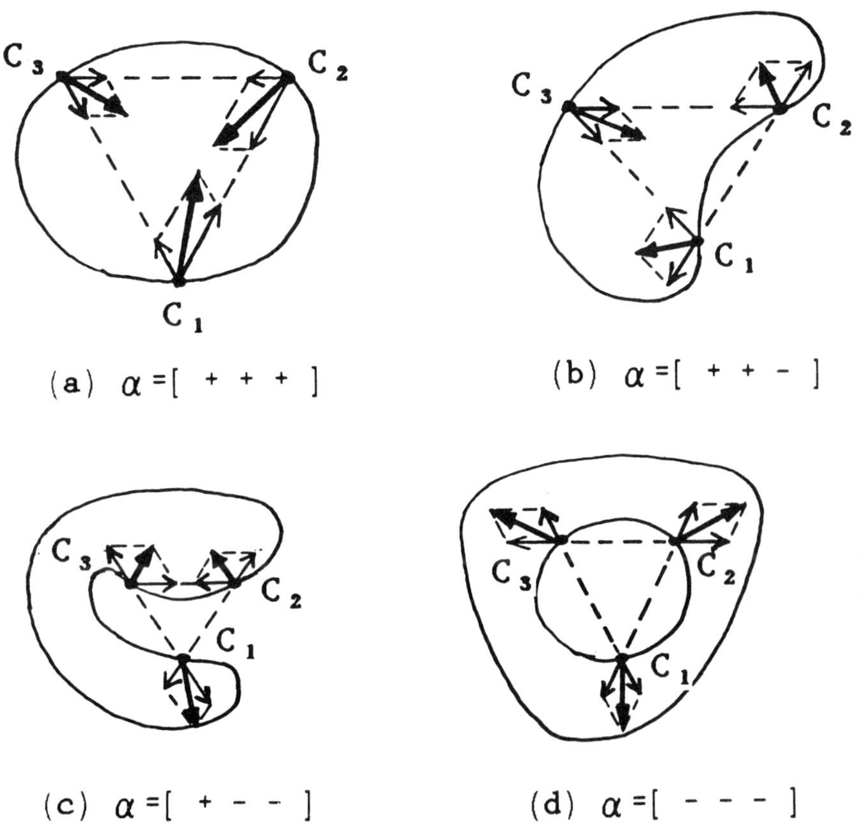

Figure 9.7: Four grasp modes.

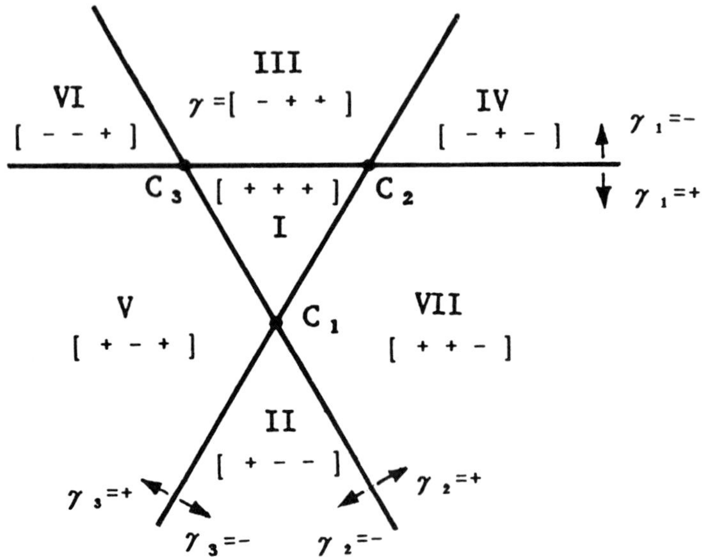

Figure 9.8: Seven regions on plane Q and their codes (©IEEE, Yoshikawa and Nagai 1987).

$$\beta \triangleq [sgn(e_{\rho 1}^T a_1), sgn(e_{\rho 2}^T a_2), sgn(e_{\rho 3}^T a_2),] \tag{9.37}$$

This parameter β represents the relation between the location of focus P and the shape of the object. Define the regions I through VII on the plane Q and their code $\gamma = [\gamma_1, \gamma_2, \gamma_3]$ ($\gamma_i = +1$ or -1) as shown in Figure 9.8. Then we can easily show that for an internal force to be a grasping force, one of the relationships among, α, β and the regions I -VII listed in Table 9.1 should hold. These relationships can further be condensed to the following relation.

$$\beta = \gamma \text{ or } -\gamma \tag{9.38}$$

$$\alpha = (\gamma_1 \cdot \gamma_2 \cdot \gamma_3)\beta \tag{9.39}$$

Variables α, β and γ will be denoted by only signs + and - hereafter. For example, [+1, +1, -1] will be expressed as [+ + -]. Note that equation 9.38 is a condition on the relation between the location of P and the shape of the object with respect to P, and equation 9.39 tells the grasp mode realizable under the given location of P and shape of object. Relation in equation 9.39 is schematically shown in Figure 9.9. Summarizing the argument, we have

Proposition 2 Suppose that the grasping positions r_i, the inward normal directions a_i and the static friction coefficients μ_i are given. Then a grasping force can exist if there is a focus P satisfying the following two conditions.

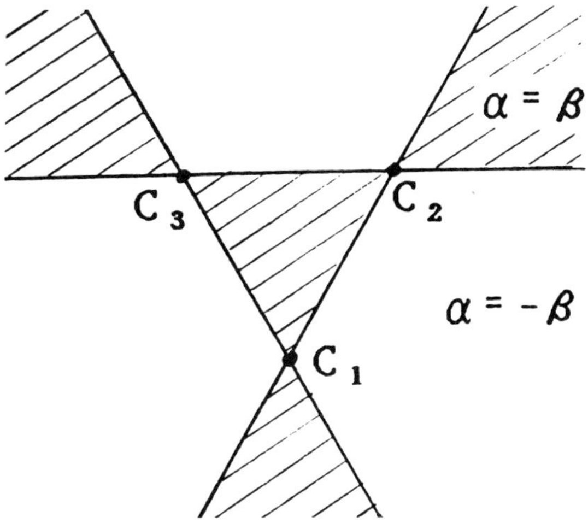

Figure 9.9: Relation 36.

1. a_i, e_{pi}, and μ_i satisfy the constraint equation 9.34
2. β and γ satisfy condition in equation 9.38

Using the expression 9.22 for the internal force and the mode α (i.e., the signs of z_{ij}), expression of the grasping force $F_g \triangleq [f_{g1}^T, f_{g2}^T, f_{g3}^T]^T \epsilon R^3$ is given by

$$F_g = B_g h_g \tag{9.40}$$

where

$$B_g \triangleq \begin{bmatrix} 0 & \hat{e}_{13} & \hat{e}_{12} \\ \hat{e}_{23} & 0 & \hat{e}_{21} \\ \hat{e}_{32} & \hat{e}_{31} & 0 \end{bmatrix} \tag{9.41}$$

$$\hat{e}_{i(i+1)} \triangleq \alpha_{1+2} \cdot e_{i(i+1)}, i = 1,2,3 \tag{9.42}$$

$$\hat{e}_{i(i+2)} \triangleq -\hat{e}_{i(i+2)} i, i = 1,2,3 \tag{9.43}$$

$$h_g \triangleq [h_{g1} \; h_{g2} \; h_{g3}]^T, h_{gi} \geq 0 \tag{9.44}$$

Note that in equations 9.42 and 9.43 the first subscript i of $\hat{e}_{ij}$ implies the finger i. Vectors $\hat{e}_{ij}$ play the role of unit vectors for the grasping force.

Although there is a close relation between the shape of the object and the mode α of realizable grasping force, there are cases where two or more modes are possible for a given shape of an object. Figure 9.10 shows the realizable modes of grasping force for an cylindrical object. The plane Q is assumed to be perpendicular to the axis of the object, so that the vectors α

Region	γ (code of region)	β (shape of object)	α (grasp mode)
I	[+ + +]	[+ + +]	[+ + +]
		[- - -]	[- - -]
II	[+ - -]	[+ - -]	[+ - -]
		[- + +]	[- + +]
III	[- + +]	[- + +]	[+ - -]
		[+ - -]	[- + +]
IV	[- + -]	[- + -]	[- + -]
		[+ - +]	[+ - +]
V	[+ - +]	[+ - +]	[- + -]
		[- + -]	[+ - +]
VI	[- - +]	[- - +]	[- - +]
		[+ + -]	[+ + -]
VII	[+ + -]	[+ + -]	[- - +]
		[- - +]	[+ + -]

Table 9.1: Relation among α, β, and γ (adapted from [Yoshikawa and Nagai 1987], ©IEEE).

are in the plane Q. It is also assumed that $\mu_i = 0.4$. Figure 9.10(a) and (b) are the cases where there is only one mode $\alpha = [+++]$ and $\alpha = [-++]$ are realizable, while 9.10(c) is the case where two modes $\alpha = [+++]$ and $\alpha = [-++]$ are realizable. When there is more than one realizable mode, some additional consideration for selecting one mode will be necessary.

9.2.3 Manipulating Force

Manipulating force is defined and its relation to the grasping force introduced in the previous section is discussed in this section.

Definition 2: For a given resultant force T, the fingertip force $F = [f_1^T f_2^T f_3^T]^T$ is called the manipulating force of mode α if F satisfies the following three conditions.

$$T = AF \tag{9.45}$$

$$f_i^T \hat{e}_{ij} \geq 0, \; i = 1,2,3; \; j = (i+1),(i+2) \tag{9.46}$$

$$(f_i^T \hat{e}_{i(i+1)})(f_{i+1}^T \hat{e}_{(i+1)i}) = 0, \; i = 1,2,3 \tag{9.47}$$

Note that $\hat{e}_{ij}$'s are defined by equations 9.42 and 9.43, and so are dependent on mode α. Equation 9.45 implies that F produces the resultant force T. Equation 9.46 means that the manipulating force is not in the inverse direction of the grasping force. Equation 9.47 implies that the grasping force components between C_1 and C_2, C_2 and C_3, and C_1 are zero.

In order to give an explicit expression for the manipulating force, we first introduce the following matrix.

$$B_m \triangleq \begin{bmatrix} l\bar{e}_{13} & 0 & (1-n)\bar{e}_{13} & \bar{e}_{10} & 0 & 0 \\ (1-l)\bar{e}_{21} & m\bar{e}_{23} & 0 & 0 & \bar{e}_{20} & 0 \\ 0 & (1-m)\bar{e}_{32} & n\bar{e}_{31} & 0 & 0 & \bar{e}_{30} \end{bmatrix} \tag{9.48}$$

where

$$\bar{e}_{oi} = \frac{\hat{e}_{i(i+1)} \times \hat{e}_{i(i+2)}}{\|\hat{e}_{i(i+1)} \times \hat{e}_{i(i+2)}\|} \tag{9.49}$$

$$\bar{e}_{i(i+1)} = \hat{e}_{i(i+2)} \times \bar{e}_{i0} \tag{9.50}$$

$$\bar{e}_{i(i+2)} = \bar{e}_{i0} \times \bar{e}_{i(i+1)} \tag{9.51}$$

and l, m and n are parameters which can take 1 or 0. Vectors $\bar{e}_{ij}$ given by equations 9.49, 9.50, and 9.51, play the role of unit vectors for manipulating force.

Figure 9.11 shows the vectors $\hat{e}_{ij}$ and $\bar{e}_{ij}$ schematically for several typical cases. Note that $\bar{e}_{i0}$'s are normal to the grasp plane Q and

$$\bar{e}_{ij}^T \hat{e}_{ij} \geq 0, \; i = 1,2,3, j = (i+1),(i+2) \tag{9.52}$$

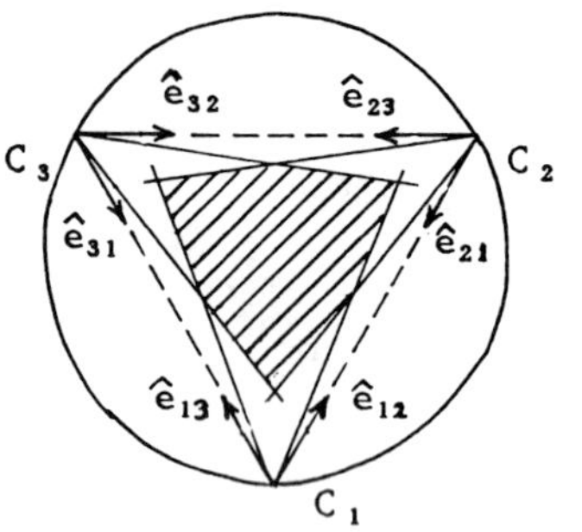

(a) Only α =[+ + +] is realizable.

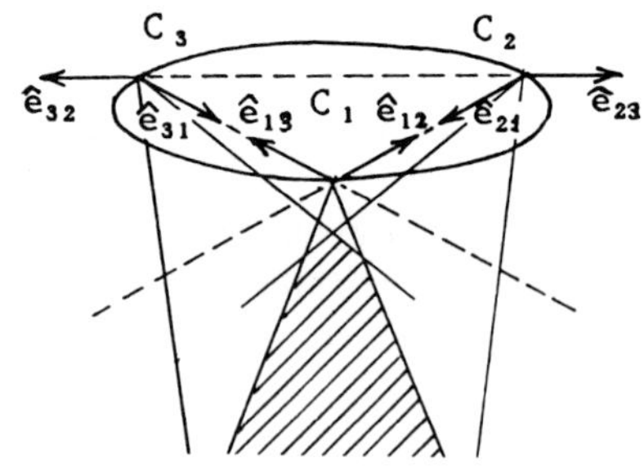

(b) Only α =[-+ +] in Region II is realizable.

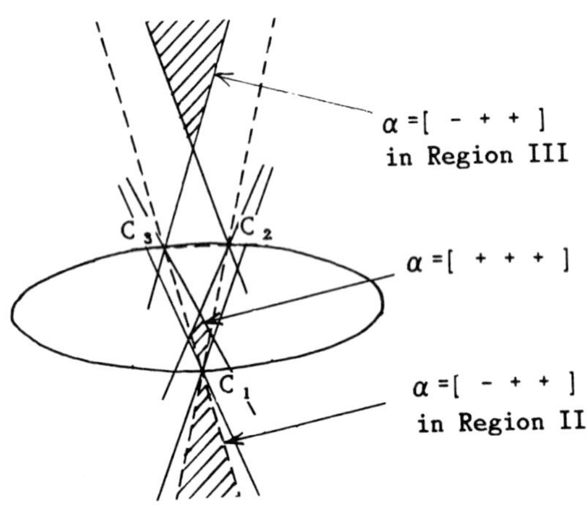

(c) α =[+ + +] and α =[- + +] in Region II and III are realizable.

Figure 9.10: Examples of realizable mode of grasping force (©IEEE, Yoshikawa and Nagai 1987).

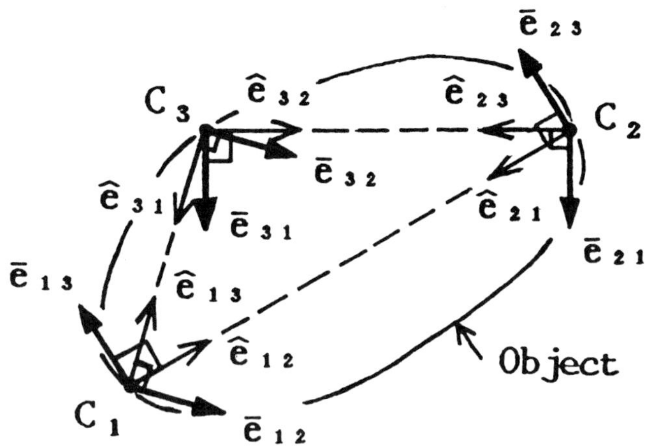

Figure 9.11: Relation between $\hat{e}_{ij}$ and $\bar{e}_{ij}$ ($\bar{e}_{i0}$: Upward and Normal to the Plane Q).

The matrix B_m is a function of α and (l, m, n). The parameter l (m, n) determines which of the two vectors $[\bar{e}_{12}, \bar{e}_{21}]$ $([\bar{e}_{23}, \bar{e}_{32}]$, $[\bar{e}_{31}, \bar{e}_{13}])$ is included in the matrix B_m. With these preparations, we state the following proposition.

Proposition 3: A fingertip force $\bar{F}$ which produces the resultant force T, is a manipulating force of mode α if it satisfies

$$\bar{F} = B_m h_m \tag{9.53}$$

for some α, (l,m,n) and $h_m \stackrel{\Delta}{=} [h_{m1}, h_{m2}, ..., h_{m3}]^T$ with $h_{mi} \geq 0, i = 1, 2, 3$.
Proof: Suppose a fingertip force $\bar{F} \stackrel{\Delta}{=} [\bar{f}_1^T, \bar{f}_1^T, \bar{f}_1^T]^T$ satisfies equation 9.53, i.e.,

$$\bar{f}_1 \stackrel{\Delta}{=} l h_{m1} \bar{e}_{12} + (1 - n) h_{m3} \bar{e}_{13} + h_{m4} \bar{e}_{10} \tag{9.54}$$

$$\bar{f}_2 \stackrel{\Delta}{=} m h_{m2} \bar{e}_{23} + (1 - l) h_{m1} \bar{e}_{21} + h_{m5} \bar{e}_{20} \tag{9.55}$$

$$\bar{f}_3 \stackrel{\Delta}{=} n h_{m3} \bar{e}_{31} + (1 - m) h_{m2} \bar{e}_{32} + h_{m5} \bar{e}_{30} \tag{9.56}$$

It is then enough to show that these $\bar{f}_i$ satisfy equations 9.46 and 9.47. From equation 9.49, 9.50 and 9.51 we have

$$\hat{e}_{i(i+1)}^T \bar{e}_{i(i+2)} = \hat{e}_{i(i+1)}^T \bar{e}_{10} = 0 \tag{9.57}$$

$$\hat{e}_{i(i+2)}^T \bar{e}_{i(i+1)} = \hat{e}_{i(i+2)}^T \bar{e}_{10} = 0 \tag{9.58}$$

Therefore, from equations 9.57, 9.58, 9.54, 9.55, 9.56 and 9.52,

$$\bar{f}_1^T \hat{e}_{12} = l h_{m1} \bar{e}_{12}^T \hat{e}_{12} \geq 0 \quad (9.59)$$

$$\bar{f}_1^T \hat{e}_{13} = (1-n) h_{m3} \bar{e}_{13}^T \hat{e}_{13} \geq 0 \quad (9.60)$$

and,

$$\left(\bar{f}_1^T \hat{e}_{12}\right)\left(\bar{f}_2^T \hat{e}_{21}\right) = l(1-l)h_{m1}^2 \left(\bar{e}_{12}^T \hat{e}_{12}\right)\left(\bar{e}_{21}^T \hat{e}_{21}\right) \quad (9.61)$$

$$= 0 \quad (9.62)$$

Hence equations 9.46 and 9.47 for i=1 are satisfied. It can also be shown by similar arguments that these equations satisfy for i=2 and 3 also by $\bar{F}\cdot$. The manipulating force will be expressed by $F_m \triangleq [f_{m1}^T, f_{m2}^T, f_{m3}^T]^T$ hereafter.

Proposition 4: For any given manipulating force F_m of mode α

$$||F_m|| \leq ||F_m + F_g|| \quad (9.63)$$

for any grasping force F_g of the same mode α.
Proof: From equations 9.40 and 9.46

$$F_m^T F_g = \sum_{i=1}^{3} (f_{mi}^T f_{gi}) \geq 0 \quad (9.64)$$

Therefore,

$$||F_m + F_g||^2 = ||F_m||^2 + 2F_m^T F_g + ||F_g||^2 \geq ||F_m||^2 \quad (9.65)$$

A manipulating force F_m of mode α which produces the resultant force T, satisfies

$$T = AF_m = AB_m h_m \quad (9.66)$$

Hence for obtaining an F_m for a given T, we first calculate $\hat{h}_m \triangleq [\hat{h}_{m1} \hat{h}_{m2} ... \hat{h}_{m3}]^T \epsilon R^3$ from

$$\hat{h}_m = (AB_m)^{-1} T \quad (9.67)$$

for eight triples (l,m,n) (l,m,n = 1 or 0). We then select the one for which $\hat{h}_{mi} \leq$ $(i = 1, 2, 3)$ as the vector h_m. Then $F_m = b_m h_m$ with the above selected values h_m and (l,m,n) is the desired manipulating force. The triple (l,m,n) which satisfies $\hat{h}_{mi} \leq 0 (i = 1, 2, 3)$ is uniquely determined except for degenerate cases where at least one of the three $\hat{h}_{mi}$ is zero. If we further wish to determine a fingertip force F which produces T, we only need to select a grasping force F_g such that $F = F_g + F_m$ satisfies the friction constraint. It is of course possible to select F_g uniquely by using some performance criterion for F_g [Yoshikawa and Nagai 1988].

9.2.4 Decomposition of Fingertip Force into Manipulating and Grasping Force

A method for decomposing a given fingertip force into a manipulating force and a grasping force will be given. Suppose that $r_i, a_i, \mu_i (i = 1, 2, 3)$, and a fingertip force F given, and that this F can be expressed as a sum of a manipulating force F_m and a grasping force F_g for some mode α. Then

$$F = F_g + F_m = BH \tag{9.68}$$

where

$$B \triangleq [B_g, B_m] \tag{9.69}$$

$$H \triangleq [h_g^T, h_m^T]^T \tag{9.70}$$

An algorithm for decomposition of F is as follows.

1. Find the set of all realizable modes α from $r_i, a_i, \alpha_i, (i = 1, 2, 3)$.

2. Pick up one realizable mode α from the set obtained in (1). Calculate $\hat{H} \triangleq [\hat{h}_g^T \hat{h}_m^T]^T$

$$\hat{H} = B^{-1}F \tag{9.71}$$

   for the eight triples (l,m,n) (l,m,n=1 or 0).

3. Select the triple (l,m,n) for which $\hat{h}_{gi} \geq (i = 1, 2, 3)$ and $\hat{h}_{mi} \geq (i = 1, 2, 3)$. Let $\hat{H}$ for this triple be H. Then $(F_m = B_m h_m, F_g = B_g h_g)$ is a decomposition of F.

4. Repeat the steps (2) and (3) until all the realizable modes are checked.

Note that the decomposition may not be unique because of the existence of multiple grasp modes. Also there are cases where such decomposition does not exist.

A simple example of decomposition will be given in the following. The object is a cylinder of radius 5 with its axis coinciding with the z axis of the object coordinate frame $\sum_0$. Suppose that the three contact points and the fingertip forces are given by

$$r_1 = [0, -5, 0]^T, \quad f_1 = [1, 8, 0]^T \tag{9.72}$$

$$r_2 = [5\sqrt{3/2}, 5/2, 0]^T, \quad f_2 = [-3, -2, 0]^T \tag{9.73}$$

$$r_3 = [-5\sqrt{3/2}, 5/2, 0]^T, \quad f_3 = [6, -2, 0]^T \tag{9.74}$$

Therefore, the plane Q is given by the x-y plane of $\sum_0$, and the a_i's are

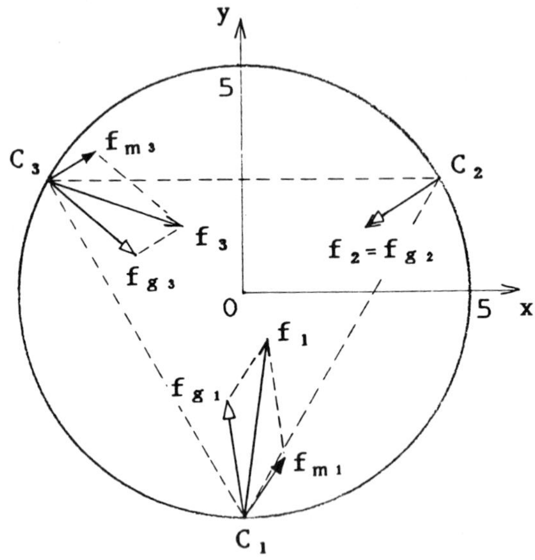

Figure 9.12: An example of decomposition of fingertip force into manipulating and grasping forces (the scale of forces is x0.5) (adapted from [Yoshikawa and Nagai 1987], ©IEEE).

$$a_1 = [0, 1, 0]^T \tag{9.75}$$
$$a_2 = [-\sqrt{3/2}, -1/2, 0]^T \tag{9.76}$$
$$a_3 = [\sqrt{3/2}, -1/2, 0]^T \tag{9.77}$$

Also suppose that $\mu_i = 0.4, i = 1, 2, 3$. Then there is one mode [+++] by Proposition 2, and using the above algorithm we obtain the following manipulating and grasping forces of mode [+++].

$$f_{m1} = [7/4, 4 - 3\sqrt{3/4}, 0]^T \quad f_{g1} = [-3/4, 4 + 3\sqrt{3/4}, 0]^T$$
$$f_{m2} = [0, 0, 0]^T \quad f_{g2} = [-3, -2, 0]^T$$
$$f_{m1} = [9/4, 3\sqrt{3/4}, 0]^T \quad f_{g3} = [15/4, -2 - 3\sqrt{3/4}, 0]^T$$

This is shown in Figure 9.12. It can be easily seen that these forces satisfy Definition 1 and 2.

Note that expressions 9.53 and 9.40 are the three-fingered hand version of two-fingered hand 9.9 and 9.10. In fact, equations 9.9 and 9.10 may be obtained by

1. Similar to equations 9.53 and 9.40, let

$$\begin{bmatrix} f_{m1} \\ f_{m2} \end{bmatrix} = \begin{bmatrix} l \\ -(1-l) \end{bmatrix} h_m, \; h_m \epsilon R^1 \tag{9.78}$$

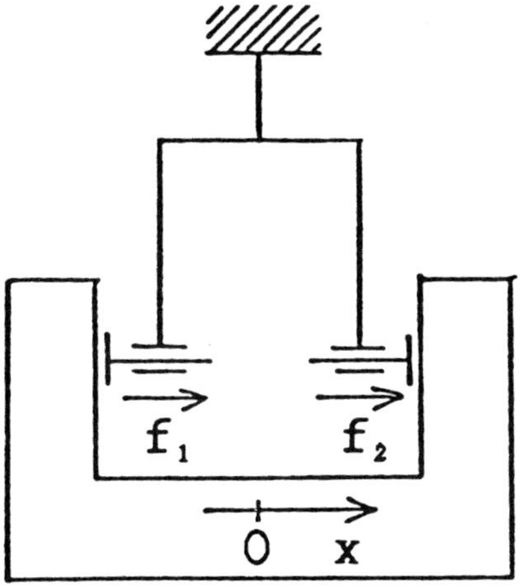

Figure 9.13: Two-fingered hand and concave object.

$$\begin{bmatrix} f_{g1} \\ f_{g2} \end{bmatrix} = \begin{bmatrix} 1 \\ -1 \end{bmatrix} h_g, \quad h_g \epsilon R^1 \tag{9.79}$$

2. Obtained l, h_m and h_g such that $h_m \geq 0$, $h_g \geq 0$.

3. Substitute h_m and h_g into equations 9.78 and 9.79.

Note also that equations 9.78 and 9.79 represent one grasp mode and there is another grasp mode which appears in the case of Figure 9.13. For the latter, we have (instead of equations 9.78 and 9.79)

$$\begin{bmatrix} f_{m1} \\ f_{m2} \end{bmatrix} = \begin{bmatrix} -l \\ 1-l \end{bmatrix} h_m, \quad h_m \epsilon R^1 \tag{9.80}$$

$$\begin{bmatrix} f_{g1} \\ f_{g2} \end{bmatrix} = \begin{bmatrix} -1 \\ 1 \end{bmatrix} h_g, \quad h_g \epsilon R^1 \tag{9.81}$$

9.3 Conclusion

The concepts of the manipulating force and grasping force, which are often used for describing some kinematic aspects of manipulation of object by human hands, have been studied for three fingered robot hands. The main results obtained in this chapter are summarized as follows.

1. A new representation of the internal force whose physical meaning is very clear, has been given.

2. Grasping force has been defined as an internal force which satisfies static friction constraints. The concept of grasp mode is also introduced.

3. The manipulating force has been defined as a fingertip force satisfying the following three conditions: (1) It produces the specified resultant force. (2) It is not in the inverse direction of the grasping force. (3) It does not contain any grasping force component. An explicit expression of the manipulating force has been given.

4. An algorithm for decomposing a given fingertip force into manipulating and grasping forces has been given.

We can easily extend these results to the case of four-fingered hands. These results are expected to be useful for developing control algorithms for multifingered robot hands and cooperated manipulation of objects by multiple robots. In particular, the concepts of grasping force and grasping mode would be helpful for secure grasp of objects with various shapes. As for the utilization of the results in this chapter to the determination of fingertip forces for given manipulation tasks, see reference [Yoshikawa and Nagai 1988]. It is possible to develop control algorithms in which the grasping control and manipulation control are separately performed using the results obtained in this chapter. This is a future research topic.

Acknowledgements: The authors would like to thank Professor H. Hanafusa, Ritsumeikan University, and Dr. Y. Nakamura, University of California, Santa Barbara, for their helpful discussions.

10

Tactile Sensing for Shape Interpretation

Ronald S. Fearing[1]

ABSTRACT Dextrous robot hands require intelligent sensing systems to manipulate unknown objects reliably in a complex environment. Tactile sensing and perception are needed to provide information on contact forces and local shape properties to ensure robust manipulation. This chapter considers the problem of inferring global object properties such as size, location and orientation from only sparse local geometric information at three fingers. Tactile sensors can provide local shape information, including surface normals, contact location on the finger, and principal curvatures and their directions. This chapter assumes that an object is a cone (not necessarily circular) but is otherwise unknown. Three contacts can in some cases determine the pose of an unknown cone, but in general, more are required. A grasping system could command finger forces to control an object's position in a hand using this tactile information. The advantage of not using specific object models is that a system will be flexible enough to handle a wide range of parts without reprogramming, and will be robust to gross and small differences among objects.

10.1 Introduction

The general goal of this chapter is to give an outline of how the problem of manipulating unknown objects in a hand can be approached using tactile sensors, and general (weak) models. Consider trying to guess the shape and pose of an object from just three local touch probes. A restricted class of objects is considered, simple convex cones. By using geometric constraints for this class of objects, it is sometimes possible to determine what type of feature the fingers would be touching, and thus the pose of the cone. The chapter concludes with a proposed hand control system that could use the orientation and position information from a "weak model system" to perform dextrous manipulation of a part in a hand.

Multifinger articulated hands have been developed [Okada 1982a; Salisbury and Craig 1982; Jacobsen *et al* 1985] that provide much potential

[1] Dept. of Electrical Engineering and Computer Science, University of California, Berkeley, CA 94720

for dextrous manipulation. Systems for controlling finger forces to grasp and manipulate objects in the hands have been proposed [Salisbury and Craig 1982; Cutkosky 1985; Jameson 1985; Kerr 1985; Ji 1987], and demonstrated [Hanafusa and Asada 1977b; Okada and Tsuchiya 1977; Okada 1982a; Kobayashi 1985a; Fearing 1986a; Loucks *et al* 1987]. None of these systems appear to use tactile sensing in the control loop, although Speeter [1987] simulates closed loop grasping using tactile force information.

The need for tactile sensors for dextrous hands has been recognized for a long time [Tomovic and Boni 1962; Hollerbach 1982], but compatible sensors for dextrous fingers (which can not be planar) have just recently been developed [e.g., Fearing, Rise and Binford 1986; Begej 1988]. Tactile sensing is needed to provide contact location, forces, and basic curvature information such as distinguishing between planes, edges, and corners of objects. Because of the complicated compliance, friction, rolling, and slipping effects that occur at finger-object interfaces, it is generally not possible to predict object location exactly from *a priori* object models and finger joint angle measurements.

Most manipulation methods have assumed that a complete object model is available. A more dextrous system should be able to handle a variety of unknown objects in unknown positions. To pick up an unknown object, 4 stages were identified in [Fearing 1986b]; the approach, initial touch, initial grasp, and stable grasp phases. During the initial touch phase, the fingers first make contact with the object. The goal of this chapter is to make as efficient use as possible of this first contact information from 3 fingers, so that in many cases enough can be inferred about the object to make a reasonable grasp attempt in the initial grasp phase. If we are lucky, this initial grasp could result in a stable grasp. Even if the grasp attempt fails, more tactile information should be available to interpret the object's shape. To make an initial grasp, the minimum representation of the object should include its location, orientation, some size constraint, and surface normal information to know whether forces will be within the friction cones at those finger sites. This chapter addresses the problem of obtaining that minimum representation.

10.2 Tactile Sensing

A tactile sensor has been packaged in a cylindrical rubber finger tip for the Stanford/JPL hand. The finger tip sensor uses an array of capacitors formed at the junctions of perpendicular copper strips [Siegel 1986], spaced at 3.3 mm along the length and 18° around the circumference of the finger. Only an 8×8 subset of the 8×20 elements is used. 3.8 mm of rubber covers the core and is essential to increase contact areas and reduce aliasing. Details of finger construction are in [Fearing, Rise and Binford 1986; Fearing 1987b].

After calibration, the sensor output is normalized to determine equivalent strain at each tactile element (tactel). The mean sensitivity of the tactels is 0.4 gram with a 3 mm diameter probe, and they are very linear up to 50 grams.

When the finger touches an unknown convex surface, principal curvatures, normal force and location are determined from a 4 by 4 window of strain measurements as described in [Fearing and Binford 1988]. Sensor strains are predicted by convolving the spatial impulse response of the rubber skin with the assumed surface pressure distribution derived from a Hertz contact model. Gradient search finds the parameters of the convex second-order shape and the force that best fit the sensor data. Experiments show radius estimation within 10%, orientation within 2.5 degrees, and sub-tactel localization 3% of the element spacing.

The curvature algorithm can also be used to distinguish between contact features such as spheres, planes, cylinders, edges and vertices. The use of these features to interpret object shape will be discussed in the next section, assuming ideal curvature and orientation data.

10.3 Shape Interpretation Without Specific Models

There have been two extremes in determining the shape of an object from tactile sensing. Allen and Bajcsy [1985] build up a map of the entire surface of unknown objects from local measurements. Gaston and Lozano-Pérez [1983] and Faugeras and Hebert [1986] assume known objects, and determine object shape by matching features in the world to specific object models, keeping all consistent matches. To manipulate unknown objects, it is desired to determine as much about the global properties of the object, such as its size and orientation, from as little local sensing as is possible. Shape classification experiments with tactile sensors for very regular objects such as circular cylinders and cones have been performed [Gurfinkel *et al* 1974; Kinoshita 1977]. These techniques lack descriptive capability and generality for more complicated shapes.

We distinguish here among object classification, model matching, and shape description without specific models. Classification typically uses pattern recognition methods, and does not answer the question of where or what the object is, only which category of object it is, for example, round or square categories. Gurfinkel *et al* [1974] used tactile curvature information obtained from a 3×3 array on a two-finger parallel-jaw gripper to classify simple shapes such as a circular cylinder, block, and sphere. Thresholds on curvature were used to classify plane, edge, vertex, spherical and planar points. The object surface was explored, and objects were classified by local curvature measurements. Kinoshita [1977; Kinoshita, Aida and Mohri 1975] demonstrated discrimination of circular, square and triangular cylinders based on the total number of sensing sites activated, using 384 binary sensors on a 5 finger hand.

Hillis [1981] proposed to classify six small parts based on 3 "features"; the shape (object long or round), bumps, and whether or not the part rolled with finger motion. Ozaki *et al* [1982] used the pattern of surface normals around the object circumference to classify its shape. For example, a quadrilateral will have all surface normals in just four directions. One problem with the classification schemes is limited reliability, for example, Gurfinkel correctly classified a cube from among 5 different objects in less than half the trials. Other problems are the lack of generality for complicated objects with slightly different features, and not determining the pose of the objects.

Unlike classification, model matching can localize and identify specific objects from a set of known objects, instead of only categorizing them. Model matching compares relations between sensed features and features on particular object models, for example, distances between points, and angles between surface normals. A consistent combination of features in the world and in the model determines the object and its position, but not necessarily uniquely. Faugeras and Hebert [1986] used surface primitive matching from range data. They minimized a "distance" measure between primitives in the model and in the sensed data to solve for object type and position. Gaston and Lozano-Pérez [1983] and Grimson and Lozano-Pérez [1984] showed rapid pruning of an interpretation tree of feature combinations (which is basically an exhaustive search of all possible feature matches). Ellis [1987] has developed a planning system to choose sensor paths that will prune the interpretation tree more efficiently using prior sensed data. The principal objection to these methods is that they need a specific model for each object that will be touched.

Shape description uses measurements and geometric constraints to derive a representation of the object, and does not require prior object models. Brady *et al* [1985] demonstrate using range data to extract surface curvatures patches. Allen and Bajcsy [1985] and Allen [1986] examined building up surface patches with a tactile sensor, in combination with vision, by exploration over the whole surface. Another type of shape description, which is used in vision, starts with all the edges in a scene linked into a line drawing, and determines valid interpretations of the edges (for example, Malik [1985]). The objects in the scene are unknown, so *a priori* geometric constraints are used, for example that the objects in the scene are bounded by piecewise smooth surfaces.

These shape description methods have available rich global data on the object. Although it is possible to scan a tactile sensor over all the surfaces of an object, it will be interesting and more efficient to use the sparse local 3D tactile data that is available from the first touches on an object. This partial preliminary information can be used to direct tactile exploration.

10.4 LSHGC Interpretation

A simplified form of the shape interpretation problem is considered here. Given three contact features on an unknown cone (not usually of circular cross-section), geometric considerations are derived to determine if the features are necessary and sufficient to determine the origin, orientation, and scale of a bounding right circular cone to the sensed cone. These contact features, which are convex elliptic, parabolic, and planar points and their sub-case vertices and edges, are determinable by the perception algorithms of [Fearing and Binford 1988]. These three contact features could come from an attempted grasp on a cone with three tactile-sensor equipped fingers. This section deals with the questions: What can be determined about an object from this initial grasp, and is there sufficient information?

Objects need a representation constrained enough that global properties can be inferred from local measurements. A reasonable class constraint is the *generalized cylinder* [Binford 1971], which is the volume generated by an arbitrary cross-section translated with or without deformation along an arbitrary space curve. This representation gives an analytic expression for the whole body, and has well defined properties. The particular class constraint that is used in this chapter is the convex Right Linear Straight Homogeneous Generalized Cylinder (RLSHGC), otherwise known as a simple convex cone. These objects have a convex constant cross-section, which is translated along an orthogonal straight axis and scaled by a linear sweeping rule. The ends of the cone are parallel. Shafer and Kanade [1983] have shown that an axis orthogonal to the cross-section exists for every LSHGC. For any arbitrary sweeping rule, a straight homogeneous generalized cylinder was shown by Ponce, Chelberg and Mann [in press] to have a unique axis if there are 2 zeros of curvature in both the cross-section and the sweeping rule. That paper also shows that tangents at parabolic points at the same height on the cone intersect on the cone's axis. The cone's axis is a global property that can be determined from global image measurements. We desire to find similar global properties for the LSHGC that can be determined from only local measurement.

We start by re-deriving the expressions for tangents and normals on the LSHGC. From Ponce *et al* [in press], a point on the LSHGC can be given in vector form:

$$\underline{x}(z,\theta) = (Az + B)\rho(\theta)\,(\cos\theta\hat{i} + \sin\theta\hat{j}) + z\hat{k} \tag{8.1}$$

where the LSHGC is aligned with the $\hat{k}$ axis, A is the scale factor, the ends of the cone are given by $z_{Top} \leq z \leq z_{Bottom}$, and $0 \leq \rho(\theta) \leq 1$ is the reference cross-section given in polar form. z is defined ≥ 0, and is equal to zero only if the cone's apex is not truncated. The cross-section closed curve represented by $\rho(\theta)$ is a star-shaped function, which has all convex cross-sections as a subset [Ponce *et al*, in press]. We consider only convex cross-sections.

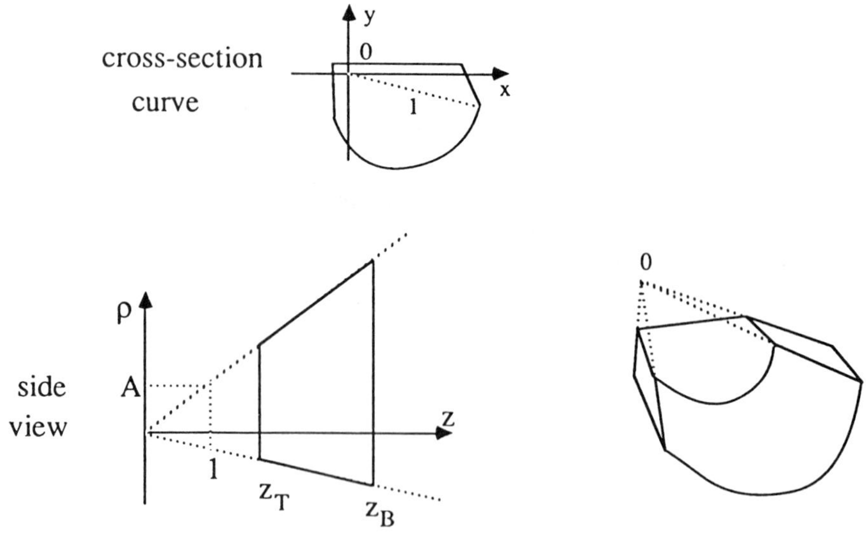

Figure 10.1: LSHGC Example. A convex Linear Straight Homogeneous Generalized Cylinder is the volume generated by sweeping a convex cross-section along the z axis with a linear expansion rule.

The convention used here is that $A = 0$ for a *regular cylinder*, and for a *cone*, $B = 0$. Both cases are subsets of the LSHGC family. The end of the cone is parallel to the $x-y$ plane, and because $B = 0$, the apex of the cone is at the origin. The maximum *radius* of the cone is $Az\rho_{max}$ (where $\rho_{max} = 1$), evaluated at $z = 1$, so the "radius" of the cone is given by A. The length of the cone is given by the distance between the bottom and top of the cone ($z_B - z_T$). The radius and length parameters give a bounding right circular cone that contains the LSHGC. There is an equivalent expression with the LSHGC cross-section in Cartesian coordinates,

$$\underline{x}(z, t) = (Az + B)x(t)\hat{i} + (Az + B)y(t)\hat{j} + z\hat{k}. \qquad (8.2)$$

An example cone with $B = 0$ is shown in Figure 10.1 with cross-section and scale factor A indicated. Note that the origin does not need to be included inside the cross-section, so cones may slanted.

To predict what the tactile sensor will feel on this class of object, we need to determine expressions for surface normals and curvatures on the cone. These expressions will be used in later sections to solve for cone parameters. We will use Cartesian coordinates to describe straight line segments, and polar coordinates for curved arcs in the cross-section. Using simple differential geometry, the tangential vectors on the side of the cone are given by

$$\underline{x}_z = \frac{\partial \underline{x}}{\partial z} = \begin{pmatrix} Ax \\ Ay \\ 1 \end{pmatrix} = \begin{pmatrix} A\rho\cos\theta \\ A\rho\sin\theta \\ 1 \end{pmatrix} \tag{8.3}$$

and

$$\underline{x}_t = \frac{\partial \underline{x}}{\partial t} = (Az + B)\begin{pmatrix} x\prime \\ y\prime \\ 0 \end{pmatrix} \tag{8.4}$$

or in polar coordinates

$$\underline{x}_\theta = \frac{\partial \underline{x}}{\partial \theta} = (Az + B)\begin{pmatrix} -\rho\sin\theta + \rho\prime\cos\theta \\ \rho\cos\theta + \rho\prime\sin\theta \\ 0 \end{pmatrix}, \tag{8.5}$$

where $x\prime = \frac{\partial x}{\partial t}$ and $\rho\prime = \frac{\partial \rho}{\partial \theta}$.

By convention, the unit normal vector points out of the body. The unit normal vector for a cross-section described in Cartesian coordinates is

$$\hat{n} = \frac{\underline{x}_t \times \underline{x}_z}{|\underline{x}_t \times \underline{x}_z|} = \frac{\begin{pmatrix} y\prime \\ -x\prime \\ A(yx\prime - xy\prime) \end{pmatrix}}{\sqrt{y\prime^2 + x\prime^2 + A^2(yx\prime - xy\prime)^2}}, \tag{8.6}$$

and for the star-shaped cross-section (which we choose to be convex) described in polar coordinates

$$\hat{n} = \frac{\underline{x}_\theta \times \underline{x}_z}{|\underline{x}_\theta \times \underline{x}_z|} = \frac{\begin{pmatrix} \rho\cos\theta + \rho\prime\sin\theta \\ \rho\sin\theta - \rho\prime\cos\theta \\ -A\rho^2 \end{pmatrix}}{\sqrt{\rho^2 + \rho\prime^2 + A^2\rho^4}}. \tag{8.7}$$

The principal directions are the two perpendicular directions for which the normal curvatures take on maximum and minimum values [Lipschutz 1969]. The first principal curvature is given by $\hat{k}_1 = \hat{x}_z$. Since the principal curvatures are orthogonal to each other and both lie in the tangent plane, the unit vector in the direction of the second principal curvature is:

$$\hat{k}_2 = \frac{\hat{k}_1 \times \hat{n}}{|\hat{k}_1 \times \hat{n}|} = \frac{\begin{pmatrix} -(A^2\rho^2 + 1)\rho\sin\theta + \rho\prime\cos\theta \\ (A^2\rho^2 + 1)\rho\cos\theta + \rho\prime\sin\theta \\ -A\rho\rho\prime \end{pmatrix}}{\sqrt{(A^2 + 1)^2\rho^2 + A^2\rho^2\rho\prime^2 + \rho^2}}. \tag{8.8}$$

It is important to note that $\hat{k}_2$ and the surface normal are generally not in the cross-section plane unless $A = 0$, that is, the object is a cylinder. The principal direction $\hat{k}_2$ will be in the cross-section plane for a circular cross-section ($\rho\prime = 0$) and when the cross-section curve goes through the origin ($\rho = 0$).

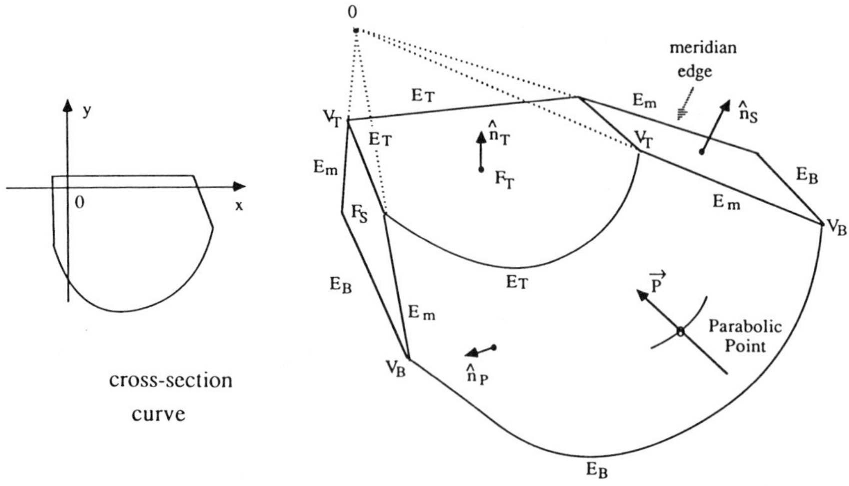

Figure 10.2: Labelled Contact Features on LSHGC. The tactile sensor can distinguish features of vertices, edges, faces, and parabolic points.

When the cross-section is curved, all points on the side of the cone will be parabolic. As expected for a parabolic point, κ_1 (curvature in the direction of $\hat{x}_z$) is zero. The other curvature (with surface normal pointing out of the body) is:

$$\kappa_2 = -\frac{(A^2\rho^2 + 1)(\rho\rho'' - \rho^2 - 2\rho'^2)}{(Az + B)(\rho^2 + \rho'^2 + A^2\rho^4)^{\frac{3}{2}}} \tag{8.9}$$

The tactile sensor is able to measure κ_1, κ_2, $\hat{n}$, and $\hat{k}_1 = \hat{x}_z$.

10.5 Contact Features for LSHGC

An LSHGC has a limited set of contact features, which can be simply classified into point/vertex, plane/face, edge, and parabolic points. This section gives the relevant equations for these feature types. Then given a set of measurements, the equations can be solved simultaneously to find the pose of the unknown object. Each contact has a central location measurement, which corresponds to the center of pressure on the finger sensor. In addition, an edge has a direction vector (two components), a plane has a surface normal, and a parabolic point has a direction vector, a surface normal and a curvature κ_2. Since these features will be mentioned frequently, they will be abbreviated as **V** for vertex (an elliptic point with very high curvature), **E** for edge, (a parabolic point with zero curvature in one direction, and very high curvature in the other), **P** for a parabolic point, (one curvature approximately zero) and **F** for a planar face, (a planar point with very small curvatures in all directions). We will classify any planar point, which corresponds to a zero of curvature in the cross-section, as a face. The tactile sensor can not distinguish between a planar point and a plane such as the face of a polyhedron.

An LSHGC with labelled contact features is shown in Figure 10.2. The subscripts to the feature labels are T for Top, B for Bottom, S for side, and m for meridian. Expressions for these contact features are derived in this section.

An edge of a cylinder end is not described well by a second order surface with $\kappa_1 \to \infty$ and $\kappa_2 \gg 0$. For a small radius of curvature, this contact can look like a vertex. Although this junction is not a C^2 surface it will be assumed that it will appear as an edge in one direction. The curvature algorithm of [Fearing and Binford 1988] finds the best fitting paraboloid to the contact shape. The curvature algorithm is not sophisticated enough to distinguish between the line and cylinder edge types.

Let z_T and z_B be the height of the top and bottom of the LSHGC respectively as measured from the origin. A vertex can only occur at the junction of the top or bottom face and a side edge, thus the contact location is:

$$\underline{V}_T = \begin{pmatrix} (Az_T + B)\rho\cos\theta \\ (Az_T + B)\rho\sin\theta \\ z_T \end{pmatrix} \text{ or } \underline{V}_B = \begin{pmatrix} (Az_B + B)\rho\cos\theta \\ (Az_B + B)\rho\sin\theta \\ z_B \end{pmatrix} . \quad (8.10)$$

There are two types of edges, one corresponding to the discontinuities of the sweeping rule which occur at the top and bottom ends, designated as E_T and E_B respectively, and one corresponding to discontinuities in the slope of the cross-section curve, designated as E_m. The edges E_m correspond to meridians on the surface. Consider a straight line segment in the cross-section. It can be parameterized by

$$x(t) = d\sin\alpha + t\cos\alpha \ , \quad y(t) = -d\cos\alpha + t\sin\alpha \qquad (8.11)$$

where d is the normal distance of the line segment from the origin. The derivatives are

$$x\prime = \cos\alpha \ , \quad y\prime = \sin\alpha \ . \qquad (8.12)$$

The top edge line is given by

$$\vec{E_T} = \underline{E}_{T_o} + t\,(\cos\alpha, \sin\alpha, 0)^T \qquad (8.13)$$

with t a free parameter. The meridian edge line is given by

$$\vec{E}_m = \underline{E}_{mo} + c\hat{x}_z = \underline{E}_{mo} + c\frac{(Ax,\, Ay,\, 1)^T}{\sqrt{1 + A^2(x^2 + y^2)}} \tag{8.14}$$

where c is the parameterization of the line.

Using the surface normal expression from eq. (8.6) and knowing that the cross-section must be linear for a planar face, the surface normal for a planar contact on the side is

$$\hat{n}_S = (\sin\alpha,\, -\cos\alpha,\, -Ad)^T\,(A^2d^2 + 1)^{-1/2}\,. \tag{8.15}$$

The parabolic point has the same location and direction as E_m, the meridian edge, and the surface normal $\hat{n}_P$ is given by eq. (8.6):

$$\hat{n}_P = \frac{\begin{pmatrix} \rho\cos\theta + \rho\prime\sin\theta \\ \rho\sin\theta - \rho\prime\cos\theta \\ -A\rho^2 \end{pmatrix}}{\sqrt{\rho^2 + \rho\prime^2 + A^2\rho^4}}\,. \tag{8.16}$$

In addition, κ_2 is measured by the tactile sensor, and its expression is given by eq. (8.9).

10.6 Feature Measurement

Consider an LSHGC that has an unknown orientation and origin in the world. This can be described by rotation about the origin by an unknown rotation matrix $\mathbf{R}$ and translation from the origin by an unknown vector $\underline{x}_o$. Thus an object feature will be subject to the transformation

$$\underline{x}_S = \mathbf{R}\underline{x} + \underline{x}_o \tag{8.17}$$

where $\underline{x}$ is a point in object coordinates, and $\underline{x}_S$ is the sensed location in world coordinates. Since the reference cross-section is not known *a priori*, any point on the cross-section can be arbitrarily chosen as $\theta = 0$. A rotation of the cross-section in the $x - y$ plane will not be distinguishable, thus only 2 rotations need to be considered. Letting ϕ be the rotation around $\hat{z}$, and ψ the rotation about the $\hat{y}$ axis (i.e. spherical coordinates), the rotation matrix is:

$$\mathbf{R} = \begin{pmatrix} \cos\phi\cos\psi & -\sin\phi & \cos\phi\sin\psi \\ \sin\phi\cos\psi & \cos\phi & \sin\phi\sin\psi \\ -\sin\psi & 0 & \cos\psi \end{pmatrix}. \tag{8.18}$$

The global unknowns for a cone, that is the parameters that affect every local measurement, are seen to be A, $\underline{x}_o$, ϕ, and ψ, for a total of 6 unidentified parameters. Given that the object is an LSHGC, with unknown cross-section, we want to localize it and bound its maximum radius. The scaling parameter A determines a right circular cone which bounds the LSHGC which should aid in planning an effective object grasping strategy. We shall show that it is not possible to determine the scaling parameter exactly with only local measurements.

10.7 Solving for Cone Parameters

Consider the cone with $B = 0$, that is, exclude the cylindrical case. (The set of feature measurements can be checked for consistency with a cone as in [Fearing 1987b], to treat the cylindrical case separately.) We will show that sensed features can be combined to provide constraints on object orientation and scaling.

Examples of the parameters obtained from a contact feature are given in Figure 10.3 for an edge on the top end of the cone, and a parabolic point on the side of the cone. The top edge line has a contact location which is specified by three equations, and a direction in space which is specified by two independent equations (8.13), for a total of 5 constraint equations. These 5 equations involve 4 new unknowns, ρ, θ, z_T, and α, in addition to the 6 previous unidentified global parameters A, $\underline{x}_o$, ϕ, and ψ.

A meridian edge has 3 equations to specify contact location, and two independent equations to specify the line orientation (8.14). The edge has 3 new unknowns (ρ, θ, z) in addition to the 6 global unknowns. A parabolic point, also shown in Figure 10.3, is similar to the meridian edge, but also determines a surface normal (8.16) and one curvature κ_2 (8.9). The surface normal is specified by one angle with respect to the meridian line $\vec{P}$ through the contact location, thus (8.16) provides one additional equation. With the curvature equation, there are 7 independent equations for the parabolic point. The surface normal and curvature equations introduce two new variables $\rho\prime$ and $\rho\prime\prime$, the first and second derivatives of the cross-section with respect to θ. There are thus 5 new unknowns at a parabolic point (ρ, $\rho\prime$, $\rho\prime\prime$, θ, and z).

Parabolic points provide two more equations than meridian edges, but have two more unknowns, thus there is no net gain in constraining equations. If $\rho\prime$ is known, say equal to zero for a circular cross-section, then a parabolic point would provide two more constraints than a meridian edge, providing 7 equations with 3 new unknowns.

Combinations of features are required to solve for unknowns; a single feature is not sufficient. For example, with a meridian edge there are 5 equations and 3 + 6 unknowns. When the number of equations is less than the number of unknowns, a unique solution is not obtainable.

The remainder of this section assumes that we have correctly interpreted the sensed features, that is, we know that a sensed face is an end face. From combinations of the feature equations, we can solve for some of the global cone parameters. A later section discusses how to interpret or label each feature, based on LSHGC geometric constraints.

10.7.1 Solving for Cone Origin, $\underline{x_o}$

Two meridians of a cone always intersect at the apex, which is the cone origin (see Figure 10.3). Consider a combination of two non-colinear merid-

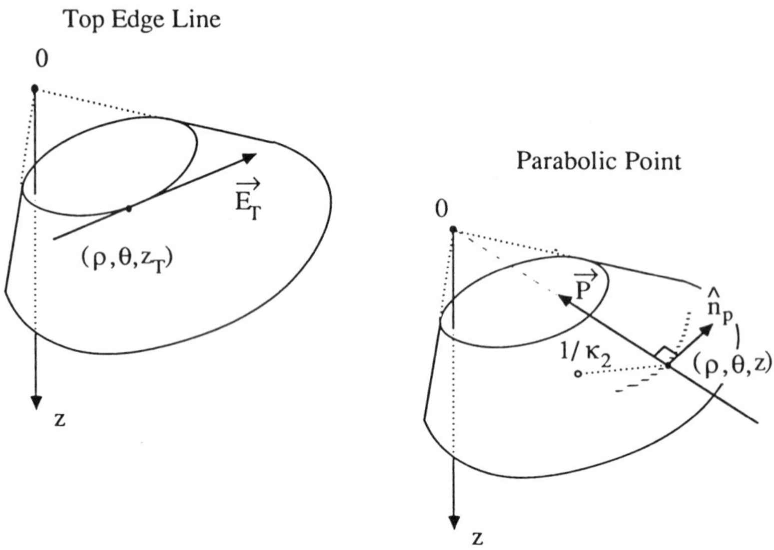

Figure 10.3: Contact Parameters for Top Edge and Parabolic Point. An edge contact determines 3 parameters for the contact location, and two parameters for the edge direction. A parabolic contact gives 3 parameters for location, 2 for meridian direction, 1 for surface normal direction, and 1 parameter for the curvature magnitude. To solve for the cone parameters, at least as many equations from the sensed parameters are required as there are unknowns.

ian edges or parabolic points. Then the sensed contact points $\tilde{\underline{P}}_1$, $\tilde{\underline{P}}_2$ are given by

$$\begin{aligned} \tilde{\underline{P}}_1 &= \mathbf{R}\underline{P}_1 + \underline{x}_o \\ \tilde{\underline{P}}_2 &= \mathbf{R}\underline{P}_2 + \underline{x}_o \end{aligned} \tag{8.19}$$

where $\underline{P}_1$, $\underline{P}_2$ are in the object frame. The two meridian lines through the contact points (from eq. 8.14) can be shown to intersect at a point, say $\underline{P}_{12}$:

$$\underline{P}_{12} = \mathbf{R}\begin{pmatrix} A\rho_1\cos\theta_1 \\ A\rho_1\sin\theta_1 \\ 1 \end{pmatrix}(z+c_1)+\underline{x}_o = \mathbf{R}\begin{pmatrix} A\rho_2\cos\theta_2 \\ A\rho_2\sin\theta_2 \\ 1 \end{pmatrix}(z+c_2)+\underline{x}_o \tag{8.20}$$

where c_1, c_2 are the free parameters of the lines. Since $\theta_1 \neq \theta_2$, the only solution is $z = -c_1 = -c_2$, so the intersection of the two lines gives the origin $\underline{P}_{12} = \underline{x}_o$.

10.7.2 Solving for Orientation, ψ and ϕ

To specify the cone orientation to one of two directions, it is sufficient to know the surface normal of an end. Then the sensed unit surface normal $\hat{s}$ at the top end of the cone is obtained from:

$$\hat{s} = \begin{pmatrix} s_x \\ s_y \\ s_z \end{pmatrix} = \mathbf{R}\hat{n}_T = -\begin{pmatrix} \cos\phi\sin\psi \\ \sin\phi\sin\psi \\ \cos\psi \end{pmatrix}. \tag{8.21}$$

Thus the orientation is solved simply from

$$\tan\phi = \frac{s_y}{s_x} \quad \text{and} \quad \cos\psi = -s_z . \tag{8.22}$$

10.7.3 Solving for the Scaling Factor A

The orientation and origin of the cone need to be found before the scaling parameter, because A is defined on the $x-y$ plane through $z = 1$ as the maximum distance of any point on the cross-section curve from the origin. We have for any (x, y) on or inside the cross-section:

$$\sqrt{x^2+y^2} \leq A \cdot z \cdot \rho_{max} . \tag{8.23}$$

But the cross-section is scaled such that $\rho_{max} = 1$, so

$$A = \max_{i,\ z=1}\left\{\sqrt{x_i^2+y_i^2}\right\}, \tag{8.24}$$

where the maximum is taken over all (x, y) in the cross-section. (Equivalently, this is the maximum $\rho(\theta)$ for $0 \leq \theta \leq 2\pi$.)

Once the orientation and origin have been found, one can determine the x and y location for any z location on any line through the origin from a meridian edge, side face, or parabolic patch. That is, any side contact

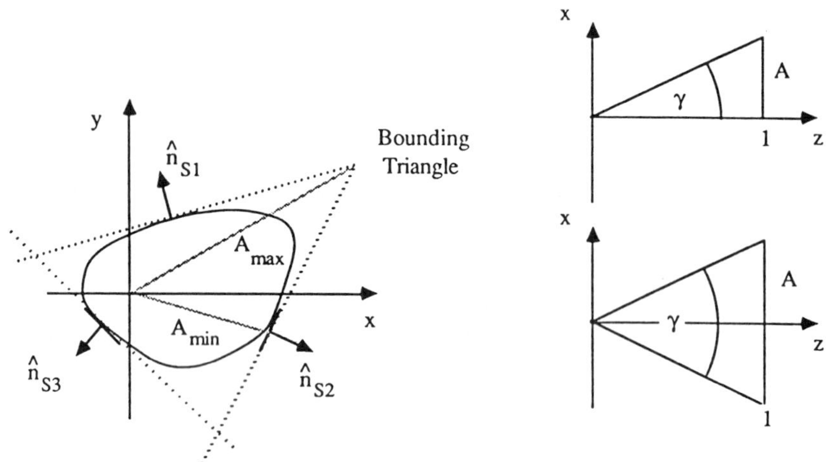

Cross Section Curve at z= 1 Side View Of LSHGC

Figure 10.4: Constraints on Scale Parameter. The minimum and maximum values of the scale parameter A ca sometimes be bounded by 3 side contacts for a convex cross-section. The scale parameter is important, since a cone with a large included angle γ will be hard to grasp on its sides.

location can be projected into the plane $z = 1$. Given only local information, not the whole contour, the maximum must be taken over the available points in the cross-section, thus only a lower bound on A can be found.

Using the convex cross-section assumption, 3 surface normals defined at side faces or parabolic points put an upper bound on A. Figure 10.4 shows a bounding triangle that must contain the cross-section. The three side contacts will not necessarily form a triangle, in which case the maximum A is unbounded in some directions. The maximum radius of the bounding circle to the cross-section must be $\geq A_{min}$, the furthest contact from the origin, and $\leq A_{max}$, the distance of the extremal vertex of the bounding polygon from the origin.

The included angle γ (at the vertex of the cone) is bounded by

$$\tan^{-1} A \leq \gamma \leq 2 \tan^{-1} A \tag{8.25}$$

and has the maximum angle for a fixed A if the origin is in the "center" of the cross-section.

Solve for ϕ and ψ	Restrictions
F_T or F_B	
$E_{T1}E_{T2}$ or $E_{B1}E_{B2}$	
$E_T E_B$	not parallel
$V_{T1}V_{T2}V_{T3}$ or $V_{B1}V_{B2}V_{B3}$	
$E_T V_T$ or $E_B V_B$	not coincident

Table 10.1: Features for Orientation Solution. Some of the sufficient feature sets that can be used to solve for the cone orientation.

10.7.4 SUFFICIENT FEATURE COMBINATIONS

It was shown in the previous section that a combination of 2 meridian edges and a surface normal on an end face allows solution for the cone origin and orientation. Other combinations of features will also give a solution. This section considers which sets of features are sufficient. The origin and orientation solutions require different sets of features. (We assume here that the features have been properly identified.)

Table 10.1 gives combinations of features that are sufficient to determine the end face and thus the orientation. For example, three vertices V_{T1}, V_{T2}, V_{T3} define the top end. Table 10.1 is not exhaustive. Two vertices on the top end can define an equivalent edge (E_T), which could be used to solve for the end face in combination with a bottom edge (E_B). A top and bottom edge will not always be sufficient. The restriction that the edges are not parallel is noted in the table. These two parallel lines do not define a unique plane. However, if two edges are parallel, and on the same end plane ($E_{T1}E_{T2}$), there is a unique plane that contains both edges.

Meridian edges are needed to solve for the origin of the cone. As for the orientation solution, features can be combined to give equivalent meridian edges. Since a side face corresponds to a straight line segment in the cross-section, and a meridian edge corresponds to a vertex, it is apparent that the intersection line of two side planes F_{S1} and F_{S2} must also be a meridian, even if the sides are not adjacent. Therefore, three independent planar contacts on the sides of the LSHGC will define three meridian lines which must intersect at the origin. Table 10.2 gives combinations of features which can be used to solve for the origin. Table 10.2 is not exhaustive. For example, corresponding vertices on the top and base of the cone define an equivalent meridian edge. The restrictions in Table 10.2 ensure the independence of

Solve for $\underline{x}_o = (x_o, y_o, z_o)^T$	Restrictions
$P_1 P_2$ or $E_{m1} E_{m2}$ or PE_m	not colinear
$E_m F_S$ or PF_S	not coplanar
$F_{S1} F_{S2} F_{S3}$	none coplanar

Table 10.2: Features for Origin Solution. Some of the sufficient feature sets that can be used to solve for the location of the cone apex.

the feature equations.

A solution for the cone orientation and origin requires one set of features from Table 10.1 and one set from Table 10.2. The minimum set of features is two meridian or parabolic edges or a meridian edge and a side face for the origin solution, and the top or bottom face for orientation. We are particularly interested in the minimum sets of features, because these are the only sufficient sets of features for a 3 finger hand with a single touch at each finger. If we are not fortunate enough to touch the cone and obtain one of the minimum sets of features, additional features will need to be identified until the feature combinations of Tables 10.1 and 10.2 are satisfied.

10.7.5 Necessary Feature Combinations

Previous sections have shown that 2 meridian edges and an end face are sufficient for cone origin and orientation determination, but are they necessary? Can orientation be determined or at least constrained without an end face or feature combination that gives an end face? We can show that no combination of parabolic points, side faces, or meridian edge points can determine orientation.

A side face, parabolic point, and meridian contain equivalent information about cone orientation. The surface normal at a parabolic point, eq. (8.16), gives one more equation than is obtained from a meridian edge, eq. (8.14). This equation specifies the orientation of the normal with respect to the meridian line. The parabolic point has 2 more equations and 2 more unknowns than the meridian edge, so there is no net gain in equations over the meridian edge. The surface normal orientation is a function of the derivative of the radius, $\rho\prime$. Without knowing $\rho\prime$, which is not measured directly, the surface normal does not constrain the cone orientation.

A side plane provides similar information to an edge. A line coincident with this plane through the point of contact goes through the origin, just like a meridian edge. The surface normal of the plane is perpendicular to this line. Only one free parameter is required to specify orientation of a surface normal with respect to this line, hence a side face provides one more equation than a meridian edge. This surface normal is generally not in the cross-section plane, and does not provide any orientation constraint on the cone.

By the above discussion, we have shown an equivalence between meridian edges and parabolic lines in relation to the number of independent equations and parameters they provide. Fearing [1987b] shows that no number of side edge contacts will allow the cone orientation to be determined.

Thus each meridian edge gives just 3 independent equations, but introduces three new unknowns ρ_i, θ_i, and z_i, in addition to the unknown orientation parameters ϕ and ψ. For N contacts, we have $3N$ equations, but $3N+2$ unknowns. Therefore, no unique orientation is obtainable. Similarly for parabolic points and side faces, the orientation parameter can not be determined. This result agrees with Shafer and Kanade's Pivot Theorem [Shafer and Kanade 1983]:

> A non-degenerate SHGC can be described as another SHGC with a different, intersecting axis, and the same cross-section planes if and only if it is Linear.

In other words, the cross-section can not uniquely specify the cone axis; a linear SHGC can have an alternate axis with the same cross-section. Shafer's equivalent LSHGC's can have different beveled ends, which implies that one can't specify a unique right LSHGC until the end is sensed.

We have shown that side contacts can not specify the cone orientation. The end faces specify the cone orientation. Can end faces determine anything about the cone origin? Consider the orientation determined by a contact on F_T, and additional contacts at vertices on the top end, given by:

$$\tilde{\underline{V}}_{Ti} = z_T \mathbf{R}(Ax_i, Ay_i, 1)^T + \underline{x}_o \tag{8.26}$$

where $\mathbf{R}$ is known from the surface normal direction, and $\tilde{\underline{V}}_{Ti}$ is the sensed vertex position. There are 2N + 1 equations from N vertices (all z_T coordinates are the same), but 2N + 4 unknowns, Ax_i, Ay_i, z_T and $\underline{x}_o$. Thus contacts on just one end cannot be used to determine anything about the origin, other than that it must be above F_T. The top contacts also provide no constraint on the scale parameter A, since the cross-section scale depends on the distance of the top end from the origin. Vertices corresponding to the same point in the cross-section on top and bottom ends of the cone define side edges, and two side edges will define the origin.

10.8 Feature Consistency Matching/Labelling for LSHGC

We proceed in this section on the assumption that we are touching an object that is not a cylinder. The fingers are in random positions in contact with the object; the cone's pose is still unknown. (Feature constraints for distinguishing a cone from a cylinder are given in [Fearing 1987b].) This section examines combinations of one face and two edge contacts, and combinations of two faces and one edge. The edges are considered to be independent of the faces, that is, not coplanar. These are the minimum sets of three features, not including parabolic points, for determining the origin and orientation of the LSHGC. In section 10.7, we assumed that the feature set had been classified. We now attempt to determine whether an edge is a meridian or end edge, and whether a face is a side or an end. This is not always possible with only 3 contacts. By using LSHGC geometric constraints, the possible interpretations of sensed features is reduced to a small set. It is possible to uniquely label certain sets of features.

If we have 2 parabolic points and one face, it is simple to determine whether the face is an end or side. The meridian lines at the parabolic points intersect at the origin. If the face plane intersects the origin, the set of features must be $F_S P_1 P_2$ and will not give an orientation solution. Otherwise we must have $F_T P_1 P_2$ or $F_B P_1 P_2$ which gives both the origin and the orientation of the cone.

10.8.1 Interpretation of Face and 2 Edges

Consider three finger sensors which give simultaneous contact measurements on the object, and these contacts are classified by curvature into a planar contact (F) and two edge contacts (E_1, E_2). We now use the geometric constraints of the cone to determine whether these features are consistent with an end face and two meridian edges, which is a sufficient condition for determining the LSHGC. The measured feature set is checked for consistency with $F_E E_m E_m$, where F_E is an end face. For simplicity here, we won't try to label all the features, we can just check to see if it is the set of sufficient features we desire.

There are three face types (F_T, F_S, or F_B) and 3 possible edge types (E_T, E_B, or E_m). The total possible number of combinations of 1 face and 2 edges is:

$$\begin{pmatrix} \text{3 face types} \\ \text{1 sensed face} \end{pmatrix} \begin{pmatrix} \text{3 edge types } + 2 - 1 \\ \text{2 sensed edges} \end{pmatrix} = 3\frac{4!}{2!2!} = 18 \qquad (8.27)$$

Lacking a more elegant approach, the 18 possible combinations are examined individually. Is the sensed set of features an end face and two meridian edges, or some other combination that will not be sufficient? There is an initial reduction in combinations because we need only consider combina-

Case	Combination	Violates Constraint
I	$F_T E_B E_B$ $F_T E_B E_m$ $F_B E_T E_T$ $F_B E_T E_m$	meridian edge can not be parallel to end face
II	$F_S E_{m1} E_{m2}$	apex can not be on plane F_T
III	$F_S E_m E_T$	contact on E_m must be below apex
IV	$F_S E_{T1} E_{T2}$ $F_T E_{m1} E_{m2}$	sensed face could be side or base (or bottom face if cross section contains origin)
V	$F_B E_{m1} E_{m2}$ $F_S E_B E_m$ $F_S E_B E_B$ $F_S E_B P$	face could be side or base

Table 10.3: Consistency of FEE with $F_{End} E_{m1} E_{m2}$.

tions with intersecting edges. All meridian edges of a cone must intersect as was shown in Section 10.7.1. If the edges do not intersect, we have an insufficient measurement set, and need to explore the object further. An example of this is a set of features $F_T E_B E_T$, $F_S E_B E_T$, or $F_B E_B E_T$. If an edge is coincident with a face, there will be an insufficient number of equations for pose solution, so 4 more combinations can be ignored: $F_T E_T E_T$, $F_T E_T E_m$, $F_B E_B E_B$, and $F_B E_B E_M$.

Table 10.3 lists the remaining possible combinations of two edges and a face. A single parabolic point is treated as a meridian edge. We sense a face and two edges, where the two edges intersect, but are not coplanar with the face. We have not yet properly labeled the sensed features. Let us suppose that one of the edges is perpendicular to the sensed surface normal of the face. From Case I of Table 10.3, this violates the constraint that a meridian edge is not parallel to an end face. We now know that we do not have two meridian edges, and need to probe the object to get more. But we have found the top and bottom of the cone, because we know that the face and edge correspond to $F_T E_B$ or $F_B E_T$.

Cases I through III are determined by simple geometric considerations;

the details can be found in [Fearing 1987b]. All but 5 edge-face combinations are eliminated by the constraints. There are only two interpretations for the face; if we were lucky with our contact locations, we have the object pose. Otherwise we have a useful indication of what direction to make the next contact from.

Case I:

A meridian edge E_m can not be parallel to an end face, which can be seen from:

$$\hat{n}_T \cdot \hat{E}_m = \frac{\begin{pmatrix} 0 \\ 0 \\ -1 \end{pmatrix} \cdot \begin{pmatrix} Ax \\ Ay \\ 1 \end{pmatrix}}{\sqrt{1 + A^2(x^2 + y^2)}} = \frac{-1}{\sqrt{1 + A^2(x^2 + y^2)}} \neq 0. \tag{8.28}$$

More generally, a meridian edge is never parallel to any face on a cone unless the edge is coincident with that face:

$$\hat{n}_S \cdot \hat{E}_m = \frac{A(x \sin\alpha - y\cos\alpha - d)}{\sqrt{(1 + A^2d^2)[1 + A^2(x^2 + y^2)]}} \neq 0, \tag{8.29}$$

where the surface normal is defined in eq. (8.15). Now since the object is not a cylinder, $A \neq 0$, the edge can be perpendicular to the surface normal only if

$$x \sin\alpha - y\cos\alpha - d = 0. \tag{8.30}$$

Any vertex in the cross-section plane (which generates the meridian edge) which satisfies the above equation is on the edge generating the side face, thus violating the condition that the meridian edge is not coplanar with the side face. Since $\hat{n}_T \cdot E_B$ is identically zero, any of the feature sets under Case I can be classified as having at least one end edge and one end face, not a sufficient set of features.

Case II:

For a side face and 2 meridian edges ($F_S E_{m1} E_{m2}$), the two edges intersect the face at a common point, so the origin $\underline{x}_o$ (from eq. 8.19), is coplanar with the face. By the LSHGC definition, this can never be the case. The scaling rule is zero at the origin, so there will be a vertex at the apex of the cone; there can not be a planar patch there. Thus this set of edges and face is distinguishable from the desired set.

Case III:

Any contacts on the cone must be below the apex of the cone (i.e. have $z > 0$). For Case III, we sense a set of features $F_S E_T E_m$, where E_T and E_m intersect, call the intersection point $\underline{x}_o$. (The case when the edges do not intersect can not be meridian edges, so there will be no ambiguity). Now if the sensed face were actually an end face (as we are testing for), it would have to be the base end, because the supposed apex of the cone $\underline{X}_o$ is on the inside of the surface normal.

A contact point on the meridian edge, say $\underline{E}_{mo}$ must be below the apex, that is

$$\hat{n}_B \cdot (\underline{0} - \underline{E}_{mo}) = \begin{pmatrix} 0 \\ 0 \\ 1 \end{pmatrix} \cdot -z_o \begin{pmatrix} Ax_o \\ Ay_o \\ 1 \end{pmatrix} = -z_o < 0. \tag{8.31}$$

But this is not the case when the side face is interpreted as the base. Without loss of generality, assume that the top edge of the face is in the $\hat{y}$ direction, with the normal along the $\hat{x}$ direction. Note that the intersection of $\underline{E}_m$ and $\underline{E}_T$ must occur at $z = z_T$. We have

$$\underline{x}_o = z_T \begin{pmatrix} Ax_o \\ Ay_o \\ 1 \end{pmatrix}, \underline{E}_{mo} = z \begin{pmatrix} Ax_o \\ Ay_o \\ 1 \end{pmatrix}, \text{ and } \hat{n}_S = \frac{\begin{pmatrix} 1 \\ 0 \\ -Ad \end{pmatrix}}{\sqrt{1 + A^2 d^2}}. \tag{8.32}$$

Thus

$$\hat{n}_S \cdot (\underline{x}_o - \underline{E}_{mo}) = \frac{\begin{pmatrix} 1 \\ 0 \\ -Ad \end{pmatrix}}{\sqrt{1 + A^2 d^2}} \cdot (z_T - z) \begin{pmatrix} Ax_o \\ Ay_o \\ 1 \end{pmatrix} \tag{8.33}$$

$$= (z_T - z)A(x_o - d) < 0.$$

This quantity is greater than zero because $z_T < z$, $A > 0$, and x_o must be less than d since the point must be on the interior of the side face since the cross-section is convex. Therefore the point $\underline{E}_{mo}$ is above the assumed apex, a violation of the definition of the LSHGC. The same result holds even when the origin is not contained within the cone cross-section (a tilted cone). Therefore a side face, meridian edge, and top edge can be distinguished from the desired combination $F_B E_{m1} E_{m2}$.

Case IV:

There are five possible interpretations of one face and two edges (out of 18 total) that satisfy all the LSHGC constraints for $F_E E_m E_m$, the desired set. Two of these, $F_T E_{m1} E_{m2}$ and $F_S E_{T1} E_{T2}$ (when the edges intersect above the side face) are equivalent. These two combinations are separable because the supposed origin (determined from the intersection of the edges) is above the surface normal of the face. This can only happen if the face is the top of the cone. The other 3 feature sets have edge intersections below the face: if the edges were meridian edges, the face would be the base of the cone. Unfortunately, there are two sets of measurements that can sometimes be indistinguishable from the set $F_B E_{m1} E_{m2}$. Figure 10.5 shows how $F_S E_B E_m$ or $F_S E_B E_B$ can sometimes be indistinguishable from the desired set of $F_B E_{m1} E_{m2}$. A parabolic point in the set $F_S E_B P$ does not prevent misinterpretation. F_S may still look like F_B if only local information is available.

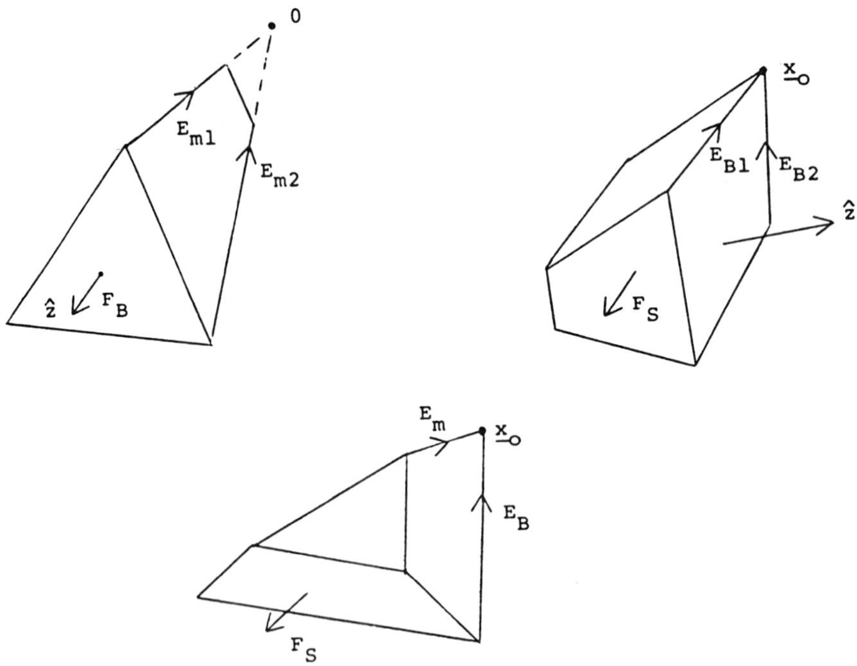

Figure 10.5: Multiple Cones for $F_B E_{m1} E_{m2}$. These three contact features on these cones can all be interpreted as the base end and two meridian edges. There are no further geometric constraints that can be applied to these three features to distinguish among these cones. Thus there are three valid interpretations of the contact data.

Case	Combination	Violates Constraint
I	$F_T F_S E_B$ $F_B F_S E_T$	$E \cdot \hat{n} \neq 0$ meridian edge can not be parallel to end face
II	$F_S F_S E_m$	apex can not be on plane F_T
III	$F_T F_B E_m$	end and side planes are never parallel
IV	$F_T F_S E_m$	satisfies all requirements
V	$F_B F_S E_m$ $F_S F_S E_T$ $F_S F_S E_B$	face could be side or base

Table 10.4: Consistency of FFE with $F_{End} F_S F_m$.

10.8.2 Interpretation of 2 Faces and an Edge

We consider now the combination of 2 faces and an edge. There are twelve possible combinations of two faces and one edge, but fewer combinations with the edge independent of the face (combinations with coincident edges $F_T F_S E_T$ and $F_B F_S E_B$ can be ignored). Combinations with parallel faces and coincident edges $F_T F_B E_T$, and $F_T F_B E_B$ can also be ignored; they are not sufficient for determining cone origin, although they do provide the orientation of the cone. To solve for the origin and orientation of the LSHGC, the desired sets of sensed features are $F_T F_S E_m$ or $F_B F_S E_m$. For the desired features, the edge intersects both faces, and the edge is not coplanar to a face.

As in the previous section, we measure two faces and one edge, and test this set for consistency with the desired set of features using the LSHGC constraints. Table 10.4 summarizes the constraints that can be used to distinguish a set of $F_E F_S E_m$ from the insufficient sets of two faces and one edge. Cases I and II are analagous to the same cases for combinations of a face and two edges.

The combination $F_T F_S E_m$ of Case IV is the only combination of features where the intersection of the face and edge is above the sensed end face. Thus the end face must be the top face. The combinations $F_S F_S E_T$ and $F_S F_S E_B$ can have a side face interpreted as the base. Since the plane corresponding to a face and an edge will intersect unless they are parallel, there is a much looser constraint from a face and an edge than from two faces. For example, the contact points will not be above the apex in the combination $F_S F_S E_T$ as they are in $F_S E_m E_T$, which is the roughly corresponding feature combination.

10.8.3 Conclusions for Shape Interpretation

If the first 3 contacts are not adequate, we want to know where to move the fingers. The convexity assumption can be used in combination with the partial constraints to choose finger paths that are very likely to make further contact with the cone. An insufficient set of features must be correctly labelled to use partial constraints. For example, with an edge parallel to a face and an object that is not a cylinder, we know from case I of Table 10.3 that the face is an end, and the edge must be on the other end. In this case we have the orientation as a consolation prize, but the length of the cone as a bonus. Tables 10.1 and 10.2 give necessary features for partial solutions to orientation or the origin of the cone. For example, given $F_S F_S F_T$, one obtains a line of points for the origin location.

Because of the cylindrical finger shape, we do not get a constraint at vertices or edges that can tell us which half space the object is in. A half-space constraint would help to localize the object with the three fingers. Without higher order edge sensing, to get the angles of planes that join to form an edge, or the shape of two surfaces intersecting, we cannot determine the attitude of the local surface from single contacts at curvature discontinuities.

Contrary to expectations, the LSHGC (a cone) is harder to characterize from local sensing with three contacts than was expected. There are only a few combinations of features that can be unambiguously identified and uniquely specify the cone's orientation and origin. The only unique solutions require sensing 1) the top or bottom end of the cone and two parabolic points, or 2) the top end of the cone, a side face, and a meridian edge. The scale parameter is unobtainable without sensing over the whole cross-section.

A similar analysis could be used to find the axis of a surface of revolution with local tactile sensing. In some ways it may be simpler to identify features on an SHGC that was not generated by a linear sweeping rule: faces are unique, and must occur on the ends.

10.9 A Tactile Servo System

In principle, the results of the previous section allow us to determine the orientation, origin and partially constrain the size of an object with a three fingered hand, given at least 3 suitable contact locations. While a vision

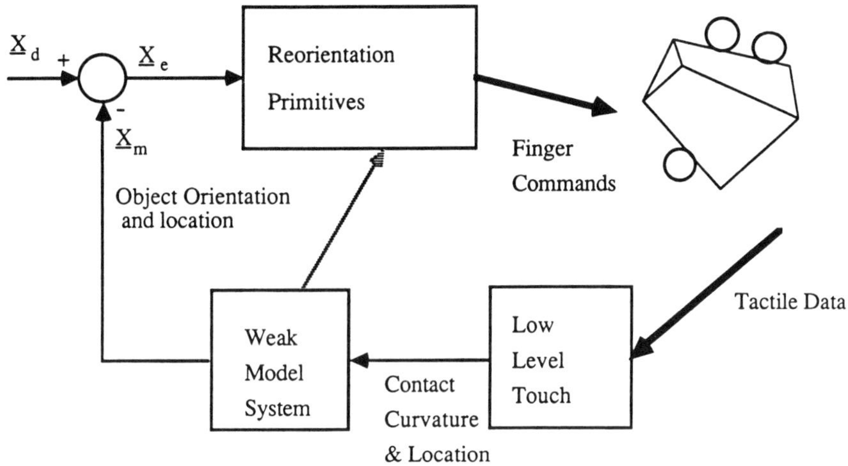

Figure 10.6: Proposed Dextrous Manipulation Servo System. Tactile data from finger contacts could be processed and interpreted by a weak model system to determine the axis and origin of a grasped LSHGC. Commands to the system could specify desired pose for the object in the hand. Reorientation primitives can be combined to regrasp the object if required.

system would be quite helpful during the grasp acquisition process, when the object is held by the hand, the fingers would tend to obscure the object. Thus this chapter considers the simpler problem of what can be done with only tactile information. This section considers the simplified condition where an object has been previously grasped by a three finger hand, and the goal is to reorient this object in the hand. This capability is required when changing from an acquisition grasp to a grasp used for insertion, or a grasp used for stable transport as the hand moves through space. A control system using tactile feedback to reorient unknown objects reliably in the hand is outlined. This example demonstrates directions for future work, and shows how the work in this chapter contributes to the dextrous manipulation problem.

Figure 10.6 shows a high level block diagram for a proposed dextrous manipulation system. Consider an unknown cone (of LSHGC form from Section 10.3) stably grasped by a three finger hand. Input commands to the system may specify regrasping the object so that the axis of the LSHGC is normal to the palm of the hand. There are many common industrial objects that can use this axis description. For example, in attaching a nut to a bolt, the orientation of the nut could be specified without prior knowledge of the nut's diameter or length. This ability would give a flexible assembly system that would not need specific modeling or reprogramming for different nuts.

Tactile information from the fingers of a dextrous hand can be interpreted to yield contact curvature and location. Important object features such as vertices, edges, planes, and curved surfaces can be characterized by curvatures, surface normals, orientations and locations. At a lower level than is shown in this figure, contact force information from the fingers could be used in a faster servo to control finger and object forces during contact.

Contact features from multiple fingers can be interpreted in a Weak Model System to determine cone orientation and origin, as hinted at in Section 10.3. The weak model system has two functions, to continuously update the estimate of the position of the object with respect to the hand, and to refine estimates of the global properties of the object such as its size, ends, and cross-section. With multiple continuous measurements, the data set is no longer sparse, and sensor uncertainty models can be used to get more reliable model estimates. For handling objects, the orientation of the cross-section curve can be important, thus the sensed object position is specified by a six vector $\underline{X}_m$. The orientation about the cone axis can be specified relative to some sensed feature. In Section 10.3, this angle was unspecified, because the cross-section was not predetermined. The weak model system could also be used in developing exploration strategies for determining object properties before the object is grasped.

The input to this dextrous manipulation system would be $\underline{X}_d$, the desired location of the end of the cone, and its axis orientation in the hand. The system should also control the forces of objects in contact, such as during insertion. Relative rotations should also be specifiable, such as turning a nut on a bolt.

The relative position error $\underline{X}_e$ is input to the reorientation subsystem, which needs to plan a sequence of object rotations and translations to reposition the object. A successful plan will depend on coefficients of friction, finger workspace limitations, the angle of the apex of the cone, the size and ends of the object, and the shape of the cross-section. This plan requires the global object properties determined by the weak model system.

Finger force strategies depend on the object shape. For example, in regrasping a circular cone with a large apex angle with a three finger hand, the object must be stably grasped with two fingers while the third finger is repositioned to a new grasp location. Finger force strategies need to include gravitational force, which can sometimes be used as a "fourth finger". It is important to position the two fingers along a diameter to maintain grasp stability with low friction coefficients. The weak model information may also indicate that better grasp points are available to improve reorientation.

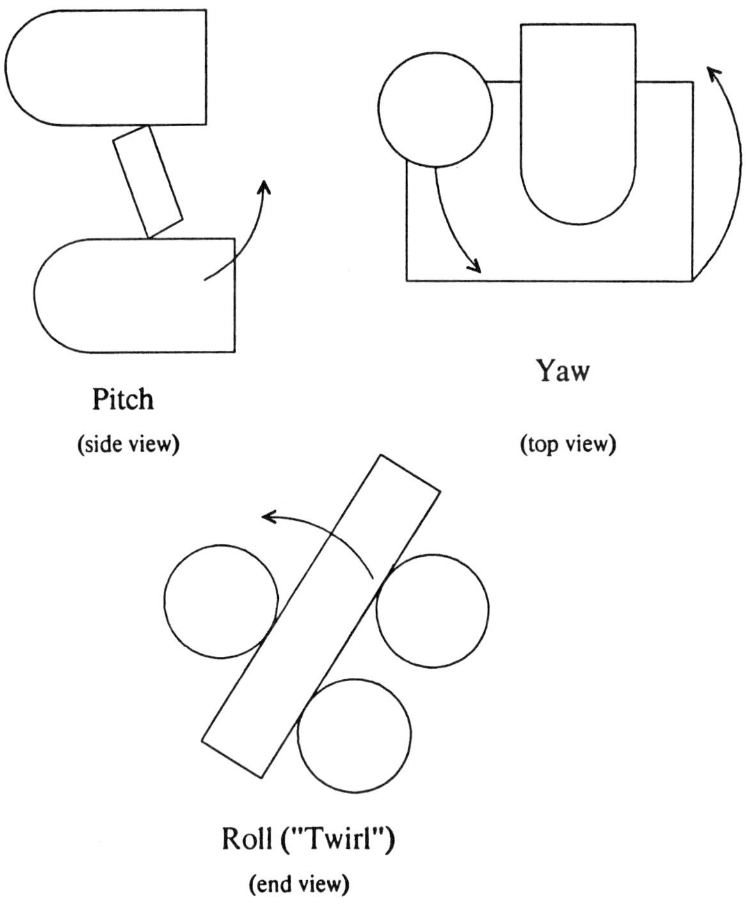

Figure 10.7: Three Basic Reorientations. For rectangular objects, roll, pitch, and yaw motions can form a basis for controlling object position in a hand. (©IEEE, Fearing 1986a).

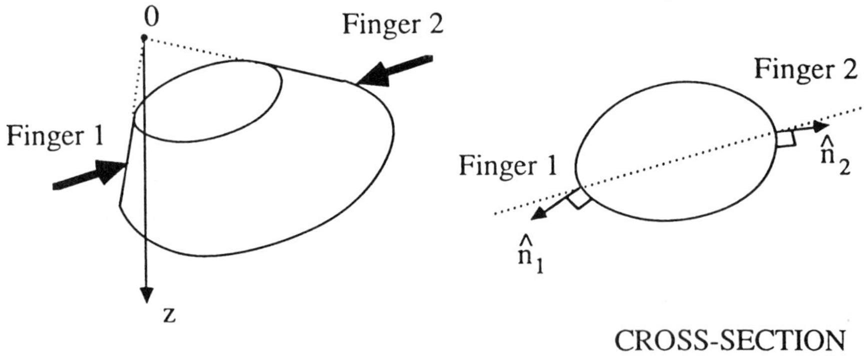

Figure 10.8: Grasping Cone with Two Fingers. The most stable grasp locations will be on one of the two pairs of double normals.

There has been little implemented work on motion operations with dextrous hands. Kobayashi [1985a] used a force control system to control local object motions with fingers in constant grasping locations. Reorientations where objects need to be regrasped, have been shown for rotating a bar through many revolutions using preprogrammed finger position commands [Okada 1982a], and by [Fearing 1986a] using a sequence of applied forces at each finger. A useful set of re-orientation primitives for simple objects like blocks, is shown in Figure 10.7. Regrasping objects with a simple parallel jaw gripper has been demonstrated by [Tournassoud, Lozano-Peréz and Mazer 1987]. Clearly, much work remains in developing regrasping/reorientation operations.

The reorientation sub-system would command finger force, stiffness and position. High speed tactile or force sensing can monitor the grasping operation for unexpected forces, contact slip, and loss of contact. The reorientation planner may also continuously monitor object position during reorientation, and develop error recovery strategies if unexpected object loads are encountered.

The right circular cone is the most difficult kind of LSHGC to grasp. Figure 10.8 illustrates the problem of grasping a non-circular cone. With two fingers, contacts must be placed nearly diametrically opposite to each other to get colinear surface normals in the cross-section plane, and thus finger forces within the friction cones. There is high slip potential for this shape; as the fingers slip, surface normals at the finger contacts get further away from colinear, causing greater slip. For a non-circular cross-section, there are always at least 2 pairs of colinear surface normals [Jameson 1985]. One of these pairs will give a more stable grasp with two fingers. A circular

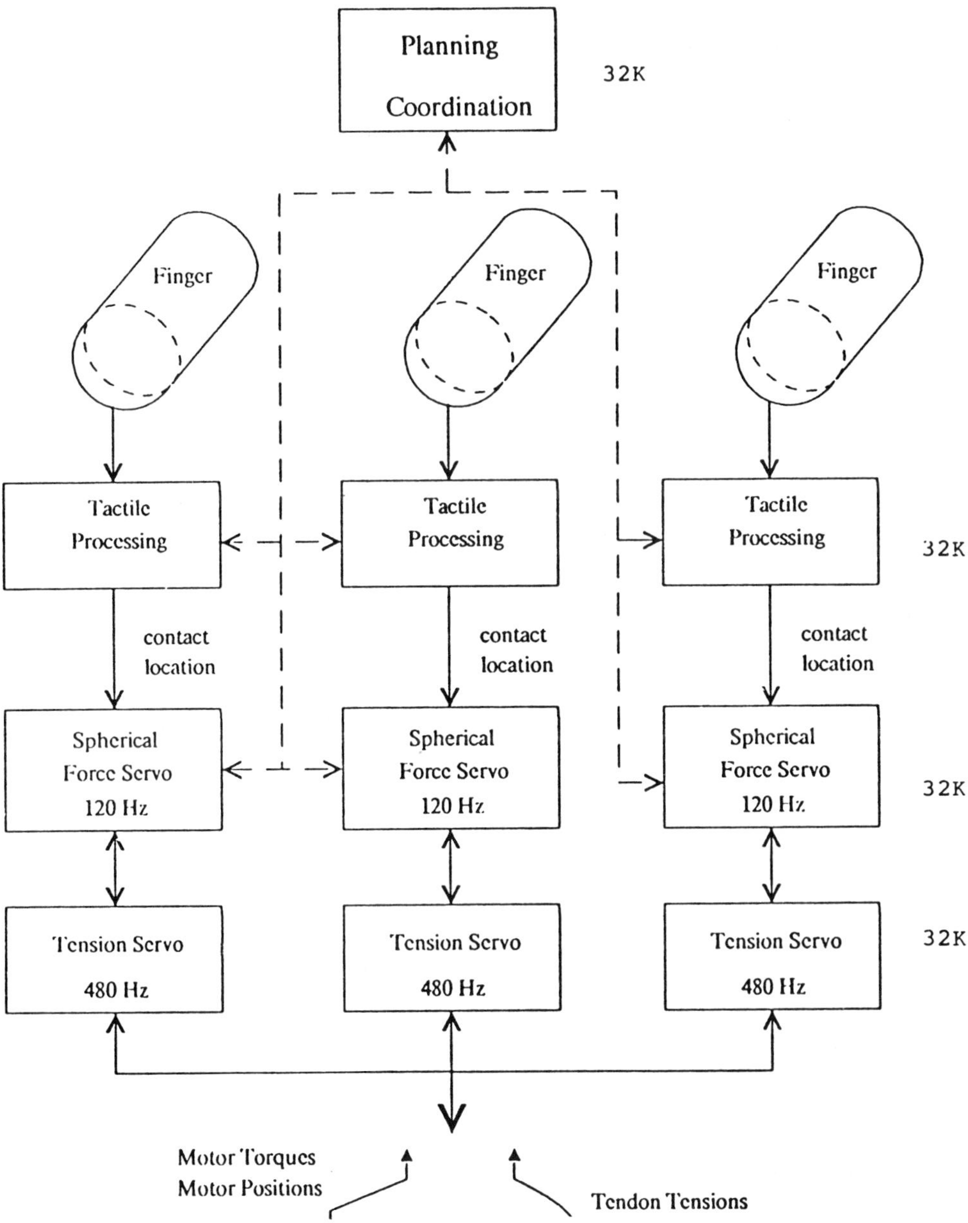

Figure 10.9: Proposed Controller Architecture for Dextrous Manipulation System. (©IEEE, Chen *et. al* 1986).

cross-section has equally unstable grasping for every diameter.

The proposed high level dextrous manipulation system of Figure 10.6 requires a large amount of computational power both for force control for fingers, and sensor processing. The NYMPH multiprocessor system has been implemented at Stanford for this task [Chen *et al* 1986]. Figure 10.9 shows a proposed processor arrangement using 10 National 32016 32-bit computers. A 7-processor version for hand control without tactile processing has been tested for reorienting objects in one plane ("TWIRL"), and work is in progress at incorporating three additional processors for tactile-based object control. At the lowest level, two processors are used for each finger for a force control system in spherical coordinates [Fearing 1986a]. At the highest level, a planning and coordination processor directs the sequencing of finger force commands.

10.10 Conclusion

We have shown that only a few specific combinations of three features on an LSHGC are sufficient to determine the cone's orientation and origin. Some combinations of sensed features will partially constrain the object's origin. Other combinations of features will constrain the object to be in one of three distinct orientations. Features such as edges and faces are ambiguous; they could be on the sides or ends of the cone. It is possible to determine the labellings of these features that are consistent with the geometric constraints. To simplify the labelling problem, only sets of contact features that are sufficient to determine the cone attitude and location were considered. Expressions were derived for various surface features such as edges, vertices, planes, and curved regions. The origin and orientation of the cone can be found from a set of necessary and sufficient combinations of labelled features.

If three contacts are not consistent with touching a cylinder, the features are examined for consistency with the desired set of contact features, using geometric constraints for the cone class of objects. When a unique labelling is possible, the cone origin and orientation will be determined.

A dextrous manipulation system that could manipulate a priori unknown objects was described as a motivating reason for using weak models for shape interpretation.

Acknowledgements: Many thanks to Jean Ponce and Tom Binford for many discussions and insightful comments on LSHGCs and shape interpretation problems. This work was supported at Stanford University by DARPA contract MDA903-86-K-0002.

11

Tactile Sensing and Control for the Utah/MIT Hand

Ian D. McCammon[1]
Steve C. Jacobsen [1]

ABSTRACT Despite recent advances in robot hand design and tactile sensing technology, dextrous robot hands with a comprehensive sense of touch are rare. The few existing dextrous touch systems universally exhibit problems of reliability, sensory coverage, bandwidth, or included volume. In an effort to establish a design methodology for dextrous tactile systems, this chapter will explore the obstacles that must be overcome in order to reliably acquire data from a large number of sensors on a dextrous robot hand. This work emphasizes the often overlooked difficulties of collecting tactile data in a practical and reliable way. After reviewing fundamental features of tactile systems for dextrous hands, we will discuss the ways in which these features affect hand performance, and which methods of addressing tactile sensors are effective and which are not. The design of a robust sensory architecture will be described, and progress on the development of a tactile sensing system for the Utah/MIT Dextrous Hand will be reported.

11.1 Introduction

Mankind's desire to extend his presence into complex environments has engendered interest in machines that emulate the functionality of the human hand. A broad spectrum of "dextrous" machines has been developed, each emerging from a particular set of assumptions about the nature of grasping and manipulation. Some designs, evolving from biological biases, take approximately anthropomorphic form while others, motivated by task oriented considerations, follow radically functional lines. But regardless of the approach taken, researchers universally agree that a variety of senses are essential for machine dexterity. Sensing of a hand's internal position (joint angles) and internal forces (joint torques) has been implemented on a majority of robotic hands [Hollerbach 1987; Grupen, Henderson and McCammon 1989], but imparting a sense of touch to these hands remains

---

[1]Center for Engineering Design, University of Utah, Salt Lake City, Utah 84112

somewhat enigmatic. No consensus exists among researchers as to how many or what type of tactile sensors are necessary, and although numerous tactile arrays have been developed, no rigorous theory of manipulation exists which explicitly relates tactile sensing function to manipulator performance. One of the primary obstacles to the development of such a theory is a lack of experimental hardware.

Substantial advances have been made in fabricating tactile sensors, but an emphasis on devising and refining transducer arrays has diverted attention away from how these sensors will perform in the context of a dextrous hand. The result is that the majority of tactile arrays remain on the lab bench, and the few that make their way onto multifingered hands often do not have the functional richness or reliability that is needed for serious machine manipulation research. This chapter is aimed at the portion of this problem that is concerned with acquiring tactile data in a practical and reliable manner. After presenting some background material on the Utah/MIT Dextrous Hand and related research, we will review three critical characteristics of tactile systems: data bandwidth, large sensor populations, and reliability. We will describe how each of these characteristics affects the overall performance of the hand. A discussion of methods for modeling tactile systems follows, where we compare different system architectures in terms of these three characteristics. With our design constraints clearly defined, we conclude the article with a description of the tactile sensing system currently under development for the Utah/MIT Dextrous Hand.

11.1.1 THE UTAH/MIT DEXTROUS HAND

In 1982 the Center for Engineering Design at the University of Utah and the Artificial Intelligence Laboratory at the Massachusets Institute of Technology began development of the Utah/MIT Dextrous Hand. Intended for experimental investigation into machine manipulation, the dextrous hand has evolved from early prototypes into the current Version IV System (Figure 11.1). Because the dextrous hand is an early tool in machine manipulation science, there are relatively few guidelines for the construction of such a device. General attitudes have played an important role in the design process, and three underlying assumptions have dominated the development effort. The first assumption is that the hand must exist as a reliable, integrated system in order to be fully useful to researchers in machine manipulation. Despite the convenience of building the hand from off-the-shelf parts, components such as valves, actuators, and supporting structures were specifically designed to achieve the speed and reliability sought by the design. So often, an amalgamation of mechanical and electronic parts simply does not provide the high performance needed by new areas of experimentation. A second assumption is that the hand should incorporate a rich diversity of sensors and the performance needed by experimenters. Joint angle and joint torque sensors were included in the Version IV System, and

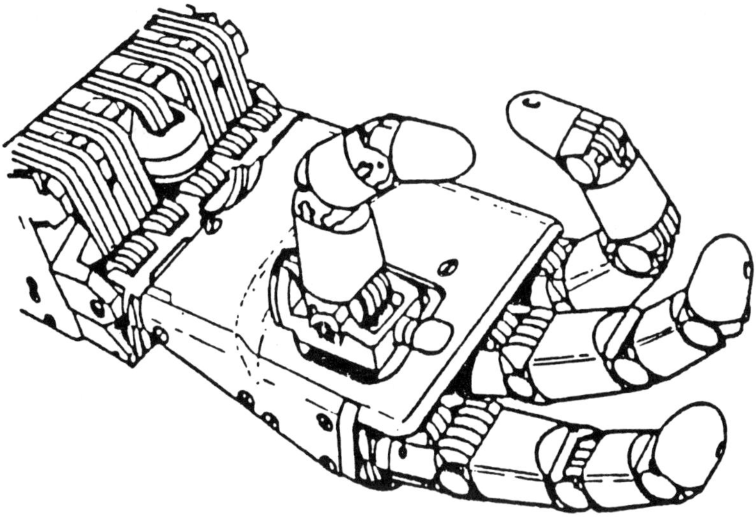

Figure 11.1: The Version IV Utah/MIT Dextrous Hand. The system also includes remotized actuators and a controller that responds to digital or analog commands. (©IEEE, Jacobsen *et al* 1986).

tactile sensors, as described here, are under development. By contemporary standards, the hand is both fast and strong, while supporting the sensory systems needed to study machine grasp and manipulation. The final assumption is that control of the hand must be accessible at a variety of levels. At the lowest level, specific joint angles and torques can be specified while at the highest level, grasping primitives can be invoked. After extensive testing, a robust and powerful controller has evolved. Throughout the development process, an emphasis on the design of a complete system, rather than an assembly of functioning components, has fostered the growth of a working, high performance device.

The Utah/MIT Dextrous Hand incorporates six specific subsystems within its structure. These subsystems are:

1. **Hand Structures**: The links and joints of the dextrous hand are roughly anthropomorphic in form, a geometry that was chosen for three primary reasons. First, the human hand stands as an existence proof that an anthropomorphic hand geometry is capable of refined manipulation, although all of the reasons for this are not fully understood. Second, the anthropomorphic form permits experimenters to more easily visualize new paradigms for machine grasping and manipulation. Finally, the approximate kinematic equivalence of the dextrous hand to the human hand simplifies its use as a teleoperated

slave. Each of the four fingers exhibits four degrees of freedom resulting in a sixteen degree-of-freedom hand. Fingers are modular and can be reconfigured as new theories of manipulation develop.

2. **Internal Sensors**: To obtain accurate joint angle measurements, rotary Hall effect sensors are mounted within each joint. Each sensor provides an output corresponding to an angular deflection of 0 to 95 degrees, which is linear to within 5constraints, tendon tensions cannot be measured directly at the joints, but are instead measured at the wrist via idler pulleys and Hall effect sensors. Tendon tension sensors provide an output that corresponds to tensions up to 30 pounds. Hall effect sensors were chosen for both sensing modalities because of their simplicity, high frequency response, small size, and reliability.

3. **Covering Systems**: These protect the finger structures from ,damage and are sites for the application of experimental tactile sensors and skin materials. Cylindrical covers are approximately 1/10th of an inch thick, and are molded from compliant urethanes. All covering segments are injection molded, and can be inexpensively made from a variety of plastic and rubber materials.

4. **Actuators**: In order to achieve the desired active and passive performance, antagonistic pneumatic actuators were used in the Version IV system. The product of years of development and exhaustive testing, 32 specially designed glass and graphite pneumatic cylinders control tendons via associated jet-pipe valves. Actuators operate at relatively low pressures (50 to 100 psi), and exhibit a flat frequency response exceeding 20 Hz. Actual forces are transmitted to the hand via high strength polymeric tendons and pulleys.

5. **Power Transmission**: Antagonistic tendons convey the actuator output to the hand via a remotizing structure, and a disconnection system permits disassembly of the structure for transport or maintenance. The remotizer has been implemented in such a way that moving the hand throughout the workspace does not affect tendon lengths, and as a result does not change the joint positions.

6. **Controllers**: The low level control system for the Utah/MIT Dextrous Hand can be used alone for subsystem diagnostics or can receive inputs from higher level controllers for more complex hand motions. The system incorporates 16 variable loop gain position servos to operate finger joints and 32 variable loop gain tension servos to control joint torques. The analog and digital control system operates with five Motorola 68000 microprocessors, which are accessible via a Multibus system.

A more detailed discussion of the design of the Utah/MIT Dextrous Hand can be found in [Jacobsen *et al* 1984; Jacobsen *et al* 1985; Jacobsen *et al* 1986]. Control of the hand is the described in [Siegel *et al* 1985; Biggers *et al* 1986; Narasimhan *et al* 1986; Narasimhan *et al* 1988]. A discussion of computational architectures for the dextrous hand can also be found in [Narasimhan *et al* 1988, this volume]

11.1.2 Tactile Sensing for Dextrous Hands

Many robot hands have been developed in the past, and some of these have been furnished with a sense of touch. But not all approaches to hand based touch sensing are appropriate for use with the Utah/MIT Dextrous Hand. Some of the barriers to practical implementation of dextrous tactile sensing systems are reviewed below.

Most of the work in tactile sensing has focussed on the implementation of planar transducer arrays. Many optical, electrical, magnetic, and mechanical phenomena have been exploited as tactile sensors, some of which are reviewed by [Harmon 1980; Dario and de Rossi 1985; Pennywitt 1986; Hollerbach 1987]. Despite substantial efforts by researchers, relatively few designs that are compatible have emerged with dextrous robot hands. Brock and Chiu [1985] describe a fingertip shaped module which can resolve forces and moments which are applied to the fingertip. Consisting of eight pairs of strain gauges mounted in a Maltese cross structure, this sensor is compatible with the Stanford/JPL Hand [Mason and Salisbury 1985]. Although it is well suited as a fingertip sensor, this approach does not leave interior space for the pulleys, cables, and support structures that are found in the proximal finger segments of most robotic hands. Also, as Fearing and Hollerbach [1985] point out, multiple contacts can result in erroneous readings. Begej [1988] describes advances in optical fingertip tactile sensors based on total internal reflection. These sensors convert normal force data into an optical display, and images can be readily processed using conventional vision algorithms. Because data transduction and transmission is optical in nature, this device exhibits exceptional immunity to the electromagnetic interference that plagues other sensors. Like the Brock and Chiu sensor however, this approach leaves little room for other structures within the fingers. Abramowitz, Goodnow and Paul [1983] report the performance of a polymeric piezoelectric skin (PVF2) for a mechanical hand. Investigators reported that pyroelectric effects and drift posed serious problems in this implementation. Dario and Buttazzo [1987] have overcome these difficulties and implemented a similar design on an anthropomorphic finger. By emphasizing sensor based control in their design, the researchers were able to implement a comprehensive system for fingertip-based exploration. Fearing, Rise and Binford [1986] have developed a fingertip capacitive transducer array for the Stanford/JPL Hand, and have used it to estimate the radii of cylinders contacted by the fingertip [Fearing and Binford 1988]. Tise [1988] reports using a force sensing resistive material to devise a tactile array that is compatible with the curved surfaces of robot fingers. Clark [1988] has used solid state magnetic field sensors in a highly compliant fingertip sensor although, again, no space has been left for other internal structures. In this volume, Okada [this volume] presents a unique approach where the superstructure of the finger itself, rather than a sensing array, transduces contact forces and moments. Although multiple contacts may confuse this sensor, the author discusses its application to segments other than the fingertip. Despite the apparent success of current research in deploying fingertip sensors, problems still remain with sensory coverage, included volume, and reliability of these systems.

Early in the design of the Utah/MIT Dextrous Hand, project personnel recognized the need to incorporate tactile sensing. But because no compatible sensing system was available, customized touch sensors were built. Like other researchers, our initial efforts were directed towards the development of transducers with desirable size, resolution, and dynamic range characteristics, and a prototype tactile sensing scheme was implemented on the hand [Jacobsen *et al* 1984]. By utilizing an optical transduction method involving stress-induced birefringence, we were able to achieve excellent linearity and noise immunity in the sensors. But problems with spatial density and cable breakage led to a closer look at the problem of acquiring tactile data from the hand. It became clear to us that merely building a transducer array and providing the means for data processing was not enough. The mechanical problems of routing many data wires throughout the hand necessitated some sort of multiplexing scheme, but this involved a trade off with system bandwidth. With a few exceptions, previous investigators had virtually overlooked these practical difficulties. What was needed was a closer look at the requirements of tactile sensing for the Utah/MIT Dextrous Hand.

11.2 Design Criteria for Dextrous Tactile Systems

In order to be fully useful in dextrous machine research, multifingered robotic hands require more than a finger shaped tactile array and the software to drive a CRT display. An obvious requirement is sensory coverage over finger segments and palmar surfaces as well, since robust manipulation and powerful grasp utilize large areas of the hand [Cutkosky 1985]. A second requirement is a high tactile data bandwidth. If signal conditioning, digital conversion, and multiplexing are too time consuming, then data acquisition will be too slow to investigate the dynamic aspects of manipulation and sensing. A final and very important requirement is that a tactile system must be reliable enough to support serious experimentation. The complexity of external wiring and termination is a major factor in defining system reliability. This section quantitatively discusses each of these requirements and relates them to the performance of dextrous robot hands.

11.2.1 TACTILE DATA BANDWIDTH

When a tactile sensor is in use there is a time delay between a request for tactile data and the actual arrival of the contact information from the transducers. The inverse of this time delay is the bandwidth of the tactile data. Robot applications involving object presence or feature detection (such as workpiece recognition) have relatively low bandwidth requirements since the nature of contact is not typically changing with time. During dextrous manipulation however, a robot hand must respond to encounters with unexpected surfaces, slipping or falling objects, and complex interaction forces during assembly operations. Tactile data must be collected quickly in these situations so that the hand can respond in an appropriate manner. This section will discuss tactile data bandwidth requirements and limitations in the context of machine manipulation.

11.2.2 MANIPULATION AND ACTIVE EXPLORATION

No theory currently exists that explicitly defines the relationship between tactile data bandwidth and dextrous machine manipulation. Two areas are of interest, however; force control and feature resolution during exploration.

Researchers generally agree that information describing the location and the nature of contact is essential for object manipulation by robotic hands [Mason and Salisbury 1985; Brock 1988]. Grasp stability is a product not only of the geometry of grasp (placement of fingers on the object), but also of the control of forces within the grasp. Many robotic hands (including the Utah/MIT Dextrous Hand) incorporate sensors that provide joint torque information, from which an estimate of contact force can be derived once the location of contact is known. This approach is not entirely accurate however, since it does not take into account the effects of contact friction of forces obliquely applied to the contact surface. If manipulation is viewed as succession of stable grasps within a compliant hand (see, for example, [Fearing, Rise and Binford 1986]), high bandwidth force data is not strictly necessary, but bandwidths must still be high enough to achieve stable grasp. When tactile sensors are used to interpret contact force data, the relationships between bandwidth and dexterity remain unclear.

When a dextrous hand is in the process of actively exploring or manipulating an object, it may encounter surface features that were not previously known. The hand's ability to discriminate such features is a function of data bandwidth, since the rate at which individual sensors are sampled determines the minimum size feature that can be resolved by the sensors. For example, consider a tactile sensor located at the fingertip of the hand which is sampled at a maximum rate of f_s samples per second, as shown in Figure 11.2. The sampling theorem tells us that only features which appear to

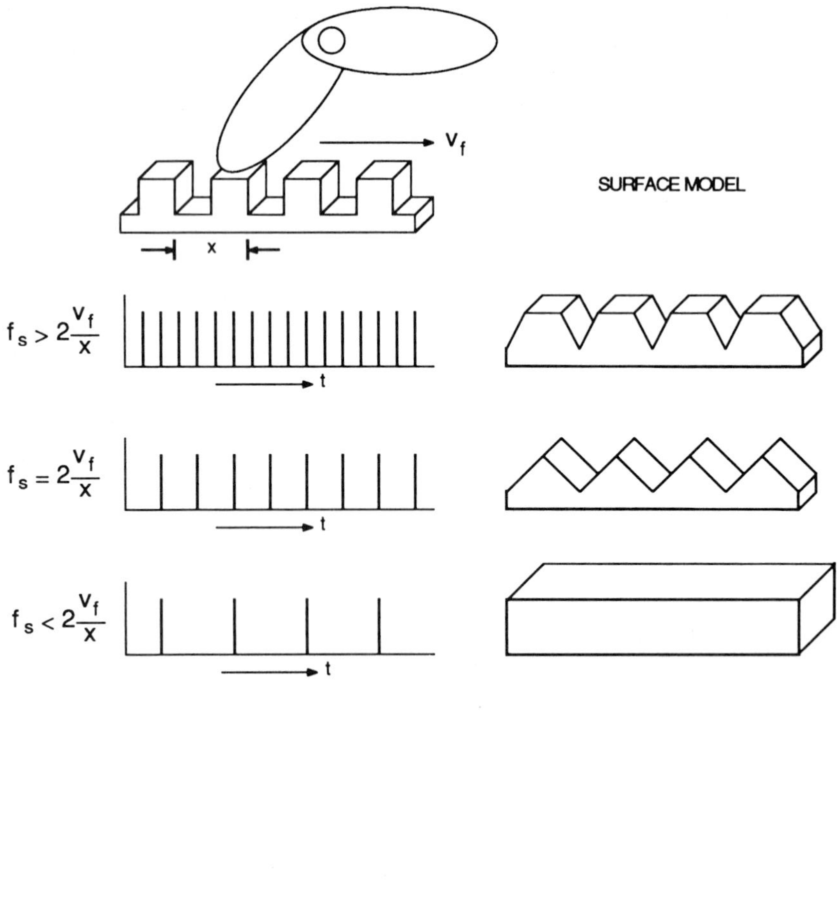

Figure 11.2: Spatial aliasing during active exploration. If the fingertip transducer is sampled at less than twice the spatial frequency of the features, incorrect models of the surface result.

the sensor at a rate less than $\frac{f_s}{2}$ features per second will be resolved by the system. If we assume that between encounters with immediately adjacent features, the finger travels a constant speed v_f, then the minimum feature size resolvable by the system is:

$$X_{min} = \frac{2\,v_f}{f_s} \tag{11.1}$$

where f_s is the bandwidth of the tactile data acquisition system. Essentially, an exploring sensor is "blind" to features that fall below the minimum size. In order to resolve very small features then, the finger must either move very slowly or the tactile data bandwidth must be very high.

11.2.3 Bandlimiting factors

High tactile data bandwidth is desirable for dextrous manipulation research, but there are a number of obstacles to achieving this ideal. The first of these concerns the tactile sensor itself. Most tactile transduction schemes have inherently high bandwidths (capacitive, piezoresistive, piezoelectric, and optical methods are typical examples), but when arrays of sensors are built it is common for many transducers to share a single amplification stage. Because of variations between transducers and problems with noise, shared signal processing is not generally optimized for each transducers, and bandwidth may be reduced by bandlimited methods of amplification. For example, a piezoelectric film commonly used for tactile transducers (PVF2) has a flat frequency response (depending on film thickness) of up to 10 GHz [Chatigny 1984]. In a recent application however, output drift and noise required amplification and filtering which reduced individual transducer bandwidths to less than 7KHz [Dario and Buttazzo 1987]. Bandlimiting methods are commonly used in tactile sensing, and are in fact necessitated by the presence of electromagnetic noise.

A second source of bandlimited tactile data arises from the mechanical response of the robot skin above the transducer. Compliant skins are generally desirable in robot hands because they contribute to the stability of grasp [Cutkosky and Wright 1986a], and protect the contact sensors from damage. These skins are by their nature quite hysteretic, and take some time to return to their unstressed state after being subjected to high contact forces. One implementation reports a mechanical bandlimit of 15 Hz, even after selecting a skin material based on its low hysteretic properties [Fearing, Rise and Binford 1986].

A third source of bandlimiting is the multiplexing of multiple transducer outputs through a transmission medium or signal processing module. In general, the bandwidth of each sensor is reduced proportionally to the number of sensors scanned. Bandlimiting due to multiplexing is described in further detail later in this chapter.

11.2.4 Tactile Sensors

It is tempting to judge tactile sensing devices by their most obvious physical characteristics. Sensor spacing, linearity, force resolution, and hysteresis all provide handy figures of merit for determining the relative worth of a particular design. But no experimental work has yet been done that conclusively shows how these characteristics relate to dextrous manipulation. For example, tactile arrays with very high spatial densities have been implemented (see, for example, [Mott *et al* 1986]), but no research shows that such densities are actually required for robot grasping and manipulation. With no empirical results to draw on then, how do tactile sensor designers identify their requirements?

Tactile sensor design objectives typically emerge from one of two commonly held perspectives. The first is based on imitating a supremely efficient model - the human hand. The indisputable ability of the hand to perform highly dextrous tasks has led some investigators to examine the characteristics of the human tactile system and by extension, tentatively establish sensory requirements for dextrous robotic hands (see, for example, [Lederman and Browse 1988]). The underlying assumption of this approach is that if critical functional elements of the hand can somehow be replicated the machine will exhibit human like dexterity. Although this approach is often criticized, it is nevertheless generally dominant and is implicitly assumed in a majority of the literature.The second perspective is based on analyzing the physics of multifingered grasp and contact. Not convinced that solutions to the problems of robotic manipulation should necessarily take an anthropomorphic form, some investigators have explored analytical methods for specifying sensor number, spacing and function [Fearing and Hollerbach 1985]. But rigorous analysis, hampered by a lack of experimental hardware to verify results, is less common. This section reviews the performance of dextrous robot hands in terms of these two perspectives, and how they relate to sensor modality, spacing, and resolution. what should tactile sensors measure?

Despite much discussion among researchers in the field of machine manipulation, there seems to be no consensus about the information that a tactile transducer must provide. Arrays of simple binary sensors were among the first to be successfully implemented on multifingered hands [Kato *et al* 1973], and have the attractive feature of being fairly simple to interface with digital electronics. A rough estimate of contact normal forces (neglecting friction) can be obtained from measurements of the joint torques and joint angles. But many researchers feel that binary touch is insufficient for refined machine manipulation and suggest that, as a minimum, tactile sensors must measure the magnitude of contact forces normal to the contact surface [Fearing and Hollerbach 1985]. Arrays of normal force detectors represent the majority of research and development efforts in tactile sensing, and have proliferated both in research laboratories and industrial settings. Because tactile data in this form typically consists of a two dimensional array of intensities and bears a striking resemblance to visual data, some researchers have successfully applied image processing algorithms for edge detection and pattern recognition [Ellis 1987]. Such sensors measure absolute contact force and so the problem of mechanical hysteresis, where the sensor's output is a function of its recent history of activation, has been the focus of much concern. Highly hysteretic sensors, if subjected to large or prolonged contact forces, retain the impression of the contact and report contact when no contact is in fact occurring. Studies of the human hand, however, indicate that the majority of the hand's mechanoreceptors are not sensitive to absolute force, but to changes over time in applied force [Vallbo and Johansson 1978]. This probably explains why the skin of the hand can be so hysteretic, and yet still provide very accurate information about the condition of contact. Noting this situation, some investigators have advocated the use of on-site processing to provide the first and second time derivatives of surface deflection, and thus eliminate hysteresis problems [Raibert and Tanner 1982]. Although this approach may improve accuracy, associated processing may reduce transducer bandwidth and will certainly increase system cost. The need for sensing other forces on the contact surface, such as shear and moments, has been indicated by researchers in machine manipulation [Mason and Salisbury 1985]. The sensation of slip in the human hand in fact suggests that this modality might be crucial to dexterity [Hollerbach 1984]. Tactile transducers that measure forces and torques at the point of contact have been developed (see, for example, [Hackwood *et al* 1983; Brock and Chiu 1985]), but the amount of information that is required from such sensors remains unclear.

11.2.5 Spatial resolution

No comprehensive theory currently exists which relates tactile sensor spacing to dextrous manipulation tasks. In the absence of such a theory, what guidelines can a designer use to determine a practical sensor density in an experimental array for a dextrous robot hand? From a biological perspective, the human hand provides a variety of clues. Measurements of the hand's ability to statically distinguish between two points placed on the adult fingertip skin (simultaneous spatial threshold) yield resolutions of between 1.5 and 2.2 mm [Vallbo and Johansson 1978]. Meissner corpuscles, which correspond roughly to contact velocity detectors and are the most densely packed mechanoreceptor in the fingertip, have a center to center spacing of approximately 0.9 mm [Schmidt 1986, p. 44]. Other receptors are present in the skin, of course, and have complex interactions that give rise to the hand's spatial resolution [Lederman and Browse 1988]. Based solely on physiological considerations then, a spacing of 1.0 mm between sensing elements would approximate the sensory abilities of the human hand [Hollerbach 1984].

Efforts to analytically correlate sensor spacing with contact feature resolution have focussed on the avoidance of spatial aliasing. Simply stated, spatial aliasing arises from the sampling theorem, which states that a time varying signal must be sampled no less than twice its highest frequency in order to be reconstructed. In the context of tactile sensing, the time varying signal is replaced by contact profile, and the sampling intervals become sensor position. So, in order to resolve features of a pattern that are equally spaced, a touch array's sensor spacing must be greater than twice the spatial frequency of the pattern. For example, if a pattern of bars occurring at a frequency of 4.0 mm is to be sensed, then sensors must be placed a minimum of 2.0 mm apart. In actual use however, the sensors are covered by some sort of compliant skin to improve the stability and compliance between the object and the fingertips [Cutkosky and Wright 1986a]. Fearing and Hollerbach [1985] recommend that a practical spacing for tactile sensors is twice their depth beneath the skin, but in an actual implementation, Fearing *et al* [1986] point out that a practical compromise between sensitivity and spatial aliasing occurs when sensor spacing is approximately equal to sensor depth. Nonhomogeneous distributions of sensors [Dario *et al* 1986], and relationships between sensor spacing and object discrimination [Togai *et al* 1984] have also been explored.

11.2.6 Number of Wires

State of the art tactile sensor arrays embody substantial advances in transducer miniaturization and performance. Ironically, it is the failures of mundane sensor components such as wires and connectors that seriously impair the utility of these arrays in dextrous robot hands. Large bundles of wires impede finger mobility and add to manipulator weight, but most importantly, they are extremely vulnerable to breakage. Reviews of early work in tactile sensing [Bejczy 1977], show that this problem plagued the first prototypes, and even contemporary researchers point out the need to ameliorate wiring problems [Goldwasser 1984; Raibert and Tanner 1982]. The authors' own experience with prosthetic devices [Jacobsen *et al* 1982], dextrous robot hands [Jacobsen *et al* 1986], teleoperated machines [Jacobsen *et al* 1989], and in entertainment robotics has taught us not to underestimate the importance of wire breakage and connector failure. Although somewhat inglorious, the mechanical issues surrounding cable routing and termination demand close attention if a tactile sensing system is to be robust enough for more than occasional use.

Given the many interconnection schemes that are commercially available, it is hardly possible to derive a rigorous theory that specifies optimal wiring and connection configurations. It is possible however, to make some rough estimates of how bundle size affects system reliability. Consider a situation where a bundle of wires carries tactile data across the joints of a dextrous robot finger. As the finger performs manipulation tasks the wires flex and collide with surfaces and edges in the environment. As long as the wiring is reasonably robust to begin with, we can assume that failure of electrical conductors will occur randomly within the bundle. If we also assume that breakage or connector failure occurs at regular intervals (α), then we can write a simple relation for the mean time between failures (MTBF) of the tactile data acquisition system [Amerasekera and Campbell 1987, pp. 6-7]:

$$MTBF = \frac{\alpha}{W} \tag{11.2}$$

where W is the number of wires in the bundle. Clearly, in designs that contain a large number of wires, the system failure interval may be so small as to render the system impractical for serious experimentation.

There are three ways to increase the reliability of a tactile data transmission system. The first is to reduce the number of wires (W) through the use of time division multiplexing. Although this approach is initially unappealing because of the complexity and expense of locating multiplexing electronics at the fingertip, compact hybrid and VLSI technology exhibit reliability while tolerating complexity very well. Data reduction methods have been implemented to reduce the number of wires [Raibert and Tanner 1982], but these approaches are probably most useful when the exact nature of tactile data for manipulation is known. A second option for improving system reliability is to provide mechanically and electrically robust conductor and connector components (large a). Careful design of these components has alleviated reliability problems in similar areas [Gordon and Johnston 1985]. Finally, redundancy can be built into the system by having one signal on two or more wires.

11.3 Modeling Tactile Sensing Architectures

The preceding section discussed how the performance of robot hands is affected by tactile data bandwidth, the number of sensors, and the number of wires transmitting tactile data from the hand. In this section, we will explore models that relate these parameters at the level of transducer arrays and at the system level. Based on a performance comparison, we will show how the constraints of dextrous tactile sensing favor one particular sensory architecture. Although these results are be fairly obvious on an intuitive level, the models allow us to quantify the reasons for choosing a particular design, and to more fully understand the trade offs of the design process.

Before we begin, it is convenient to make several assumptions that simplify the analysis. First, we consider here only those tactile data acquisition systems that are synchronous. This simply means that all logical activity within the system is synchronized by a clock signal that is simultaneously present throughout the system. Synchronous architectures offer the advantage of simplicity since no transmission protocols are required for data transmission (as with asynchronous communication) and hardware is relatively straightforward to build (no small consideration in view of space constraints!). A second assumption we make is that all addressing is binary. Commercial availability of components and high signal to noise ratios are obvious advantages. Finally, we assume that wire reduction is achieved through time division multiplexing. High bandwidths and good signal to noise ratios favor this approach over other methods of multiplexing [Tanenbaum 1981, pp. 104-108]. In this section, we are only considering the number of wires needed to address these arrays of sensors; the wires for sensor output and electrical power are not included in wire counts. Note that we have placed no restrictions on the tactile transducers themselves; for our purposes here they are merely "black boxes" containing contact data. A more general discussion of sensory system modeling can be found in [McCammon and Jacobsen 1988].

11.3.1 Array-level Addressing

At the array level our primary concern is the addressing of individual sensors so that the contact profile can be determined. Depending on how addresses are formatted and what multiplexing hardware is locally present, addressing can take a variety of forms. Typically, sensors are arranged in a rectangular array (X by Y) as shown in Figure 11.3(a). Individual sensors are positioned at the junction of two conductors; one supplying an enable or carrier signal, and the other providing a conduction path for the sensor's output. The state of any sensor can thus be determined by specifying the column (i) and row (j) in which the sensor resides. The output appears on the DATA line and may be in analog or digital form. This multiplexing scheme is prevalent among tactile sensing devices [Beni *et al* 1983], and is particularly useful with robotic hands since it can be applied to cylindrical finger geometries.

There are several methods for addressing the sensor array shown in Figure 11.3(a). Five of these methods are summarized in Table 11.1 and are analytically developed in the Appendix; a qualitative description is presented here.

1. The first method assumes that i and j addresses can be simultaneously supplied via two parallel address busses. Since all of the address bits for a single sensor arrive together, one complete address can be transmitted during a single clock cycle. The access time for the entire array then, is simply the clock period times the number of sensors in the array. For equal numbers of sensors, this approach is equivalent to the configuration shown in Figure 11.3(b) in terms of access time and wire count. This scheme is functionally identical to the random access of data in a computer memory.

2. A second approach involves the use of a single serial line to transmit

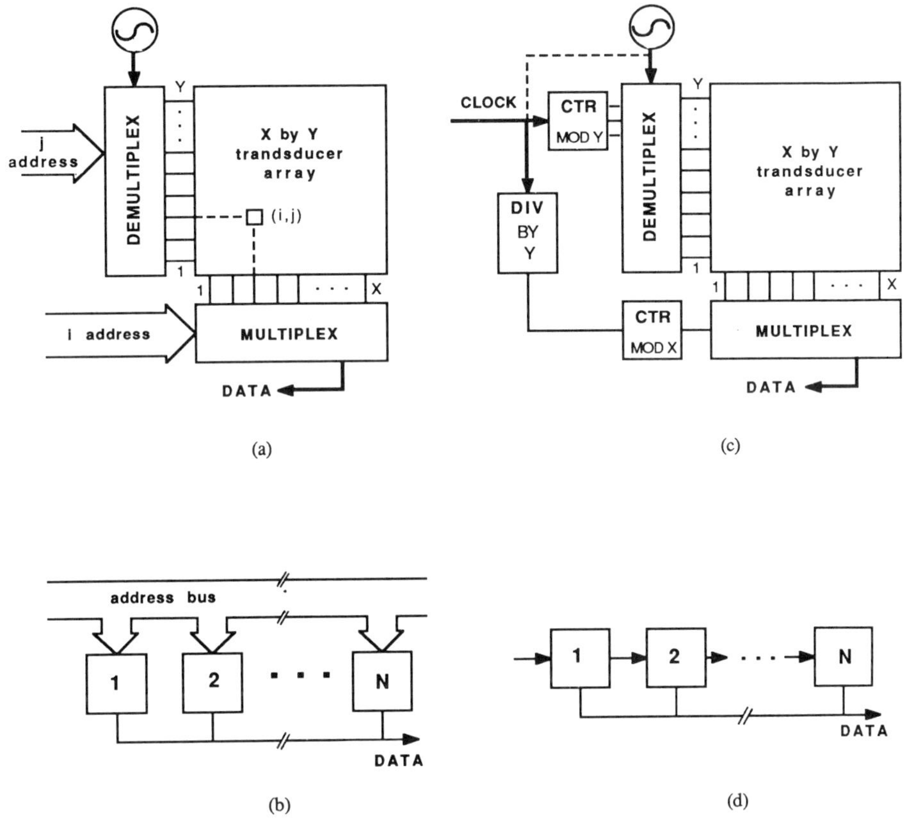

Figure 11.3: Architectures for array-level tactile data acquisition. (a) Sensor arrays are typically addressed in a row-column fashion. (b) A common address bus can randomly address distributed sensors. (c) By incorporating counters into the multiplexing hardware, an array of sensors can be collectively addressed. (d) Collective addressing of distributed sensors. In the case where X*Y=N, (a) is equivalent to (b) and (c) to (d).

	addressing scheme	access time	wires needed
	random:		
1	parallel	nt_c	$\log_2 N$
2	serial	$nt_c \log_2 N$	1
	matrix:		
3	parallel	$(x-1)(X+1)t_c + (y+1)t_c$	$\log_2 N$
4	serial	$(x-1)(X+1)t_c \log_2 X +$ $(y+1)t_c \log_2 Y$	1
5	*collective*	Nt_c	2

n=number of observed sensors
x=number of observed columns
y=number of observed rows
t_c=clock period

N=total number of sensors
X=total number of columns
Y=total number of rows

Table 11.1: Access times & number of address wires *vs* schemes

i and j sensor addresses. The access time for each sensor is somewhat longer in this case since the address transmission requires a number clock cycles equal to the number of address bits. Although this approach has the desirable effect of reducing the wire count, it requires increased hardware complexity (not shown in the figure) within the demultiplexing and multiplexing stages to de-serialize the incoming address.

3. Another way to read the array arises when the addresses i and j are on parallel busses but are transmitted during separate clock cycles. Rather than randomly send addresses of sensors within the array, a single column (i) can be transmitted and latched followed by a scan of all the rows (j varies from 1 to Y), thereby reading the entire sensor column. By advancing through the columns (i varies from 1 to X) and repeating the process, the entire array can be read. The primary drawback of this approach is that it sacrifices the individual addressability of sensors within a column. Note that if addresses i and j can be simultaneously transmitted, this method offers no advantages.

4. The same procedure can be performed in serial form. Again, longer transmission times are required due to the serial nature of the data but addressing takes place over a single wire.

5. The last approach involves mandatory scanning of the entire array. As shown in Figure 11.3(c), including counters in the array hardware enables a single clock signal to scan all of the sensors. Each clock cycle advances the reading to the next sensor, and sensor outputs appear serially on the DATA line out of phase with the clock signal by a constant amount. Figure 11.3(d) illustrates a functionally identical situation where sensors are linearly arranged. The obvious disadvantage of this approach is that individual sensors are no longer immediately accessible; the entire array has to be scanned before any one sensor can be read a second time. Also, a synch (reset) pulse must be provided to synchronize data storage with data acquisition (hence two wires are indicated in Table 11.1). Although this approach may at first seem unwieldly, it is quite efficient when large amounts of data must be acquired quickly, as in machine vision systems [Snyder 1985, pp. 259-261].

Given the options for addressing shown in Table 11.1, which particular scheme is the most efficient way to read an array of tactile sensors? The answer to this question depends not only on the number of sensors that are of interest, but also on the total number of sensors in the array. Figure 11.4 depicts the rise in access time for each addressing scheme as the number of observed (addressed) sensors increases. The array contains 64 sensors and is supplied with a 10KHz clock. The shortest access time for any given number of observed sensors is obtained by randomly accessing the array via a parallel addressing bus (curve 1). The biggest disadvantage of this approach is that the number of wires (six for addressing alone) increases the likelihood of system failure. Placing all address transmission on a single wire improves reliability but increases access time by a factor of six (curve 2 - note the vertical logarithmic scale), drastically lowering system bandwidth when many sensors must be read. Improving this situation by matrix addressing only partially increases performance (curve 4) by lowering access time. Matrix addressing of tactile sensors with a parallel bus (curve 3) gives improved performance but again, the number of wires (six) is a detriment to reliability (most tactile arrays have been implemented using the scheme of which curve 3 is an example, which might explain some of the reliability problems researchers have had). In collective addressing (curve 5) the access time is invariant of the number of sensors observed. This means that when data from only a few sensors is needed, the performance is poor compared to other methods of obtaining the same data. When many sensors must be read however, the performance of this method is quite good.

The "best" method for addressing a tactile array then, depends on the average fraction of sensors that will be observed during a typical manipulation task. If only a few sensors will be of interest and their locations (addresses) are readily available, then random access over a serial line (curve 2) provides a flexible balance between reliability and speed. On the other hand, if a large fraction of the sensors must be observed (for example, when the point of an imminent contact is not precisely known), then collective addressing (curve 5) is more effective. The reliability of two wires, relatively fast access, and a simple implementation are distinct advantages of collective addressing in this case. To quantify the trade off we need only find the intersection of these two functions (see Appendix), which occurs at

$$F = (log_2 N)^{-1} \tag{11.3}$$

where F is the fraction of the sensor array that is to be observed and N is the total number of sensors in the array. In the 64 element array that is represented by Figure 11.5, the break point between random addressing and collective addressing occurs at about 17 per cent of the total array. This tells us that if less than 17 percent of the tactile sensors will be needed during a typical manipulation task, then random addressing of the array over a single serial line is a good choice for an implementation. But if more than 17 percent of the array must be monitored, collective addressing is required to keep the data bandwidth high and maintain reliability.

The nature of dextrous manipulation suggests that collective addressing of tactile sensors may be desirable in robot hands. During object acquisition procedures, the exact points of contact with the hand may not be knowable a priori due to inaccurate object models, imperfect hand positioning,

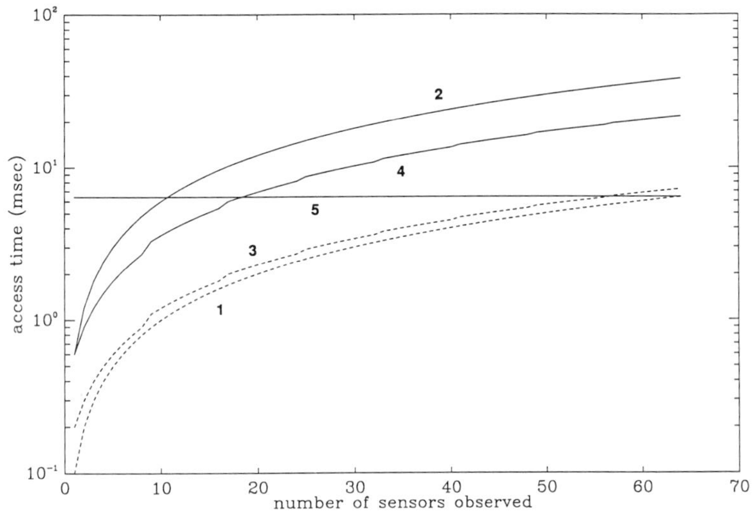

Figure 11.4: Access time versus the number of sensor observed (addressed) for a 64 element tactile array. Numbered curves correspond to methods of addressing listed in Table 11.1. Discontinuities in curves (3) and (4) arise from column scans in matrix addressing. Clock frequency is 10KHz.

or finite sensor resolution [Fearing, this volume]. In fact only regions of probable contact, such as fingertips, may be suspected. Studies of human manual exploration show that large surfaces of the hand are used to both explore and stabilize grasped objects [Klatzky and Lederman, this volume]. The evidence indicates that general regions of the hand (such as the fingertips or palm), rather than individual sensors, should be monitored. Once obtained, of course, the tactile data must be properly sorted to accurately determine locations of contact and forces on the hand for use by higher level manipulation algorithms. But at the array level, large portions of sensors are most useful for dextrous robot grasp and manipulation.

11.3.2 System-level Addressing

At the system level, we are concerned with acquiring large amounts of tactile data from different sensor arrays on the robot hand. Data bandwidths must be high enough to allow for robust grasping and manipulation of objects, but the wire count must be low to ensure reliability. Although this situation is distinct from the previous section where we were concerned with addressing single sensors, there are some similarities. The physically distinct kinematic structures of finger segments and palmar regions are convenient divisions for tactile arrays. If collective addressing is used as described in the previous section, then each array can be treated as a "virtual" sensor containing a number of data bits equal to the number of actual transducers in the array. Thus we have a system consisting of R virtual sensors, where R is the number of physically distinct sensing regions of the hand. The models presented in the previous section are analogous in this situation. Again, there is a trade off between the immediate availability of virtual sensory data, the bandwidth at which the data can be provided, and the number of wires used for addressing.

The nature of dextrous manipulation suggests that reliability and tactile data bandwidth are primary system level design targets. As a robot hand performs manipulation tasks the relative movement of the joints promotes wire and connector failure, obviating the need for an addressing scheme that uses a minimum number of wires. One of the serial methods described in the last section would therefore be a good choice. During fingertip exploration high data bandwidth is needed on the fingertip sensing array to dynamically resolve surface features. Fingertip prehension is commonly used in machine dexterity studies [Stephanou and Erkmen 1988], so bandwidths for fingertip contact data should be fairly high to accurately monitor interaction forces. When an object has been firmly grasped however, bandwidths may be relatively low for those portions of the hand that are stabilizing the object (palm), but high for those portions that are imparting motion to the object (fingertips). These constraints seem to indicate the use of a random addressing scheme for acquiring data from each tactile region. This allows data bandwidths for individual regions to be adjusted by higher level manipulation algorithms. Access time for a number of sensing regions on the hand is simply

$$T = \Sigma_i (log_2 R + N_i) t_c \tag{11.4}$$

where R is the total number of sensing regions and Ni is the number of sensors in the ith region. During manipulation tasks where bandwidth is critical the number of sensor regions read should be kept to a minimum. To illustrate this, consider a tactile system consisting of R=16 sensing arrays, each having N=64 sensors and synchronized by a 10 KHz clock. The data bandwidth of a single array, such as a fingertip, is about 147 Hz. The bandwidth of the entire system however, is only about 9 Hz.

A practical tactile sensing system for dextrous manipulation then, would utilize a random addressing scheme where groups of sensors, rather than individual sensors, are addressable. Which sensor groups must be monitored during particular manipulation and grasping sequences must be specified by higher level algorithms, and is an interesting topic for further research.

11.4 Tactile Sensing for the Utah/MIT Hand

In previous sections of this chapter we discussed the basic requirements for dextrous tactile sensing and presented a method for reconciling their conflict in system design. In this section we will describe how we have applied these principles in the implementation of an actual system for Utah/MIT Dextrous Hand. As of this writing much of the work is still in progress, but preliminary results will be reported where possible.

The goal of our development effort has been to enervate those surfaces of the dextrous hand that are likely to be used during grasping and manipulation operations. The shaded portions of Figure 11.5(a) depict these surfaces as they appear on the hand. Although sensory coverage may appear somewhat limited, we emphasize that one of our primary objectives in this project has been to develop a minimal system with high reliability. Once the basic obstacles of packaging and interconnection have been overcome, development work can proceed to more comprehensive systems. Figure 11.5(b) shows the components of a single tactile finger segment. Hybrid preamplification and multiplexing circuits are mounted on the back of the finger, and help to reduce wiring complexity. A detailed description of tactile sensing subsystems follows, and proceeds from the low levels of transducers and preprocessing to the higher levels of data transmission and system architecture.

11.4.1 Transduction and Preprocessing

Historically, tactile sensing research has produced a wide variety of devices, yet few have found their way onto dextrous robot hands due to practical

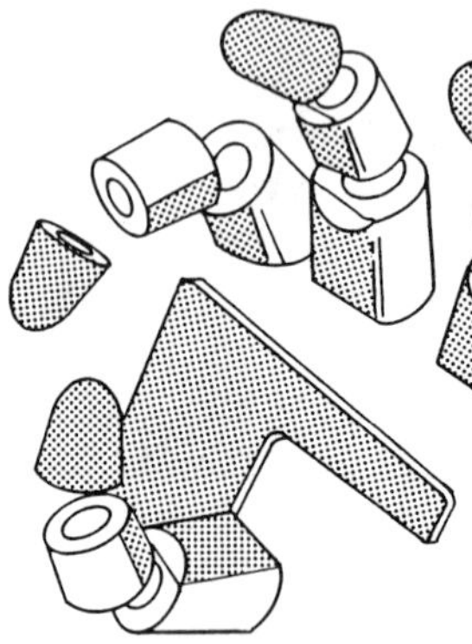

(a)

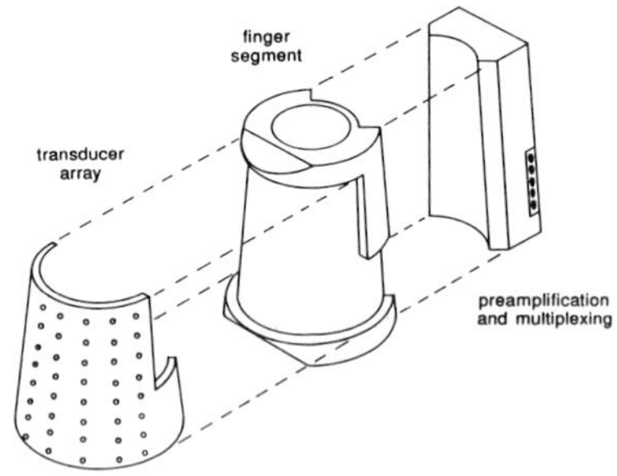

(b)

Figure 11.5: (a) Tactile surfaces of Utah/MIT Dextrous Hand Version IV. (b) Components of a tactile sensing finger segment.

difficulties. With this in mind, we have focussed our early efforts on demonstrating the feasibility of a practical distributed sensing system rather than on advanced sensors. By building on early successes with relatively unsophisticated components, we can later add degrees of complexity as new theories of tactile sensing arise.

The first stages of the project have focussed on the refinement of a capacitive force transducer. As demonstrated in [Wolffenbuttel and Regtien 1987], tactile sensors, variable gap capacitors have advantages of high inherent bandwidth, repeatability, reproducibility, stability, and high signal to noise ratio even in very small geometries. Capacitive tactile transducer arrays of the past have typically consisted of two orthogonal sets of conductors separated by a deformable dielectric. As force is applied to the array the dielectric compresses and the capacitance at the site of contact increases. Addressing is typically achieved through row and column methods (Figure 11.3a). Capacitive tactile arrays have been successfully implemented by [Boie 1984; Fearing, Rise and Binford 1986; Siegel *et al* 1987]. One of the drawbacks of using an opposed plate scheme is that the upper traces, which undergo flexing during use, experience longitudinal stresses and may fracture over time. Because the capacitor array is formed by perpendicular conductive traces, entire columns of transducers may fail as a result of a single trace fracture. Figure 11.6(a) shows a variation of the traditional capacitive sensor which overcomes this problem. By using a floating conductor that is capacitively coupled to two coplanar plates, all electrical

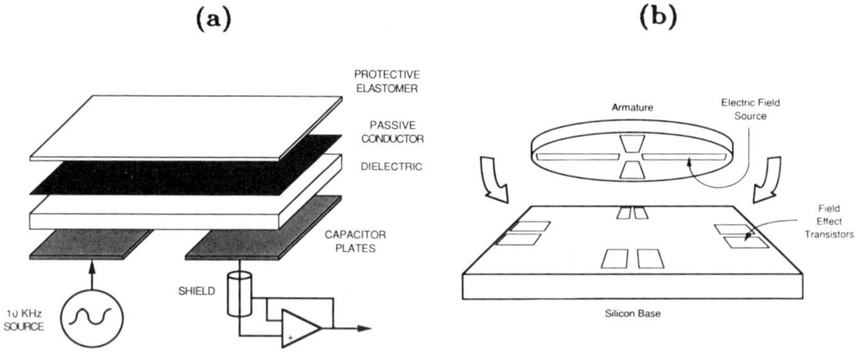

Figure 11.6: Transducer configurations. (a) A reliable capacitance based transducer which provides contact normal information. (b) Multiaxis strain microsensors have been developed as part of CED's ongoing work in micromechanical systems, and may eventually used as tactile sensors. (11.6(b) ©IEEE, Jacobsen *et al* 1987).

connections to the upper, flexing layer can be avoided. Row and column addressing of this sensor is accomplished by using a double-sided circuit board. Wolffenbuttel and Regtien [1989] have successfully demonstrated a planar capacitive array that is similar in principle to this approach. Preliminary experiments with a prototype array have shown that surface deflections as small as 50 microns are resolvable. Electromagnetic noise is minimized by careful conductor design, shielding with conductive silicones, and by placing differential charge preamplifiers near the array.

As tactile sensing research progresses and experiments are conducted that demonstrate the necessity of complex force and moment sensing arrays, multiaxis sensors can be used. These sensors, schematically illustrated in Figure 11.6(b), utilize open gated field effect transistors and are part of our ongoing research on microsensors [Jacobsen *et al* 1988]. Prototype tactile sensors have been constructed that measure six degrees of freedom, and preliminary measurements indicate that deflections on the order of 10 microns and rotations of 0.3 degrees are resolvable.

The proposed tactile sensing system for the Utah/MIT Dextrous Hand consists of 677 capacitive transducers. On the third link (the fingertip) and the second link, transducers are spaced on 2.0 mm centers, and lie under a layer of electrical and mechanical shielding at about the same depth. Fearing *et al* [1986] suggest that these conditions will yield a spatial resolution of about 2 mm features in the presence of symmetry. In other areas of the hand, spacings are somewhat greater, yielding feature resolutions of between about 2 and 3 mm on the proximal finger segment (link 1) and the palm. Table 11.2 summarizes the transducer array characteristics for the tactile sensing system. Because the hand is an experimental device, we do not intend that these spatial densities are definitive for dextrous manipulation; rather, we feel that further investigations will define more precisely the relationships between sensor spacing and manipulation tasks.

11.4.2 Multiplexing and System Architecture

In a previous section, we described why collective addressing of a sensor array is a good trade off between transmission bandwidth and reliability when many sensors must be monitored. This approach has been implemented for scanning the sensors in the tactile transducer arrays. A 10KHz clock is common to all finger segments and synchronizes addressing and data collection. A separate reset line is employed for each array to provide a synch pulse and ensure that retrieved data is in the appropriate order. In the initial phases of the design, each tactile array is served by six wires (clock, reset, data, ground, 1supply) but in later iterations the use of a customized multiplexing IC (currently under development) will permit multiplexing of the entire system on four wires.

Bandwidths for each of the tactile segments are listed in Table 11.2. If we are interested in manipulation tasks that use only one of the arrays, tactile data bandwidths can be quite high. An example of this is fingertip exploration of an object, which could be monitored at bandwidths as high as 185 Hz. Use of multiple arrays yields bandwidths that are somewhat lower. If the dextrous hand is used to manipulate an object with only three of the fingertips, monitoring of the contact forces can be achieved at bandwidths of about 60 Hz. Scanning all of the tactile sensors in the hand can be accomplished only at bandwidths of 15 Hz or lower; clearly, this is only an option when the dynamics of the hand or workpiece are minimal. An obvious way to increase bandwidth is to increase the frequency of the system clock. Experiments have shown that clock frequencies as high as 150KHz are possible, but limitations of the coherent detection electronics produce a notable decrease in the effective transducer sensitivity. A such high frequencies the amplification and filtering subsystems give little more than one part in two (binary) force resolution. Again, this represents yet another trade off between performance and speed. We are continuing with our efforts to find ways of increasing the system bandwidth in later versions without sacrificing resolution.

Once a common addressing and data bus has been installed between the tactile segments, a random addressing scheme can be employed. As we

sensor region	number of sensors	spacing (mm)	frame rate (@10 kHz clock)
fingertip	54	2.0	185 Hz
link 2	48	2.0	208 Hz
link 1	56	2.5	178 Hz
palm	45	2.5	222 Hz
system total	677	-	15 Hz

Table 11.2: Tactile sensing characteristics for the various segments of the dextrous hand shown in Figure 11.5(a). Higher frame rates can be obtained by increasing the system clock frequency, but force resolution is sacrificed.

have shown earlier in this chapter, random addressing is desirable when the data from only a few arrays is, on the average, needed. In addition, random addressing allows for expansion of the system to include three more tactile arrays (for a total of sixteen) with no changes to the multiplexing hardware. Expansion of the system might include higher density arrays or more complete sensory coverage of the hand. Random addressing allows investigators the flexibility of high bandwidth tactile data when contact surfaces of the hand are known ahead of time. An interesting area for further research would be to scan the different parts of the hand at duty cycles that are proportional to their probability of contact. Successive trials of a particular task might enable the hand to "learn" which tactile regions are involved and modulate their update rates accordingly, and so have the highest possible data bandwidth available for enhanced dexterity.

11.5 Conclusion

The objective of this chapter has been to explain the foundations of our approach to designing a tactile sensing system for the Utah/MIT Dextrous Hand. Beginning with a general description of the hand itself, we showed how the hand development effort has been aimed at providing investigators with a high performance research tool for studying machine manipulation. The dextrous hand has not been meant as an archetype for dextrous machine design, but rather represents an early step in the evolution of manipulation science. We reviewed past efforts in order to illustrate the problems that are typically encountered when robot hands are endowed with a sense of touch.

Critical features of dextrous tactile systems are their reliability, bandwidth, and the number and spacing of the sensors of which they are composed. We have quantitatively reviewed each of these features, and explored the ways in which these features interact to affect dextrous hand performance. By comparing models that relate tactile data bandwidth, the number of wires in a system, and the number of tactile sensors supported, we have suggested that dextrous touch favors an architecture where arrays of sensors, rather than individual transducers, are addressable. In view of these considerations we concluded the chapter with a discussion of the tactile sensing system for the dextrous hand that is currently under development. The practical aspects of the data acquisition system at both low and high levels was described.

Our efforts to develop a robust tactile data acquisition system for the dextrous hand have been characterized by a focus on the practical aspects of its design. Our immediate goal is demonstrating a robust and complete system using relatively simple components, rather than achieving high sensor densities or implementing novel transduction schemes. Once the obstacles to a system implementation have been overcome, complexity can be added as it is needed. We believe that by taking a balanced and practical approach to dextrous tactile sensing, useful sensory tools for the study of machine manipulation will emerge.

11.6 Appendix.

The expressions for the various methods of addressing tactile sensor arrays are easily derived, and appear below. Note that this is not an exhaustive treatment of all possibilities, but simply a listing of the methods that are commonly proposed or used in actual systems. The interested reader is referred to [McCammon and Jacobsen 1988] for a more general discussion.

Random parallel addressing: Since each parallel sensor address is transmitted during a single clock cycle (t_c), any portion of n sensors from an array of size N will require n clock cycles to be synchronously addressed. The access time is simply

$$T_1 = n\, t_c$$

Where n can have any integer value between 1 and N. Note that this expression is valid regardless of the way in which the sensors are arranged.

Random serial addressing: To randomly address n sensors over a serial line requires a number of clock cycles equal to the number of address bits. The number of address bits (B) needed to address one of M sensors can be found by simply applying the definition of information [Young 1971], which states that the amount of information (I) contained in any message is related to the probability of that particular message being sent:

$$I_i = -\log_k (P_{(i)})$$

Where k is the logarithmic base that determines the units of I_i. For binary systems k=2 to yield units of bits. If M sensors are being read in sequence, the probability of any one being read is $\frac{1}{M}$, and so the number of bits needed is

$$B = \log_2 (M)$$

If N sensors are arranged in an X by Y array, both the i-address and the j-address must be transmitted in sequence for each sensor, so for n sensors the access time is

$$T_2 = n\, t_c\, (\log_2 (X) + \log_2 (Y))$$

By the multiplication property of logarithms, this reduces to an expression that is independent of the physical arrangement of the sensors

$$T_2 = n\, t_c \log_2 (N)$$

Matrix parallel addressing: When an array of sensors is scanned by reading one column after another (a common implementation) the row index increases during each clock cycle and the column index is sent only once but latched. In the case where the parallel row and column addresses arrive during successive clock cycles, one column address is sent followed by y row addresses, where y is the number of rows to be read

$$t_y = (y+1)\, t_c$$

When x columns of rows are read, the access time becomes

$$T_3 = (x-1)\,(X+1)\, t_c + (y+1)\, t_c$$

to account for all of the columns. If the row and column addresses arrive simultaneously then this case reduces to random parallel addressing.

Matrix serial addressing: This situation is similar to the previous example, except that row and column addresses are transmitted over a serial line. Again, each address requires a number of clock cycles that is equal to the number of bits to be sent

$$T_4 = (x-1)\,(X+1)\,t_c\,log_2(X) + (y+1)\,t_c\,log_2(Y)$$

Collective addressing: By including counters with the multiplexing hardware, sensors can be sequentially scanned over a single line. As with the random parallel case, the number of clock cycles is equal to the total number of sensors in the array

$$T_5 = N\,t_c$$

A second wire is required for resetting the counters. To determine the trade off point between collective addressing and random serial methods we only need to to set these two functions equal to one another

$$N\,t_c = n\,t_c\,log_2(N)$$

The fraction of the total number of sensors at which this break point occurs (F) is simply the ratio of n to N:

$$\begin{aligned} F &= \frac{n}{N} \\ &= (log_2(N))^{-1} \end{aligned}$$

In situations where the fraction of sensors observed is greater than F (array level), collective addressing is very practical. But where only a few sensors (or "virtual" sensors) are needed, random serial addressing is preferable.

Acknowledgements: The work that has been described in this chapter is being performed at the Center for Engineering Design (CED) at the University of Utah under Massachusetts Institute of Technology subcontract number GCA 802430-01. The authors would like to thank Dr. John Hollerbach, David Siegel, and Sundar Narasimhan of the MIT Artificial Intelligence Laboratory for their valuable assistance in this ongoing project. Various people at the CED have contributed to this project, most notably Klaus Biggers, Richard T. Johnson, Nelson Clayton, and Russ Wilson. The technical assistance of Chuck Phipps, Mike Bouchard, and Jeff Barrett has also been most appreciated.

12

A New Tactile Sensor Design based on Suspension-Shells

Tokuji Okada[1]

ABSTRACT Requirements of sensors for dextrous fingers have been discussed and appropriate designs to satisfy these requirements have been proposed. Miniaturization of a tactile sensor element has also been considered. However, only a limited number of such sensors might be attached to the surface of a finger body. In addition, with present day tactile sensing technology, it is difficult to make all of the surfaces sensitive or efficiently handle all the signal lines. In order to solve these problems, a new tactile sensor devoid of blind sectors is proposed. The sensor uses a suspension shell covering the surface of the finger body. Tactile sense is obtained by detecting the relative displacement of the suspension-shell with respect to the finger body.

12.1 Introduction

Proximity information is important for the control of a robot hand when positioned very close to an object [Johnston 1973; Okada 1982b; Kanade and Sommer 1983]. Tactile sensors attached to an end effector of a robot, on the other hand, are sensitive to non-visual information like pressure, force and moment which are obtained by touching object. For instance, they make the perception of whether an object is simply located on, or tightly fastened to another object possible. Specifically, in complicated assembly tasks, tactile information must supplement information about geometrical relationships among assembly parts for fine control of the hand.

In general, when a sensor detects unexpected signals during hand movements, we recognize either an obstacle invasion or a disturbance caused by abnormal task completion, and can revise or change programs to cope with such unexpected events. Tactile sensors not only gather such information for controlling the hand, but also for recognizing size and shape of an object. In the latter scenario, the hand is used like a probe and *scans* the object to be recognized [Hillis 1982]. Information from the sensor is synthesized

[1]Department of Mechanical Engineering, Niigata University, 2-8050, Ikarashi, Niigata, Japan 950-21

with control history of the hand position to search, verify and recognize the object. Object attributes such as hardness and ruggedness may also be evaluated by synthesizing sensory information. All of this is very important for the determination of grasp forces and enable manipulation of even fragil objects without either crushing or dropping.

In this chapter, we discuss tactile sensors which are required for dextrous fingers and we propose an appropriate sensor design. Kinematic analysis and measurement principles are presented.

12.2 Requirements of Sensors for Dextrous Fingers

Various tactile sensors using opto-electrical [Rolls Royce 1973; Taylor *et al* 1981; Rebman and Morris 1983], magnetic, electrostatically capacitive, mechanical and electro-conductive principles have been proposed thus far [Larcombe 1981; Harmon 1982; Dario *et al* 1983; Dario *et al* 1984; Dario and Buttazzo 1987]. Load cells consisting of strain gauges have also been suggested for measuring and resolving applied forces and moments into their orthogonal components [Edmond 1978; Salisbury 1985]. Load cells typically consist of connected spring elements, and the forces and moments they are subjected to are computed from angular and translational displacement measurements.

Generally, strain gauge sensors are fragile. An appropriate protection of the gauge from extraneous forces and moments must be provided [Wang and Will 1978]. Further, measurements are considerably influenced by temperature. While electrical compensation techniques have been applied to correct for signal variations and noise, it has been difficult to obtain reasonable signal resolution. Similar drawbacks plague load cells, since they are basically constructed using strain gauges, along with flexure beams, cantilever bars, tension traps and so on.

Multi-axis load cells through which vectors representing all force and moment components may be obtained have a relatively large amount of cross-talk or undesired co-dependence [Edmond 1978]. Some load cells also suffer from low resonant frequency. Therefore, fabrication of load devices is rather difficult and expensive, and their undesirable features make them virtually non-applicable in manufacturing and assembly.

When conventional sensor elements are employed on the finger, several difficulties arise. Most importantly, a large number of tactile elements have to be arranged in order to obtain tactile information over a wide range on the surface of the finger body [Kinoshita *et al* 1975; Raibert and Tanner 1982]. Insensitive regions inevitably arise [Tomovic and Stojiljkovic 1975], despite the use of flexible rubber sheets [Bajcsy 1984]. In addition, it is difficult to reduce the sensor size and weight as a whole since a large num-

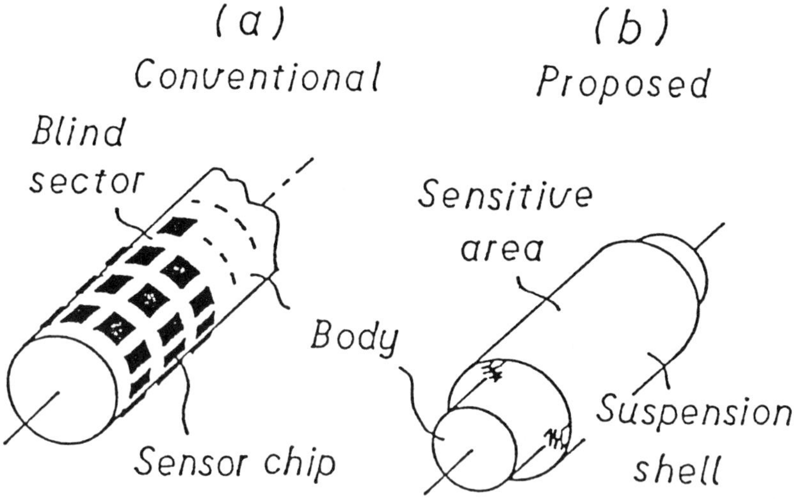

Figure 12.1: Sensor structure.

ber of elements have to be arranged in a discrete fashion [Jacobsen *et al* 1984]. Signal processing blocks increase with the number of sensor elements, resulting in increased complexity.

From the above discussion, we can extrapolate that sensors for the dextrous fingers should be

1. small and smart enough to be housed in a finger body, since the dextrous fingers can perform not only such simple motions as bending and extending but also lateral flexing motions such as adduction and abduction (signal lines are expected to be reduced in number if appropriate scanning methods are adopted).

2. capable of measuring the magnitude, direction and point of action of the external force applied on the finger body without insensitive regions (information of pressure distribution is often needed),

3. light in weight which causes no error in a fine control of finger joints,

4. high sensitivity with robustness to endure the external force,

5. low drift when subjected to a change in temperature, and,

6. good dynamic range and accurate spatial specificity.

We propose a tactile sensor that meets these requirements under the assumption that fingers are cylindrical.

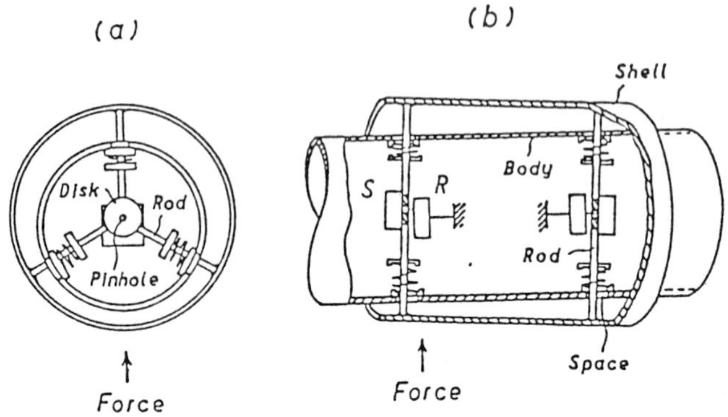

Figure 12.2: Inside view of the proposed sensor.

12.3 Structure of Proposed Sensors

Most of the fingers get tactile information from electrical or mechanical sensor elements installed on a surface of a finger body. The number, size and arrangement of elements might be considered according to accuracy and resolution demands. For instance, if the sensor is utilized to identify shape of an object, a number of sensor elements will be installed to make the resolution high. A finger with such sensors would resemble the schematic shown in Figure 12.1(a). It is obvious that surfaces of the finger become grotesque and restrict flexible motion.

Figure 12.1(b) shows a finger on which the proposed tactile sensor is mounted. The finger body is veiled in a cylinder that will be referred to as *suspension-shell.* A suspension shell is free to move axially and radially since materials connecting the shell to the finger body are elastic. A force applied on the finger will cause the shell to move. Their magnitudes and directions may be computed from measured radial, axial, and angular shell displacements. In order to make it possible to detect these displacements, we use optical means located in the central area of the finger body, by taking advantage of the fact that the finger body is hollow, in general.

Figure 12.2 shows rough sketches of the cross sectional and longitudinal views of a finger equipped with such a sensoring unit. Symbols S and R indicate a light beam source and 2-D photosensor array, respectively. Notches express an elastic material like a spring. Finger cross-section is not required to be circular. Exact measurement of the shell deviation against the finger body is described in Section 12.5.

12.3.1 Suspension Types

Three arrangements are considered for suspension of the shell coaxially around the finger body. Figure 12.3 shows these arrangements. Ends of

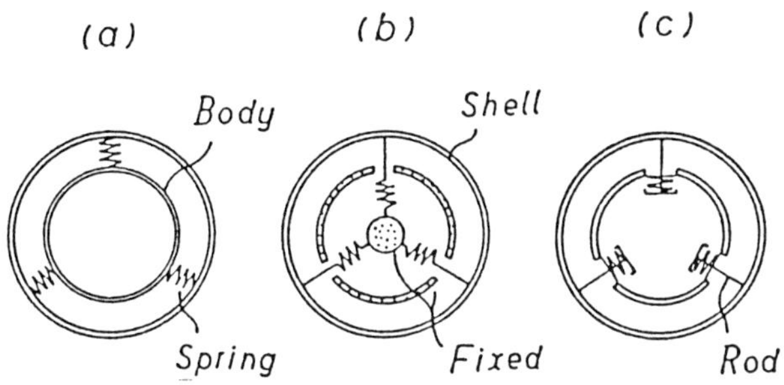

Figure 12.3: Three suspension types.

each spring are connected with the shell or the body, directly or indirectly through a rod. In Figure 12.3(a), the spring connects shell and body directly, thus the spring is housed in the gap between the body and the shell. On the contrary, in Figure 12.3(b) and 12.3(c), it is located in a space of the finger body by using the rod of connection. The dotted area in Figure 12.3(b) is a part of the finger body. Central spaces are occupied for the suspension in Figure 12.3(b) and 12.3(c). However, the central space of the body remains remains unoccupied in Figure 12.3(a).

12.3.2 Sensor Stiffness around a Finger Axis

For the proposed sensors, tactile information is available all around the finger body. Stiffness of the sensor depends on the elasticity of the springs. Let us consider a spring of stiffness k crossing the basic line x at an angle α. Then the restoring spring force in the direction of θ is expressed as $kxcos(\alpha + \theta)$ for any displacement x (see Figure 12.4). The composite stiffness k_θ coupled to a mass m is written as

$$k_\theta = \sum_{i=1}^{n} k_i \, cos^2(\alpha_i + \theta). \tag{12.1}$$

Equation (12.1) describes sensor stiffness about the finger axis. When springs are arranged so as to divide a circumference with equal angles, we obtain

$$\begin{aligned} k_\theta &= (k_1 + k_2)\cos^2\theta, \; when\, n = 2. \\ k_\theta &= \frac{1}{4}(4k_1 + k_2 + k_3)\,\cos^2\theta + \frac{3}{4}(k_2 + k_3)\sin^2\theta \\ &\quad + \sqrt{3}/4(k_2 - k_3)\sin 2\theta, \; when\, n = 3. \\ k_\theta &= k_1 + k_3 + (k_2 + k_4 - k_1 - k_3)\sin^2\theta, \; when\, n = 4. \end{aligned} \tag{12.2}$$

Left and right parts of Figure 12.5 show spring arrangements and calculated

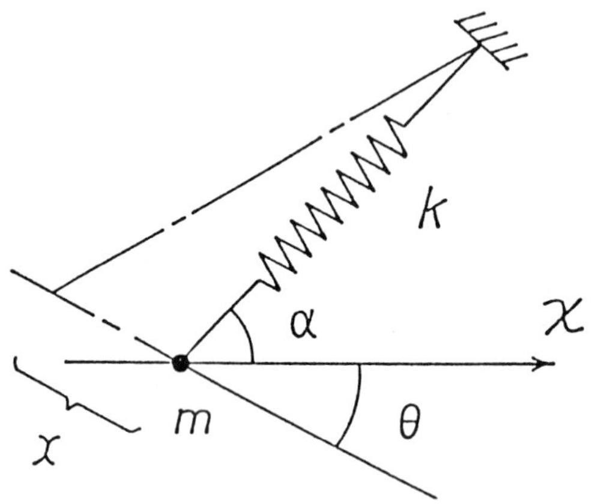

Figure 12.4: Installing condition of a spring.

stiffness patterns, respectively. Regardless of the value of θ , the composite stiffness remains constant when $n = 3$ along k_1, k_2, and k_3, and along k_1 when $n = 4$.

Above discussions allow both extension as well as compression springs. If a spring could be utilized for both extension and compression, only one spring is needed to have the stiffness pattern shown in Figure 12.5(a). When lateral stiffness of the spring is not negligible, the pattern in Figure 12.5(a) becomes a single loop.

12.4 Kinematic Analysis of the Sensor

In order to make it possible to sense tactile information from a wide area of a finger surface, we suspend a cylindrical shell at two sections. A and B which are close to the ends of the shell (see Figure 12.6). Forces acting upon the shell will result in unreliable sensor responses by causing vibrations. To avoid this, it is very important to know a behavior of the shell, especially the natural frequency. Therefore, we analyze the kinematics of the shell in this section under the assumption that the shell is circular. We neglect torsional vibrations, bending stresses, and deflections in between supports under the assumption that the suspension-shell is rigid. To begin the analysis, we define the following

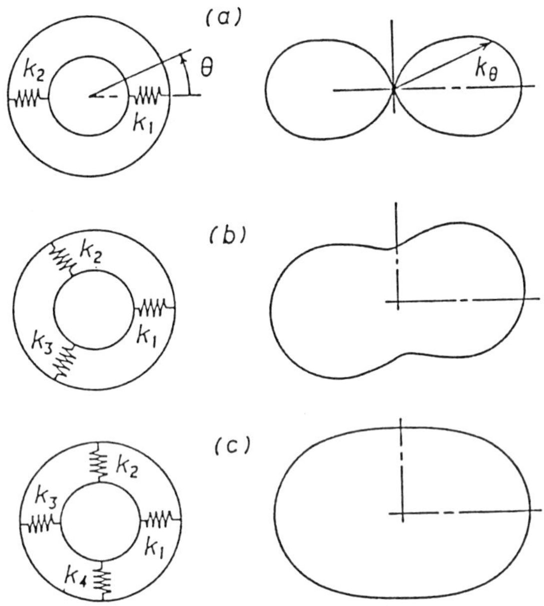

Figure 12.5: Characteristics of a stiffness around the finger axis.

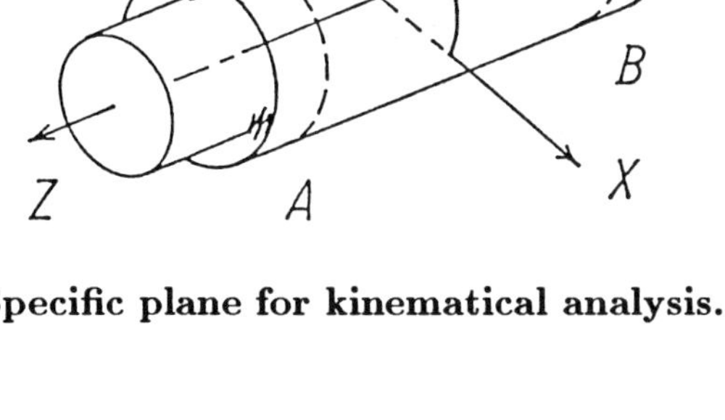

Figure 12.6: Specific plane for kinematical analysis.

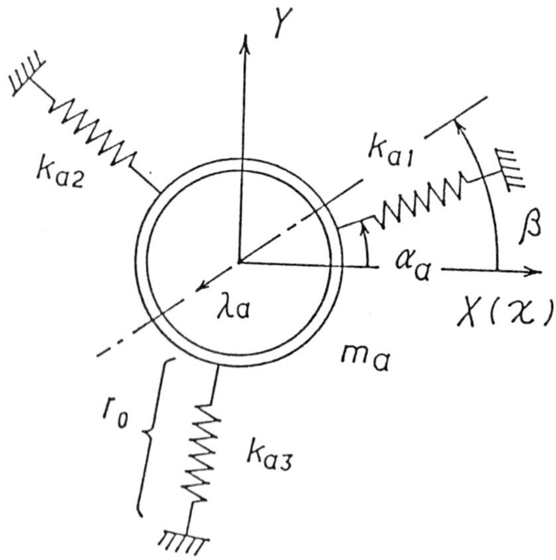

Figure 12.7: Cross sectional view of sensor in balanced condition.

12.4.1 Nomenclature

F:	external force
f_n, f_s, f_r :	rectangular components of F
a,b:	subscript for suspension sections A and B
G:	gravity center of a suspension shell
g:	gravity
h:	position of the contact point along Z axis
I:	moment of inertia
i:	subscript for discriminating springs
k:	longitudinal stiffness of a spring
k_r :	lateral stiffness of a spring
l_o :	length of the suspension shell
l_a :	distance from A to suspension end as shown in Figure 12.8
l_b :	distance of B from suspension end
l_g :	distance from G to shell edge near A
M:	mass of a suspension shell
m_a, m_b :	division of the mass M
r_o :	length of a spring
r_1 :	inner radius of the shell
r_2 :	outer radius of the shell
T:	kinetic energy

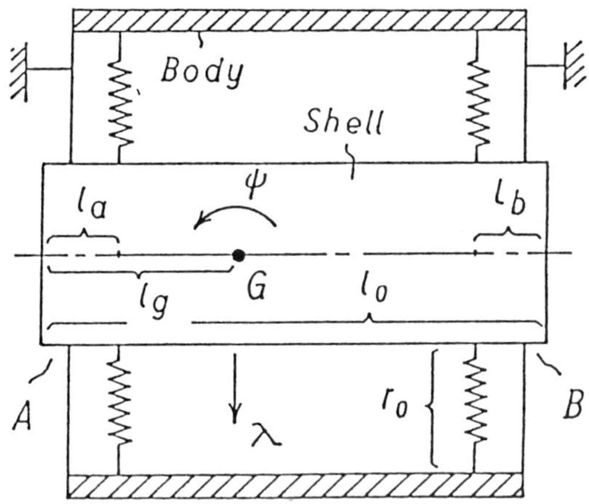

Figure 12.8: Side view of the sensor in balanced condition.

U:	potential energy
X,Y,Z:	rectangular coordinate system
α :	angle for the spring arrangement as shown in Figure 12.7
β :	rotation angle about the Z axis
δ :	spring displacement
ρ :	angular shell displacement about body axis
λ :	displacement of G from a stable equilibrium (downside positive)
λ_a :	displacement of m_a
λ_b :	displacement of m_b
ϕ :	rotation angle about axis through G and parallel to X (positive in CCW)

12.4.2 Analysis of Sensor Behavior

Since behavior of the shell in arbitrary attitude is too difficult to analyze, we assume that the finger axis is either horizontal or vertical. In this section, we consider the former case.

Let the shell be suspended by adopting the arrangement in Figure 12.3(c). Note that the analysis plane crosses X-Z at an angle β. Other relations between the shell, spring and finger body in the cross section A are illustrated in Figure 12.7.

Figure 12.8 illustrates the parameters in the analysis plane. From this, we can express the kinetic energy as

$$T = \frac{1}{2} M\dot{\lambda}^2 + \frac{1}{2} I\dot{\phi}^2 \tag{12.3}$$

When the shell is uniform, we can write

$$l_g = \frac{1}{2} l_0 \tag{12.4}$$

$$I = \frac{M}{12}\{l_0^2 + 3(r_1^2 + r_2^2)\} \tag{12.5}$$

The displacements of the mass m_a and m_b are (see Figures 12.7 and 12.8)

$$\lambda_a = \lambda + (l_g - l_a)\phi \tag{12.6}$$

$$\lambda_b = \lambda - (l_0 - l_g - l_b)\phi \tag{12.7}$$

Note that the potential energy of the spring at the section A and B are (respectively)

$$U_{ai} = \frac{1}{2}k_{ai}[\delta_{ai}^2 - \{\delta_{ai} - \sqrt{\{r_0\cos(\beta - \theta_{ai}) + \lambda_a\}^2 + \{r_0\sin(\beta - \theta_{ai})\}^2} + r_0\}^2] \tag{12.8}$$

$$U_{bi} = \frac{1}{2}k_{bi}[\delta_{bi}^2 - \{\delta_{bi} - \sqrt{\{r_0\cos(\beta - \theta_{bi}) + \lambda_b\}^2 + \{r_0\sin(\beta - \theta_{bi})\}^2} + r_0\}^2] \tag{12.9}$$

It follows that the potential energies of the mass m_a and m_b are

$$U_{ap} = -m_a g \lambda_a \sin\beta \tag{12.10}$$

$$U_{bp} = -m_b g \lambda_b \sin\beta \tag{12.11}$$

The sum of all potential energies are

$$U = \sum U_{ai} + \sum U_{bi} + U_{ap} + U_{bp} \tag{12.12}$$

For the spring to be balanced, we have

$$\sum k_{ai}\delta_{ai}\cos(\beta - \theta_{ai}) = m_a g \sin\beta \tag{12.13}$$

$$\sum k_{bi}\delta_{bi}\cos(\beta - \theta_{bi}) = m_b g \sin\beta \tag{12.14}$$

The dynamic equations of motion for the suspension system can be derived using Lagrange's equations:

$$\frac{\partial}{\partial t}\left(\frac{\partial T}{\partial \dot{q}_n}\right) - \frac{\partial T}{\partial q_n} - \frac{\partial U}{\partial q_n} = Q_n \tag{12.15}$$

where

T	total kinetic energy
U	total potential energy
Q_n	generalized external forces
q_n	generalized coordinates, and
q_n	generalized velocities

Inserting Equation (12.3) and (12.12) into Equation (12.15), we obtain

$$\begin{aligned} M\ddot{\lambda} &+ I\ddot{\phi} + \sum \frac{\partial U_{ai}}{\partial \lambda} + \sum \frac{\partial U_{ai}}{\partial \phi} \\ &+ \sum \frac{\partial U_{bi}}{\partial \lambda} + \sum \frac{\partial U_{bi}}{\partial \phi} = 0. \end{aligned} \tag{12.16}$$

Assuming that $\lambda_a/r_o << 1$, we have

$$\begin{aligned} \frac{\partial U_{ai}}{\partial \lambda} &= k_{ai}\frac{\lambda_a}{r_0}\{\delta_{ai} - (\delta_{ai} + r_0)\cos^2(\beta - \theta_{ai})\} \\ &\quad + k_{ai}\delta_{ai}\cos(\beta - \theta_{ai}) && (12.17) \\ \frac{\partial U_{ai}}{\partial \phi} &= (l_g - l_0)\frac{U_{ai}}{\partial \lambda} && (12.18) \end{aligned}$$

Assuming, in addition, that $\lambda_b/r_o << 1$, we have

$$\begin{aligned} \frac{\partial U_{bi}}{\partial \lambda} &= k_{bi}\frac{\lambda_b}{r_0}\{\delta_{bi} - (\delta_{bi} + r_0)\cos^2(\beta - \theta_{bi})\} \\ &\quad + k_{bi}\delta_{bi}\cos(\beta - \theta_{bi}) && (12.19) \\ \frac{\partial U_{bi}}{\partial \phi} &= (l_g + l_b - l_0)\frac{U_{bi}}{\partial \lambda} && (12.20) \end{aligned}$$

Substituting Equations (12.17) through (12.20) into Equation (12.16)

$$M\ddot{\lambda} + (u + v)\lambda + \{u(l_g - l_a) + v(l_g + l_b - l_0)\}\phi = 0. \tag{12.21}$$

$$\begin{aligned} I\ddot{\phi} + \{u(l_g - l_a) &+ v(l_g + l_b - l_0)\}\lambda \\ &+ \{u(l_g - l_a)^2 + v(l_g + l_b - l_0)^2\}\phi = 0 \end{aligned} \tag{12.22}$$

where

$$u = \frac{1}{r_0}\sum_{i=1}^{3} k_{ai}\{\delta_{ai} - (\delta_{ai} + r_0)\cos^2(\beta - \theta_{ai})\} \tag{12.23}$$

$$v = \frac{1}{r_0}\sum_{i=1}^{3} k_{bi}\{\delta_{bi} - (\delta_{bi} + r_0)\cos^2(\beta - \theta_{bi})\} \tag{12.24}$$

One possible solution of Equation (12.21) and (12.22) is

$$\lambda = \lambda_0 \cos \omega t \tag{12.25}$$

$$\phi = \phi_0 \cos \omega t \tag{12.26}$$

Hence, we have

$$\begin{aligned} MI\omega^4 - [(u + v)I + M\{u(l_g - l_a)^2 + \\ v(l_g + l_b - l_0)^2\}]\omega^2 + uv\{l_0 - (l_a + l_b)\}^2 &= 0 \end{aligned} \tag{12.27}$$

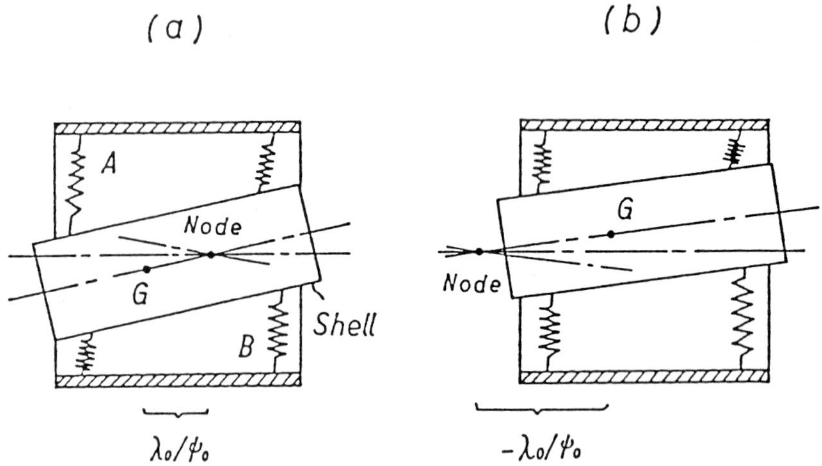

Figure 12.9: Sketches of the end of the finger body.

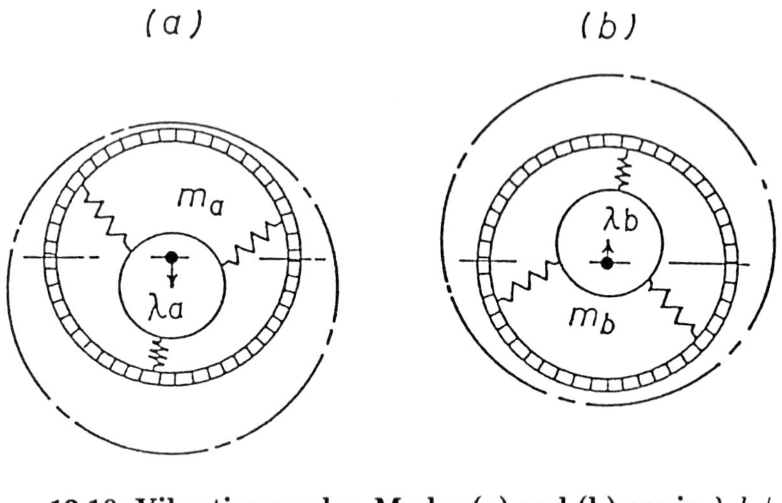

Figure 12.10: Vibration modes. Modes (a) and (b) are in $\lambda_0 l_0 \phi_0 >$ 0 and $\lambda_0 l_0 \phi_0 < 0$.

from which we may obtain the angular velocity ω. Furthermore, combining Equation (12.21) and (12.25) produces the result

$$\frac{\lambda_0}{l_0\phi_0} = \frac{u(l_g - l_a) + v(l_g + l_b - l_0)}{\{m\omega^2 - (u+v)\}l_0} \tag{12.28}$$

This result is also obtained by combining Equations (12.22) and (12.26). Vibration modes are classified into three types depending on the location of a node. Since we defined that λ and ϕ are positive in downside and counter clockwise, the node exists (1) on the point G, (2) on the right and (3) on the left, when the value of $\lambda_o/(l_o\phi_o)$ is zero, positive and negative, respectively. Deviation of the node from the point G is expressed by λ_o. Figure 12.9 shows vibration modes when $\lambda_o/(l_o\phi_o)$ is positive and negative. Cross-sectional behaviors are shown in Figure 12.10. To suppress sensor vibration effects the angular velocity ω must be many times the operating frequency of the finger. Typically, when $\omega > 30$, sensor operations are quite vibration-free.

12.5 Displacement Measurement Mechanism of the Sensor

In this section we describe displacement measurement mechanisms for two cases depending on the measurement accuracy. Characteristics of the sensor is also presented.

12.5.1 Measurement of Displacement

Figure 12.11 shows an embodiment of the tactile sensor for application at the finger tip. As shown in Figure 12.11, the cylindrical sensor body has an end plate 1 provided with a composite bearing 2 consisting of an axial slider 2a and a spherical bearing 2b at the center of the end plate. The axial slider 2a is supported by spherical bearing 2b. Sensitive part of the shell has a hemispherical form and is mounted on the finger tip with a support rod extending from its inner surface. The support rod is housed in the composite bearing 2 for axial movement and rotation.

The sensitive shell is slightly spaced apart from the end plate so that it may be stably displaced all directions with respect to the sensor body. A disk is secured to the free end of the support rod projecting from the bearing 2 into the sensor body. A plurality of uniformly spaced-apart extension spring are connected at one end of the disk and coupled at the other end to the inner periphery of the sensor body.

In the absence of any external force applied to the sensitive shell, a ring-like stopper 3 provided on an intermediate portion of the support rod comes into contact with the bearing 2, and the sensitive shell is biased such that it can balanced at a reference position and in a reference orientation. The

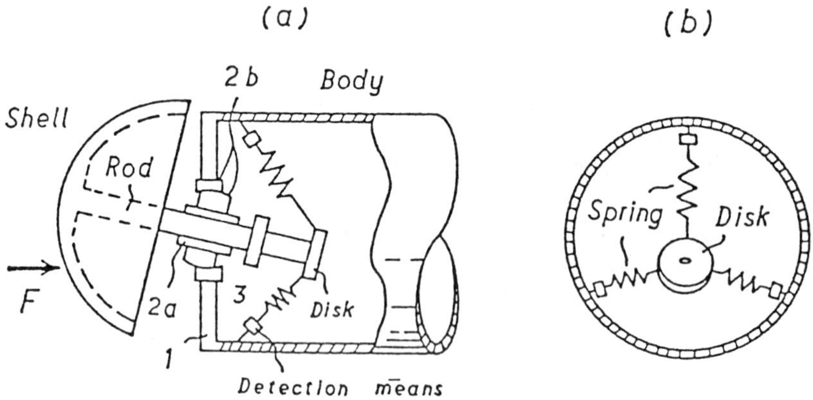

Figure 12.11: Displacement detection mechanism of a tip sensor.

displacement sensors are placed on the spring and detect deformation of the springs. Figure 12.11 shows an example of this arrangement. Displacement signals are transferred to a processing unit, and a vector of forces is calculated from the corresponding spring coefficients and angles between springs.

The basis of calculation is as follows: When the support rod is rotated about its axis, the range is generally expressed as a spherical surface. Since the distance between the ends of the springs on the disk is known and a straight line passing through the center of the disk at an angle of 90 degrees passes through the center of the composite bearing 2, the position and inclination of the disk may be determined geometrically. Using these, forces operating on the rod may be computed.

When the external force is assumed to be parallel to the axis of the finger body, the magnitude and point of action of an external force can be determined uniquely. When two or more external forces act simultaneously, the results of calculation represent a resultant force.

When the external force F is exerted on the sensitive shell in a state where the cylindrical sensor body and cylindrical sensitive shell are coaxial with each other as shown in Figure 12.2, the two cylindrical bodies become eccentric with respect to each other. More specifically, the elastic member (expressed by the spring) at positions closer to the point of the external force F are elongated, while the elastic members on the other side are contracted, causing the shell to be stabilized. When the point of action of the external force F shifts progressively to the right from the position

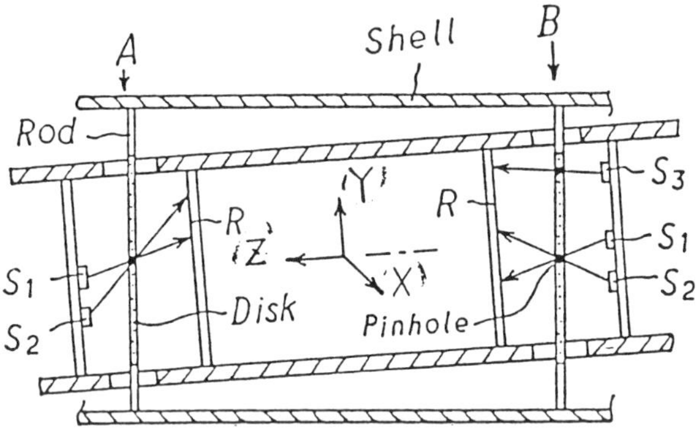

Figure 12.12: Displacement detection mechanism of a body sensor.

shown in Figure 12.2(b), the extent of deformation of elastic members on the right side is progressively increased and eventually becomes greater than that of elastic members on the left side.

The connecting disk to which the end of each support rod is connected is unified with the sensitive shell. In addition, the connecting disk may be provided with a pinhole at the center, i.e., at the position corresponding to the axis of the sensitive shell. A light-emitting element S and a position detection element R may be provided on the axis on the opposite sides of the pinhole, whereby it is possible to detect a slight displacement of the sensitive shell with respect to the sensor body. And the data obtained can be processed to calculate the magnitude and the location of the external force F applied to the sensitive shell. In this case, there is no need of providing any detection element on the elastic members. A single light-emitting element S as shown in Figure 12.2 is sufficient only for the detection of the displcement of the sensitive shell in the radial direction.

12.5.2 Measurement of the Externally Applied Force and its Location

When it is desired to obtain force detection in all directions, two light-emitting elements S_1 and S_2 may be mounted on a base unified with the sensor body as shown in Figure 12.12 from which the elastic members and detection processors have been omitted for a structural simplification.

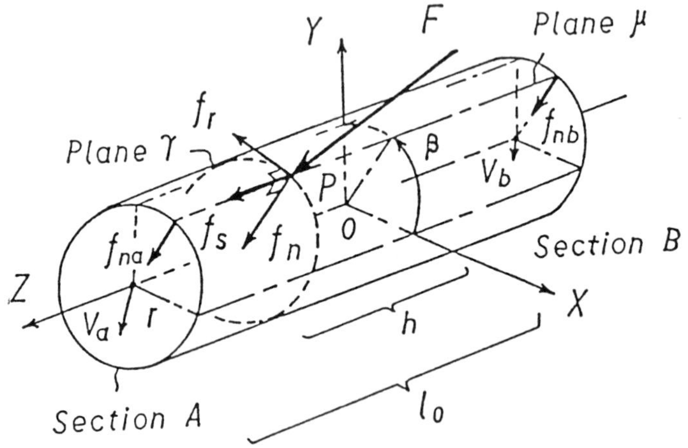

Figure 12.13: Force components of the external force F.

Positions of the light spot from the light-emitting elements are detected by a position detection element, so that it is possible to determine the position of the pinhole in the disk as an intersection of two light beam paths by a calculation unit which is not shown. However, when the detection element cannot detect simultaneously two or more light spot points in the PSD (Position Sensitive Device), it is necessary to cause flickering of the light-emitting S_1 and S_2 one after another to determine the light spot positions separately. For this purpose, light emitting sources will be scanned so that only one light source is allowed to be active at the same time.

If a light-emitting element S_3 and a decentralized pinhole are provided on either side of the sensitive shell, the rotation angle of the sensitive shell about the axis of the sensor body can be calculated, by using the positional information of a light spot on the photosensor R. When the inclination, axial deviation and rotational displacement of the sensitive shell are determined, the deformation due to the elastic members are all calculated, so that it is possible to determine the magnitude, direction and point of action of the external force on the sensitive shell in the strict sense, however the number of the points is limited to only one.

When the elastic members at two sides of the sensitive shell have the same elastic modulus, the position of the force F is determined according to the radial displacements at the two sides. If the displacements are equal, the position of action is determined to be at the middle point of the sensitive shell, and the magnitude of the force is expressed as a sum of these deformation.

Based on the above-mentioned optical measurement, we can calculate shift vectors related to the two disk centers at the cross section A and B, say $V_a(c_{1a}x + c_{2a}y + c_{3a}z)$ and $V_b(c_{1b}x + c_{2b}y + c_{3b}z)$, respectively (see Figure 12.13). Rotation angle, say ρ, related to the suspension shell is also computed. We suppose that vector V_a and V_b exist in a same place since the suspension shell is rigid and the number of the external force is only one. That is, the plane including vectors V_a and V_b also includes the axis Z and is expressed by μ in Figure 12.13. Thus, β may be expressed as

$$\beta = \cos^{-1}\{\frac{c_{1a}}{(c_{1a}^2 + c_{2a}^2)^{\frac{1}{2}}}\} \tag{12.29}$$

β may also be obtained by replacing parameters (c_{1a}, c_{2a}) with (c_{1b}, c_{2b}) in Equation (12.29). These results are used to determined the vectors of the force F and the position where the force F operates.

Let the three components of F be f_n, f_z, f_r as illustrated in Figure 12.13. The components f_s and f_r are shear forces at the contact point. Stiffnesses of suspension at cross-sections A and B may be determined by substituting β into Equation (12.1). Denote $k_{\beta a}$ and $k_{\beta b}$, for the stiffness when $\theta = \beta$ at sections A and B. Then, the normal force f_{na} and f_{nb} (see Figure 12.13) are

$$f_{na} = k_{\beta a}\sqrt{c_{1a}^2 + c_{2a}^2} \tag{12.30}$$

$$f_{nb} = k_{\beta b}\sqrt{c_{1b}^2 + c_{2b}^2} \tag{12.31}$$

Accordingly, we obtain

$$f_n = f_{na} + f_{nb} \tag{12.32}$$

Further, the unknown parameter expressing the contact position is calculated to balance out the forces such that

$$h = \frac{l_0(f_{na} - f_{nb})}{f_{na} + f_{nb}} \tag{12.33}$$

We neglected lateral stiffness of the spring so far because its effects have been relatively negligible thus far. But it is the lateral tiffness that stabilizes the sensor shell when subject to shear force f_r and f_s. It is not so easy to express the lateral stiffness since spring length affects it. In our application, the displacement of the spring is negligible in comparison with the spring length. Therefore, we use the value

$$k_r = k_\theta / J \tag{12.34}$$

where J is the constant depending on the radius, length of a spring and so on [Wahl 1966]. In general, k_r is smaller than k_{theta}. By using the stiffness k_{rai} and k_{rbi} at the sections A and B, we can approximate the force components f_s and f_r such that

$$f_s = \sum_{i=1}^{3} k_{rai}c_{3a} + \sum_{i=1}^{3} k_{rbi}c_{3b} \tag{12.35}$$

$$f_r = \sum_{i=1}^{6} (k_{rai} + k_{rbi})\rho \tag{12.36}$$

All of these calculations are valid when the suspension shell is rigid. The position parameter h will be accurate when the external force operates in between the sections A and B. Finally, the force F is written as

$$F = (f_n \cos\beta + f_r \sin\beta)x + (f_n \sin\beta + f_r \cos\beta)y + f_r z \tag{12.37}$$

Contact point is expressed by the parameter β and h. As a result, we can determine the externally applied force and its location.

12.5.3 Characteristics of the Sensor

Proposed sensor consisting of a sensitive shell and sensor body has been constructed such that there is no insensitive zone. The sensor would be most effective when utilized on artificial articulated fingers. This sensor provides an artificial finger with a sophisticated sense similar to that of human skin, though it gives no information about pressure distribution. Further, by selecting appropriate elastic members, the displacement versus force characteristics around the sensitive shell can be varied. It is possible to increase the sensitivity of the sensor in a particular direction according to the rotation angle of the sensor body about the axis. Of course this sensor has a uniform sensitivity independent of the rotation angle.

The sensor body can accommodate optical or electric components inside, so that the operation can be performed without hindrance from signal lines. Also, the entire finger can be reduced in size. Information equivalent to that obtainable when an infinite number of sensors are provided is collected with a limited number of signal lines, thus is is possible to dispense with means for scanning sensors and reduce the weight of the overall sensor device. Even if an unexpected great force should be applied to the sensitive shell, the sensor body will endure this force. Therefore, the sensitive shell needs not be made of a metal but may be made of a light material as long as the material has sufficient rigidity to maintain a shape similar to that of the sensor body. From this standpoint, the sensor body can be readily reduced in weight.

12.6 Conclusion

A tactile sensor that has the ability to detect external forces applied on its end surfaces as well as its entire outer periphery has been designed. All of electrical and mechanical elements are housed in the sensor body itself. A prototype of this sensor with $r_1 = 27.Imm^{\phi}.r_2 = 27.3mm^{\phi}, l_0 = 70mm^{\phi}, l_a = l_b = 10mm^{\phi}$, on the $23mm^{\phi}$ has already been developed and housed on a finger body. The proposed sensor does not have the ability to obtain pressure distribution directly. It is the opinion of this author that pressure distribution sensing capability is required only for a narrow set of applications, and therefore will not reduce its applicability to dextrous manipulation. The sensor may also be used as a probe. This work is definititely a great step in the direction towards the realization of sensor-integrated fine manipulations using multi-fingered hands.

Acknowledgements: The author expresses his sincere gratitude to Dr. Y. Shirai, head of the Electrotechnical Laboratory, the place of the author's previous affiliation, and to Dr. E. Takano, Professor of Niigata University, the place of his current affiliation.

Part IV

Panel Discussion

George Bekey (University of Southern California, Los Angeles, CA): I would like to open the panel discussion with the question of anthropomorphic versus non-anthropomorphic designs. How many fingers?

Ian McCammon (University of Utah, Salt Lake City, UT): Rather than emphasizing the number of fingers, I think that we should look at the results of experiments with many different designs of dextrous hands. Since these kinds of machines are new, at this point in time there is no way of knowing what particular geometry is the best. Although we do not know what the solutions may be, we have to make some initial guesses in order to build experimental hardware. I believe that only by experimenting with a number of different physical systems can we find out what we need to know to build practical robot hands.

Bekey: Thank you, that is a very good point. As you all already know, there are many other designs out there in the world. Odetics, for example, has a hand with two thumbs. A colleague of mine, Sukhan Lee, has proposed a hand that basically consists of two thumbs and one finger. With such a design, the hand is inbued with the versatility and dexterity of the two thumbs, as well as a single finger for grabbing things in the middle. In terms of designs proposed for dextrous hands, it is not clear, as yet, which of these in fact have the correct geometry.

Patrick Haggard (Haskins Laboratory, New Haven, CT): I would like to ask the robotics and psychologists if the reason why anthropomorphic geometries dominate is that many robotic models, such as the one presented by Venkataraman, have a task level. This would relate to tasks occurring at a conceptual level in psychological models. Is the issue simply that we have no understanding of what is the task space for a dextrous hand other than that we have of the way in which we perform those tasks ourselves? It seems to me that perhaps the constraints on robot geometry come in at the task level rather than at any other level.

S. T. Venkataraman (Jet Propulsion Laboratory, Pasadena, CA): I can answer that question where flexible assembly is considered as the domain of application for robotic systems. There, it is useful to look at functional specifications in terms of degrees of freedom, forces imparted, and constraints that arise from partially *affixed* components in a sub-assembly. Ken Salisbury illustrates this concept in his thesis [Salisbury 1982] by designing the Stanford/JPL Hand for stability during grasping and small manipulation, subject to the constraint that a minimum number of fingers and joints be used. I have a feeling that if one were to begin the analysis from the domain of application, and proceed to a hand design, the resulting kinematic structure of the hand may not resemble that of the human hand at all. Note that a reverse engineering process in this context would answer the question "what could one do with a given hand?". During human evolution, I feel that the path chosen combines forward and reverse engineering: the *software* in the brain performs reverse engineering, and biological evolution itself re-designs the hand.

Alan Wing (Medical Research Council, Cambridge, United Kingdom): Haggard was, I believe, making a significant point that wasn't picked up by Venkataraman quite as intended and so should be clarified. Haggard was asking whether anthropomorphic designs predominate in robotics because the designer has in mind task specifications based on high-level human characteristics. And the latter are shaped by the form of the designer's own hand. Really the issue is whether or not the way one thinks of elements of manipulation might be dependent on the form of the effector.

Ron Fearing (University of California, Berkeley): I think that task-directed design is an important issue. The Stanford/JPL hand is designed for manipulating a one-inch sphere. With a one and a half inch sphere in the grip, you are going to find a lot of places in the workspace where you cannot even grasp it. What we have now is a first pass at hand design. My uninformed opinion on the human hand is that it was designed for swinging from tree to tree. It is because of our great control system and sensing capability that we can do anything useful at all with it. It wasn't designed for performing the dextrous tasks we are now interested in.

Susan Lederman (Queens University, Kingston, Ontario): Certainly what motivates Klatzky and myself is our desire to understand both the capabilities and the limitations of the biological system we study, namely touch. One needs to obtain the complete picture when applying what we know about living perceptual systems to the design of those for sensate robots. Biological systems do a great many things very well, and as such, present an existence proof for exact replication on sensate machines. However, I think that we have moved from one extreme to the other, and I wonder if we are trying to carry this analogy too far. Robots were originally designed to perform a single task very well, which of course can be seriously limiting. But now we want robots to do everything that humans can do. This is probably just as unreasonable, and I am doubtful that we will succeed. Unlike the human hand and its sensorimotor apparatus, in robotic systems the specific hardware and processing constraints change with each new device. Those of us who are physiologists and experimental psychologists, however, can not only provide information about what specific biological systems can do well, and how they achieve it; we also may explain what each system cannot do, and again, why. With a considerable range of biological systems to study, these fields can offer informed opinions about what things might and might not effectively be incorporated into a particular robotic system. For example, in work that I conducted earlier on tactual texture perception in humans, I found that vibration is not the type of cue to which humans attend when judging surface roughness. However, vibration does seem to be one type of cue that could be usefully extracted by a robotic sensor to provide surface texture. In keeping with this, a vibration sensor, as well as the more common tactile sensor, will be placed on the UPenn hand that Dr. Bajcsy, Dr. Paul, and their associates are building, thus providing complementary information concerning surface texture. I believe that we can learn as much about robotic design from the studies on what biological systems cannot do, as from attempting to design systems completely in their likeness.

Ruzena Bajcsy (University of Pennsylvania, Philadelphia, PA): It seems to me that we have passed through one phase of research development in hand grasping and manipulation, one that was initiated and pushed forward by mechanical engineers. As one of my engineers, Nathan Ulrich, called it: the 'research hand phase'. I really applaud the speaker from Utah who said that we have to look at these machines as systems. The second evolution, as I perceive in my laboratory, starts with learning from our failures and errors from the first hands. For this, Richard Paul and Ruzena Bajcsy sat down and asked, "What do we want to use this hand for?" We both had different purposes in mind. He wanted to do hand manipulation. I am interested in collaborative research with Klatzky and Lederman in exploration and manipulation for perception. So we came up with certain specifications. For example, for feeling, I needed a particular configuration. For teaching, I need another configuration. And then for enclosing, as Lederman and Klatzky described in their paper, I need another one. I also need palming for texture and hardness exploration. We developed specifications for a certain hand design, which we passed to our mechanical engineer, Nathan Ulrich. It included tactile, force and other sensors. I perceive that, for the second phase of dextrous hand development, we will be task driven rather than hand driven. We should not, however, underestimate the evolution that we have just come through, with hands like the Utah/MIT hand and the Stanford/JPL hand. These were the pioneers who allowed us to look at how not to do things in the second evolution.

Bekey: Thank you very much, that was a very nice summary comment. A number of people have indicated that in fact dexterity of the human hand is primarily useful in highly unstructured environments – in places where the tasks are undefined – and therefore you need the kind of versatility that biological evolution has provided us with. On the contrary, for assembling widgets on an assembly line, you know the shape of the widget in advance. For this, we can generally build better special purpose devices and more useful ones. But if you are going to pick up sledge hammers, type at a terminal, and then stroke someone on the head, you need a single hand that can do all three of those things, and that's difficult.

Thea Iberall (Univ of Southern California, Los Angeles, CA): Are robot hands going to be doing these things?

Bekey: Yes, of course.

Wing: In a given skill people don't always employ the full potential for moving a hand-held object with finger movements alone. Instead the task degrees of freedom are distributed between the hand and the arm so that the arm is involved to a greater extent than is strictly necessary. An example is the way that children, in learning to write, often use a very tight grip on the pen, and form their letters with gross movements of the arm [Sasson, Nimmo-Smith and Wing 1986]. This strategy dissociates pen grip and pen movement and contrasts with skilled writers' use of finger movements to achieve pen motion in combination with stable pen grip. It appears to represent a simplification of movement control for the beginning writer. An interesting question then is how is responsibility for achieving the required movement in a given task divided between hand and arm? A complementary question, one that Sylvie Athenes has been addressing in Cambridge, is what compensatory strategies involving the hand do people employ if they don't have full freedom of arm movement, for instance, because the elbow has been immobilized?

Bekey: Yes, that's a very good point. If you have seen people with prosthetic hands, or even with a simple hook hand, it is amazing how much they can do with a limited number of degrees of freedom. I also think that we have not traded off degrees of freedom well. Rotating an object using finger motion can also be achieved by a robot hand which has a completely free-rotation wrist. How a particular functional capability is provided in a robot system has to be balanced by its cost in design complexity.

Tsuneo Yoshikawa (Kyoto University, Kyoto, Japan): I would like to make a comment about the structure of robot hands, and whether we should approach their design from our own hands or from simple mechanical hands. I think many analytical studies of hands must be done before we can answer this question. For example, we should look into the problem of degrees of freedom necessary for performing certain classes of tasks, the problem of redundancy, the problem of evaluation of manipulating ability, and so on. And then we should be able to go to the stage where the design problem or control problem could be looked at in a more systematic way. I think that hand research is still in its infancy, and should be taken much further before discussions on how hands should be built or designed can provide meaningful answers.

Venkataraman: I would like to supplement what Dr. Yoshikawa said with the following point. I think that there should also be a lot more work done on understanding the mathematics of tasks. Typically, we classify tasks as *curl* or *pinch* and decompose them manually into execution primitives such as *joint control* or *compliant tool position control* based upon our intuitive understanding of the problem, as well as how we, as human beings, would execute it. Autonomous decomposition and execution of abstract tasks in a computer is much harder, because *ad-hoc* methods may significantly reduce the generality of the solution, as well as degrade task execution *robustness.* I think that, in the context of robotics, it may be more useful if you have a notion of abstract tasks and how they get decomposed into primitive tasks for autonomous executions. This may, in fact, steer us from drawing too close a parallel with how humans perform tasks. For this, mathematical tools necessary for formal representations of robotic tasks will need to be developed.

Roberta Klatzky (University of Calif, Santa Barbara, CA): It seems to me, as I listen to this, that in robotics there is little agreement on what the criteria for judging hand effectiveness should be. A lot of them have been offered. I hear discussion of flexibility vs. specificity, or economy and efficiency vs. redundancy, user friendliness, and there is also the underlying question about esthetics, which I myself feel is almost a hidden driving force behind certain designs in robotics. And, a not so hidden force in others. As a psychologist, judgement criteria are very interesting to me. I think that perhaps what is needed is some discussion among the robotics community as to what the criteria should be. Suppose I told you that I could design a simple robotic hand that did not have any fingers at all. All it had was a sort of spatula at the end of it and it went around slapping at things. That would be a great hand for swatting mosquitoes. But I don't think that you all would be proud to bring it out and show it. On the other hand, if mosquito swatting is your goal and you have criteria such as economy and efficiency, that's the way to go. I really think some self-analysis is needed in which criteria are examined and there is explicit mention of what a goal is.

John Hollerbach (MIT, Cambridge, MA): I would like to say that there should not be an excess emphasis on the geometry of the hand. There are many other factors, like actuation, tendon routing, tactile sensing, and computational architecture to take into account. I think that focusing on geometry is emphasizing one point too much. We, the members of the Utah/MIT hand project, never claimed that the particular research goals and projects that motivated our work should be the focal point of all of hand research. We were interested in biological hand control, interested in the hand for teleoperator use, and hence the issue of similarity to the operator's hand arose. The human hand certainly is an existence proof, and we thought it easier to come up with ideas of what to do with a robot hand if we have our own hand to compare it with. So, in conclusion, there are many issues in hand control research. Geometry is just one of them. The Utah/MIT hand could be configured differently without too much trouble, and there is certainly a need to think of different hands in different contexts.

Bekey: Thank you, that was a good point related to anthropomorphic vs non-anthropomorphic designs. I was wondering if there are someone would like to comment on sensing? Dr. Wing mentioned vibration sensing. I wonder if there are other people who feel that sensing vibration is an important thing for the hand? And what are the other kinds of sensors that we should be looking at?

Greg Starr (University of New Mexico): In terms of force sensing, we found two primary sources of friction affecting fingertip force control on the Stanford/JPL hand. One is tendon conduit friction. The other is at the joints. In our experiments, if we commanded a desired fingertip force of 10 Newtons, we would measure only about 6 Newtons. So there is clearly a need to either develop feedforward friction models, or sense it at the fingertip itself and to close the loop around the actually measured contact forces. Tendons that are routed through conduits clearly suffer from friction problems. The bushings at the joints have significant friction, but cannot yet be replaced with bearings, because there are none available. As a mechanical engineer, I would much rather design to remove these effects rather than model them. There have been a few attempts at building fingertip force sensors, but much needs to be done in developing reliable and robust sensors. Prior to these experiments in our lab, I hadn't realized how critical these issues would prove to be for control.

Wing: I would like to comment briefly on vibration. In the normal hand, there are sensory receptors (Pacinian corpuscles) tuned to skin vibration whose sensitivity is maximal around 250Hz. Research by Westling and Johansson [1987] in Sweden indicates the vibration induced by slippage of an object provokes a strong signal in the corresponding afferent nerve fibers.This signal must be rapidly followed by muscle activity that arrests the slip well before it can be consciously detected. A substitute for this system has been developed for prosthetic hands. Acoustic coupling, with a microphone embedded in the fingertip, is used within a low-level negative feedback loop to increase grip force just sufficiently to prevent further slip of an object in the hand. An important direction for future research in this area will be provision of the means to override such low-level feedback in situations (such as the manipulation or release of an object) where controlled slip is desirable.

Jay Lee (US Postal Service): I would like to share some of my personal experiences in mail processing. Research and development will be more valuable if there is a clear definition of motives at the beginning of the project. We have found that many projects in adaptive robotics and computer vision apply to our problems in mail handling and mail separation. Currently we have integrated tactile sensors, force torque sensors, machine vision, and other devices into our system, but speed and time requirements make it impossible for the job to be done right. In magazine handling, a typical problem for the sensing system is the separation of magazines one at a time. If a vacuum cup is used to separate them, there is the risk of tearing the top page due to excessive forces. Two postcards stuck together can hardly be separated easily. Today's cycle time for a human is 1-2 seconds per operation, and we would like to match this performance using a mechanical hand or some high speed robotic mechanism.

Bekey: Any comments on this from the panel?

Klatzky: I hope you get rid of those postcards that always come flying out of the magazines when I get them in the mail.

Zexiang Li (University of California, Berkeley, CA): I have a question on the design of a hand. In designing manipulators, you use two kinds of criteria to guide the design. One is a kinematic one, such as what is the workspace of the manipulator. The second reflects some dynamical considerations, such as the maximum force to weight ratio. These two kinds of criteria allow us to design manipulators for a variety of optimal design criteria. For hand design, what is the corresponding optimal workspace? What kind of hand would have an optimal workspace? An answer to this question would probably help in the design of a good hand.

Bekey: That comment goes back to the statement that we should know what are our goals, purposes and criteria in advance.

Judy Franklin (GTE Laboratory, Waltham, MA): I have heard two mentions of learning so far. One was childhood learning of hand control, and the other in the work presented by George Bekey. I would like to hear some comments about what people have thought about using learning to control hands or to carry out tasks.

Iberall: From a high level black-box perspective, I think the hard issue is to figure out what it is you are suppose to learn. In terms of the inputs to the system, how do you come up with a task description that describes the inputs actually relevant to the system, as opposed to the ones a system designer thinks are relevant? In terms of the outputs, how do you accurately look at the hand, in terms of hand shapes, postures, exploratory procedures, and use of the hand? Learning involves discovering the relationship between those. There has been some work in doing this with adaptive neural networks. We are presenting a paper on this at the Neural Net Conference in San Diego in July [Liu, Iberall and Bekey 1988]. Also, I am presenting a paper at the Automatic Control Conference, showing how to use adaptive neural nets for mapping between task spaces and hand goal postures [Iberall 1988]. In our models, we start with a small training set, telling the neural network how to grasp a specific set of objects for particular tasks. Then, using supervised learning, the network generalizes these examples to other objects and tasks not in its training set. While not able, as of yet, to perform this processing in real-time due to hardware limitations, we have begun a program of research into developing grasp planners that model human performance.

Bekey: Adaptive neural nets is one way to approach it and there is also some work in machine learning. One can visualize a system in which one has a set of expectations about performance, and then evaluates the performance with respect to the expectations and adjusts on this basis. These kinds of problems are mixtures of adaptive control, automata theory, machine learning, and neural networks. There is a very interesting area for research, but we are not there yet. Of course that puts us on the interface with cognitive science. My prejudice is that the really interesting areas are the ones that are interfaces between disciplines, so as we learn more about one discipline, we learn more about the others as well.

Iberall: I have a question for the panel. For a hundred years or more, psychologists have been looking at sensorimotor integration in biological systems. We are in a new field in robotics trying to integrate sensors into dextrous hands. Do the psychologists on the panel have any comments in terms of tools that they could offer roboticists, or do the roboticists have any comments about tools that they could offer psychologists for looking at sensorimotor integration?

Wing: I think that you are wrong about the history of robotics. I used to show a slide when I was talking about the girl with the artificial hand that didn't use the thumb. There was a person named Umberto Para, who lived in the 17th century and is well known for his work on circulation. He designed an artificial hand where he had clear linkages which drove the four fingers in unison. The thumb had no linkages at all. It was drawn in clear detail, so obviously he had been thinking about the problem of how you would set up one of these systems. Robotics has been around for a long time.

Bekey: But without sophisticated sensors. Changing topics, another important issue is actuation. On the whole, as you probably realize, engineered actuators do not compare to biological actuators. Just think about the kinds of things that living organisms can do, in terms of weight to power ratios. For example, there are animals that can jump five or ten times their body size. Our actuators on a whole are not very good. Are we missing the boat?

unknown speaker: I think that our actuators are pretty good in terms of power. I think that the problem is having actuators that produce high speed and low torque. That is one of the places where the Utah/MIT hand has made a very good contribution. They introduced a pneumatic actuator that gave high holding tension forces without high speed. It has high speed, but not in comparison to DC servo motors. I don't think that it is a problem so much with the fact that the technology cannot package it small enough. It is just that we can design DC motors that move at 8,000 RPM, when our fingers want to move much slower, and so we end up using heavy-duty gear reductions.

Tokuji Okada (Niigata University): I would like to comment on the size of dextrous hands. Most hands are designed independent of the arms on which they are mounted. This implies that hand actuation systems must be placed far from the hand itself unless they are designed together. In addition for *smart* control, the hand must have an abundance of sensors. The issues of actuation and sensing are complicated by the fact that dextrous hands are expected to be small in size, yet smart in control. From my experience, finger designs should include their actuators and sensors. At present, it is difficult to find appropriate technologies for this purpose. However, we must make every effort in that direction.

Iberall: In relation to actuators, it was pointed out in [Iberall and MacKenzie, this volume] that the human hand has intrinsic and extrinsic muscles that act quite differently. I think that this is a point that roboticists should think about. Because of the control problem, only a few tendons are used in robot hands. If we can learn something from the human hand, we see that power grasping is done with the long and powerful extrinsic muscles to the fingers. For finer precision control, the intrinsic muscles in the hand are used more. If the human hand had only intrinsic muscles, it would be too fat to be dextrous; if it only had extrinsic muscles, tendon routing to support the same dexterity would create enormous problems. Another problem I see in robot hand design is that all the fingers are all the same. In the human hand, the fingers are quite different. Beyond the obvious differences between the thumb and the rest of the fingers, individual differences among the fingers include size, shape, mobility, surface characteristics, force contribution, and degrees of freedom. We don't yet understand what these asymmetries in the human hand contribute to dexterity.

Ren Luo (North Carolina State University, Raleigh, NC): I agree with Dr. Okada regarding actuators, and would like to make the following comment about sensors. In a dextrous hand, we would like to integrate lots of different sensors. Volume, bulk and heavy weight are always problems. So this is in addition to the different actuators that may be used. I want to ask the panel if they think it is possible in the future to integrate the sensor and actuator together? That means a *bidirectional* device, in which the sensor can be used as an actuator and *vice versa*. I am very frustrated with the whole situation because required micro-actuator materials are just not available commercially. I think that this is a major limitation on dextrous hands.

Bekey: Any comments? That's a challenge.

Jeff Trinkle (University of Arizona, Tucson, AR): As to actuation and sensing combined, there is a kind of material that would help materials scientists called magnetostrictive material. It has properties of constricting or extending based upon the magnetic field applied to it. It is basically a bar of metal that has a special configuration, so that when you apply a magnetic field to it, you can cause it to shrink to about 40% of its length. That seems to be a metal muscle, if you will, and offers the possibility both of low velocity actuation with high forces, as well as sensing the forces that oppose its constricting. I would like to know if there is some reason why I never see magneeto restrictive materials coming out in any research papers. Is there some limitation that prevents their actual realization in a mechanism?

Luo: Four years ago, I published a paper in SPIE on the use of magneeto restrictive strips to build tactile sensors. The strips move very little, and there were a lot of electrical problems. Another possibility is shift memory alloys. But these require a large temperatures gradients in order to obtain significant deflections.

Bekey: I remember shape memory alloys being promising also, but for relatively low power applications. I think that that is a good observation. This is an area where we are clearly not at the end, and there is a lot of exploration going on. About 15 years ago, there were a lot of experiments with pneumatically activated artificial muscles. You may have seen some of those things, they were made like tubes, and consisted of a helically woven nylon so that as you pressurized them, they shortened. They were indeed artificial muscles and they were used extensively for control of artificial limbs. But the problem was that you had to carry a tank of compressed air next to your wheelchair. These things became a lot easier as electrical motors and electrical wheelchairs came into being. People replaced all these things with electric motors, so these artificial muscles died out. But I know that at Ranch Los Amigos Hospital in California, some 50,000 of these were built in the 1960's and 70's. As technology comes and goes, it is not clear if we are going to have to rediscover some of what has been done in the past.

McCammon: There is a real problem in finding actuators that give you both speed and power. In fact, this is one of the reasons why the actuators were located far from the Utah/MIT dextrous hand. I think that this issue of integrated actuators and sensors relates to something that was said earlier about how the structure is related to function in mechanisms. Graceful behavior in machines is not just the act of performing a function efficiently, but of all system elements working together in harmony to achieve a goal. Parts of the system do not fight each other, and the grace we see is an intrinsic behavior of the system's design. For example, when we watch our hands perform some delicate task, it is difficult to see them as machines made up of linkages, sensors, and actuators that are governed by a high-level control scheme. What we see instead is a single part of ourselves that is deft and dextrous. We don't consider what each part of the machine is doing, but rather what the machine as a whole does. I think that this synergistic quality of the human hand makes it difficult to unravel the puzzle of manipulation. If we want to develop robots that are powerful and fast, we cannot talk about actuators or sensors without putting them into a system context, where they contribute with all other system elements to give a desired performance.

Bekey: This issue of grace and beauty in motion, is an important one, and I am delighted to have it mentioned. I have a colleague at USC named Margo Apostolos, who dances with robots. Some of you may have seen her work on television. She programs PUMAs so that they move gracefully, which I assure you is not an easy task, but she does it. I was wondering if Dr. Wing could make some more comments about what he feels are the possibilities of using some of the dexterity we used here for prosthetic purposes. Do you think that there is in fact a future in equiping an amputee with a dextrous hand? Or how will be get control and feedback and sensing into those kinds of hands?

Wing: With regard to higly skilled performance, *grace* probably has implications, not only for the eye of the beholder, but also for the control of movement by the actor. Speculatively, we might identify graceful hand movements with finger trajectories that minimize jerk (third derivative of position with respect to time). Certainly, a criterion of reducing jerk has proved useful in understanding hand trajectory formation [Wann, Nimmo-Smith and Wing, in press]. If this principle were applied to the design of routines used in driving prosthetic hand movements there might be cosmetic gains (from the point of view of the user, the aim of achieving natural looking movements of an artificial hand is by no means trivial). But, in addition, *natural* kinematics might facilitate control, for example, by aiding the processing of movement feedback.

Klatzky: To speak to the issue of control, it seems to me that as you get very dextrous prosthetic devices, you will have much more flexibility in how they can move. You aren't going to control them by jerking the shoulder or by activating the remaining muscles that are above the prosthesis. So one has to think in terms of preprogrammed routines. Perhaps with learning so that, for example, you could press a button and take off a jar lid. That might make them more accessible to some sort of use.

Bekey: Yes, and not ignore the question of feedback because proprioception is so important. Well if there are no further comments, let me just summarize by indicating, as Dr. Bajcsy pointed out, that is is clear that we are past the first stage in dextrous hand research. We now know that we can indeed build hands that grasp and manipulate. Now I think we are moving increasingly into questions dealing with what are the tasks for which these things are suited? And given the task and the purpose, what are the sensors we need, what are the actuators we need, and what kind of intelligence should our systems have? What should be the computational architectures? I think that we would all be wiser to move into this next phase. Thank you very much.

References

Abbott, R. A. (1970): The Comparative Myology of the Hands of the Human and the Rhesus Monkey, M.S. Thesis, College of Education, Dept Physical Education, Health and Recreation, Florida State University, FL.

Abbs, J.H., Gracco, V.L. and Cole, K.J. (1984): Control of multimovement coordination: Sensorimotor mechanisms in speech motor programming, *Journal of Motor Behavior*, 16(2): 195-231.

Abramowitz, J.D., Goodnow, J.W., and Paul, B. (1983): Pennsylvania articulated mechanical hand, *Proceedings ASME International Conference on Computers and Engineering*, Chicago, IL, August.

Ahmad, S. and Feddema, J.T. (1987): Static grip selection for robot-based automated assembly systems, *Journal of Robotics Systems*, 4(6): 687-717.

Albus, J., MacLean, C., Barbera, A., and Fitzgerald, M. (1983): Hierarchical control for robots in an automated factory, *Proceedings 13th International Symposium Industrial Robotics*, Chicago, IL, 13.29-13.43.

Allen, P. (1986): Sensing and describing 3-D structure, *Proceedings 1986 IEEE International Conference on Robotics and Automation*, San Francisco, CA, 126-131.

Allen, P. and Bajcsy, R. (1985): Object recognition using vision and touch, *Ninth International Joint Conference on Artificial Intelligence*, Los Angeles, CA, 1131-1137.

Amerasekera, E., and Campbell, D. (1987): *Failure Mechanisms in Semiconductor Devices.* NY: John Wiley and Sons.

Amis, A. A. (1987): Variation of finger forces in maximal isometric grasp tests on a range of cylinder diameters, *Journal of Biomedical Engineering*, 9: 313.

An, K. N., Chao, E.Y.S., Cooney, W.P. and Linscheid, R.L. (1985): Forces in the normal and abnormal hand, *Journal of Orthopaedic Research*, 3: 202-211.

Arbib, M. (1981): Perceptual structures and distributed motor control. In: V.B. Brooks (Ed.), *Handbook of Physiology, Section 1: The Nervous System, Volume 2: Motor Control.* Bethesda, MD: American Physiological Society, 1449-1480.

Arbib, M. A., Iberall, T. and Lyons, D.M. (1985): Coordinated control programs for movements of the hand. In: A. W. Goodwin and I. Darian-Smith (Eds.), *Hand Function and the Neocortex.* Berlin: Springer Verlag, 111-129.

Arimoto, S., Miyazaki, F. and Kawamura, S. (1987): Cooperative motion control of multirobot arms or fingers, *Proceedings 1987 International Conference on Robotics and Automation*, Raleigh, NC, 1407-1412.

Aristotle. *Parts of Animals, IV. X.* Translated by A.L. Peck. London: William Heinemann Ltd, 1968, p 373.

Athenes, S. and Wing, A.M. (1988): Knowledge directed coordination in reaching for objects in the environment. In: S.A. Wallace (Ed.), *Perspectives on the Coordination of Movement.* Amsterdam: Elsevier.

Atkeson, C.G. and Hollerbach, J.M. (1985): Kinematic features of unrestrained vertical arm movements, *Journal of Neuroscience*, 5: 2318-2330.

Bajcsy, R., (1984): What can we learn from one finger experiments? In: M. Brady and R. Paul (Eds.), *Robotics Research: The First International Symposium*, Cambridge, MA: MIT Press, 509-527.

Barber, J., Volz, R.A. and Desai, R. (1986): Automatic two-fingered grip selection, *Proceedings 1986 IEEE International Conference on Robotics and Automation*, San Francisco, CA, 890-896.

Begej, S. (1988): Planar and finger-shaped optical tactile sensors for robotic applications, *IEEE Journal of Robotics and Automation*, 4(5): 472-484.

Bejczy, A.K. (1977): Effect of hand-based sensors on manipulator control performance, *Mechanism and Machine Theory*, 12: 547-567.

Bekey, G.A. and Tomovic, R. (1986): Robot control by reflex actions, *Proceedings 1986 IEEE International Conference on Robotics and Automation*, San Francisco, CA, 240-247.

Beni, G., Hackwood, S., Hornak, L.A., and Jackel, J. (1983): Dynamic sensing for robots: An analysis and implementation, *International Journal of Robotics Research*, 2(2): 51-61.

Biggers, K.B., Jacobsen, S.C. and Gerpheide, G.E. (1986): Low-level control of the Utah/MIT dextrous hand, *Proceedings 1986 IEEE Conference on Robotics and Automation*, San Francisco, CA, 61-66.

Binford, T.O. (1971): Visual perception by computer, *Proceedings IEEE Conference on Systems and Control*, Miami, FL, Dec.

Boie, R.A. (1984): Capacitive impedance readout tactile image sensor, *Proceedings 1984 IEEE Conference on Robotics and Automation*, Atlanta, GA, 370-379.

Boullion, T.L. and Odell, P.L. (1977): *Generalized Inverse Matrices*, NY: Wiley.

Bourne, D. (1989): An overview of the intelligent machining workstation, *Proceedings NASA Conference on Space Telerobotics*, Pasadena, CA, Jan. 31-Feb. 2, to appear.

Brady, M., Ponce, J., Yuille, A. and Asada, H. (1985): Describing surfaces. In: H. Hanafusa and H. Inoue (Eds.), *Robotics Research: The Second International Symposium*, Cambridge, MA: MIT Press, 5-16.

Brock, D.L. (1988): Enhancing the dexterity of a robot hand using controlled slip, *Proceedings 1988 IEEE International Conference on Robotics and Automation*, Philadelphia, PA, 249-151.

Brock, D., and Chiu, S.L. (1985): Environmental perception of an articulated robot hand using contact sensors, *Proceedings ASME Winter Annual Meeting: Symposium on Robotics and Manufacturing Automation*, PED-15, Miami Beach, FL, 89-96.

Brooks, R.A. (1986): A robust layered control system for a mobile robot, *IEEE Journal of Robotics and Automation*, RA-2(1): 14-23.

Brost, R.C. (1988): Automatic grasp planning in the presence of uncertainty, *International Journal of Robotics Research*, 7(1): 3-17.

Buchholz, B., Frederick, L. J. and Armstrong, T. J. (1988): An investigation of human palmar skin friction and the effects of materials, pinch force, and moisture, *Ergonomics*, 31(3): 317-325.

Bunnell, S. (1944): *Surgery of the Hand.* Phila, PA: J. P. Lippincott.

Callier, F.M. and Desoer, C.A. (1982): *Multivariable Feedback Systems*, NY: Springer-Verlag.

Chao, E.Y.S., Opgrande, J. D. and Axmear, F. E. (1976): Three dimensional force analysis of finger joints in selected isometric hand functions, *Journal of Biomechanics*, 9: 387-396.

Chatigny, J.V. (1984): Piezo film yields novel transducers, *Electronics Week*, 6 August.

Chen, J.B., Fearing, R.S., Armstrong, B.S. and Burdick, J.W. (1986): NYMPH: A multiprocessor for manipulation applications, *Proceedings 1986 IEEE Conference on Robotics and Automation*, San Francisco, CA, 1731-1736.

Chiu, S.L. (1988): Task compatibility of manipulator postures, *International Journal of Robotics Research*, 7(5): 13-21.

Clark, J.J. (1988): A magnetic field based compliance matching sensor for high resolution, high compliance tactile sensing, *Proceedings 1988 IEEE*

International Conference on Robotics and Automation, Philadelphia, PA, 772-777.

Cochran, D.J. and Riley, M.W. (1986): The effects of handle shape and size on exerted force, *Human Factors*, 28(3): 253-265.

Cole, A., Hauser, J., and Sastry, S. (1988): Kinematics and control of a multifingered hand with rolling contact, *Proceedings 1988 IEEE International Conference on Robotics and Automation*, Philadelphia, PA, 228-233.

Cole, K. J. and Abbs, J. H. (1987): Kinematic and electromyographic responses to perturbation of a rapid grasp, *Journal of Neurophysiology*, 57: 1498-1510.

Cole, K. J. and Abbs, J. H. (1988): Grip force adjustments evoked by load force perturbations of a grasped object, *Journal of Neurophysiology*, 60(4): 1513-1522.

Comaish, S. and Bottoms, E. (1971): The skin and friction: Deviations from Amonton's laws, and the effects of hydration and lubrication, *British Journal of Dermotology*, 84: 37-43.

Cooney, W.P. and Chao, E.Y.S. (1977): Biomechanical analysis of static forces in the thumb during hand function, *Journal of Bone and Joint Surgery*, 59A(1): 27-36.

Cutkosky, M.R. (1985): *Robotic Grasping and Fine Manipulation*, Boston, MA: Kluwer Academic Publishers.

Cutkosky, M.R. (1989): On grasp choice, grasp models and the design of hands for manufacturing tasks, *IEEE Transactions on Robotics and Automation*, 5(3): 269-279.

Cutkosky, M.R. and Kao, I. (1989): Computing and controlling the compliance of a robotic hand, *IEEE Transactions on Robotics and Automation*, 5(2): 151-165.

Cutkosky, M.R. and Wright, P.K. (1986a): Friction, stability and the design of robotic fingers, *International Journal of Robotics Research*, 5(4): 20-37.

Cutkosky, M.R. and Wright, P.K. (1986b): Modeling manufacturing grips and correlations with the design of robotic hands, *Proceedings 1986 IEEE International Conference on Robotics and Automation*, San Francisco, CA, 1533-1539.

Dario, P., Bergamasco, M., Fiorillo, A. and DiLeonardo, R. (1986): Geometrical optimization criteria for the design of tactile sensing patterns, *Proceedings 1986 IEEE International Conference on Robotics and Automation*, San Francisco, CA, 1268-1273.

Dario, P. and Buttazzo, G. (1987): An anthropomorphic robot finger for investigating artificial tactile perception, *International Journal of Robotics*

Research, 6(3): 25-48.

Dario, P. and de Rossi, D. (1985): Tactile sensors and the gripping challenge, *IEEE Spectrum*, 22(8): 46-52.

Dario, P., de Rossi, D., Domenici, C. and Francesconi, R. (1984): Ferroelectric polymer tactile sensors with anthropomorphic features, *Proceedings 1984 IEEE International Conference on Robotics and Automation*, Atlanta, GA, 332-340.

Dario, P., Domenici, C., Bardelli, R., de Rossi, D. and Pinotti, P.C. (1983): Piezoelectric polymers: New sensor materials for robotic applications, *Proceedings 13th International Symposium on Industrial Robotics*, Chicago, IL., 14/39-14/49.

Descotte, Y. and Latombe, J.C. (1984): Making compromises among antagonistic constraints in a planner, *Artificial Intelligence.*

Dubousset, J.F. (1981): The digital joints. In: R. Tubiana (Ed.), *The Hand, Vol 1.* Philadelphia: W.B. Saunders, 191-201.

Edin, B.B. (1988): Muscle spindles influence not only the planning but also the execution of antagonistic muscle activity in humans, *Society of Neuroscience Abstracts*, 14(1): 468.

Edmond, J.J. (1978): Multi-axis load cell with arcuate flextures, United States Patent No. 4099409.

Ellis, R.E. (1987): Acquiring tactile data for the recognition of planar objects, *Proceedings 1987 IEEE International Conference on Robotics and Automation*, Raleigh, NC, 1799- 1805.

Faugeras, O.D. and Hebert, M. (1986): The representation, recognition, and locating of 3-D objects, *International Journal of Robotics Research*, 5(3): 27-52.

Fearing, R.S. (1986a): Implementing a force strategy for object re-orientation, *Proceedings 1986 IEEE International Conference on Robotics and Automation*, San Francisco, CA, 96-102.

Fearing, R.S. (1986b): Simplified grasping and manipulation with dextrous robot hands, *IEEE Journal of Robotics and Automation*, RA-2(4): 188-195.

Fearing, R.S. (1987a): Some experiments with tactile sensing during grasping. *Proceedings 1987 IEEE International Conference on Robotics and Automation*, Raleigh, NC, 1637-1643.

Fearing, R.S. (1987b): Tactile sensing, perception, and shape interpretation, Ph.D. dissertation, Dept of Electrical Engineering, Stanford University, Stanford, CA.

Fearing, R.S. and Binford, T.O. (1988): Using a cylindrical tactile sensor for determining curvature, *Proceedings 1988 IEEE International Conference on Robotics and Automation*, Philadelphia, PA, 765-771.

Fearing, R.S. and Hollerbach, J.M. (1985): Basic solid mechanics for tactile sensing, *International Journal of Robotics Research*, 4(3): 40-54.

Fearing, R.S., Rise, A. and Binford, T.O. (1986): A tactile sensing fingertip for a dextrous hand, *Proceedings SPIE Symposium on Intelligent Robotics and Computer Vision*, Cambridge, MA, 378-387.

Fitts, P. M. (1954): The information capacity of the human motor system in controlling the amplitude of movement, *Journal of Experimental Psychology*, 47: 381-391.

Fraser, C. and Wing, A.M. (1981): A case study of reaching by a user of a manually-operated artificial hand, *Prosthetics And Orthotics International*, 5: 151-156.

Garner, W. (1974): *The Processing of Information and Structure.* Hillsdale, NJ: Erlbaum.

Gaston, P.C. and Lozano-Pérez, T. (1983): Tactile recognition and localization using object models: The case of polyhedra on a plane, *Memo 705*, Artificial Intelligence Laboratory, MIT, Cambridge, MA.

Gauthier, D., Freedman, P., Carayannis, G. and Malowany, A.S. (1987): Interprocess communication for distributed robotics, *IEEE Journal of Robotics and Automation*, RA-3(6): 493-504.

Gibson, J. J. (1962): Observations on active touch, *Psychological Review*, 69: 477-490.

Glicenstein, J. and Dardour, J. C. (1981): The pulp: Anatomy and physiology. In: R. Tubiana (Ed.), *The Hand, Vol 1.* Philadelphia: W.B. Saunders, 116-120.

Goldwasser, S.M. (1984): Computer architecture for grasping, *Proceedings 1984 IEEE International Conference on Robotics and Automation*, Philadelphia, PA, 320-325.

Gordon, H.B. and Johnston, J.E. (1985): Flexible circuit design for high speed digital systems, *Connection Technology*, March/April.

Greer, K. (1984): Physiology of motor control. In: M.M. Smyth and A.M. Wing (Eds.), *The Psychology of Human Movement.* London: Academic Press, 17-46.

Grimson, W.E.L. and Lozano-Pérez, T. (1984): Model-based recognition and localization from sparse range or tactile data, *International Journal of Robotics Research*, 3(3): 3-35.

Grupen, R.A. and Henderson, T.C. (1988): A control paradigm for general purpose manipulation systems, *Proceedings 1988 IEEE International Conference on Robotics and Automation*, Philadelphia, PA, 710-715.

Grupen, R.A., Henderson, T.C. and McCammon, I.D. (1989): A survey of general-purpose manipulation, *International Journal of Robotics Research*, 8(1): 38-62.

Grusser, O.J. (1983): The multimodal structure of extrapersonal space. In: A. Hein and M. Jeannerod (Eds.), *Spatially Oriented Behavior*, NY: Springer-Verlag, 327-352.

Gurfinkel, V.S., Shneyder, A.Yu, Kanayev, Ye.M. and Gurfinkel, Ye.V. (1974): Tactile sensitizing of manipulators, *Engineering Cybernetics*, 12(6): Nov.

Hackwood, S., Beni, G., Hornak, L.A., Wolfe, R. and Nelson, T.J. (1983): A torque sensitive tactile array for robotics, *International Journal of Robotics Research*, 2(2): 46-50.

Hanafusa, H. and Asada, H. (1977a): A robot hand with elastic fingers and its application to assembly process, *Proceedings IFAC Symposium on Information and Control Problems in Manufacturing Technology*, Tokyo, 127-138.

Hanafusa, H. and Asada, H. (1977b): Stable prehension by a robot hand with elastic fingers, *Proceedings 7th International Symposium on Industrial Robots*, Tokyo, 361-368.

Hanafusa, H., Yoshikawa, T., Nakamura, Y. and Nagai, K. (1985): Structural analysis and robust prehension of robotic hand-arm system, *Proceedings 1985 International Conference on Advanced Robotics*, Tokyo, 311-318.

Harmon, L.D. (1980): Touch-sensing technology: A review, *Technical Report MSR 80-03*, Society of Manufacturing Engineers, Dearborn, Michigan.

Harmon, L.D. (1982): Automated tactile sensing, *International Journal of Robotics Research*, 1(2): 3-32.

Hay, L. (1984): The development of movement control. In: M.M. Smyth and A.M. Wing (Eds.), *The Psychology of Human Movement.* London: Academic Press, 241-267.

Hayati, S. (1986): Hybrid position/force control of multi-arm cooperating robots, *Proceedings 1986 IEEE International Conference on Robotics and Automation*, San Francisco, CA, 82-89.

Hazelton, F.T., Smidt, G.L., Flatt, A.E. and Stephens, R.I. (1975): The influence of wrist position on the force produced by the finger flexors, *Journal of Biomechanics*, 8: 301-306.

Hillis, W.D. (1981): Active touch sensing, S.M. thesis, MIT, Dept of Electrical Engineering and Computer Science, Cambridge, MA.

Hillis, W.D. (1982): A High-resolution image touch sensor, *International Journal of Robotics Research*, 1(2): 33-44.

Hogan, N. (1985): Impedance control: An approach to manipulation: Part I - Theory, Part II - Implementation, Part III - Applications, *Journal of Dynamic Systems, Measurement and Control*, 1-24.

Holder, H. (1983): Developmental constraints and the evolution of vertebrate digit patterns, *Journal of Theoretical Biology*, 104: 451-471.

Hollerbach, J.M. (1980): A recursive Lagrangian formulation of manipulator dynamics and a comparative study of dynamics formulation complexity, *IEEE Transactions on System, Man, and Cybernetics*, SMC-10(11): 730-736.

Hollerbach, J.M. (1982): Workshop on the design and control of dexterous hands, *Memo 661*, Artificial Intelligence Laboratory, MIT, Cambridge, MA.

Hollerbach, J.M. (1984): Tactile sensors and interpretation of contact features, *SRI Workshop on Intelligent Robots: Achievements and Issues*, Menlo Park, CA, 143-152.

Hollerbach, J.M. (1987): Robot hands and tactile sensing. In: W.E.L. Grimson and R.S. Patil (Eds.), *AI in the 1980's and Beyond*. Cambridge, MA: MIT Press, 317-342.

Hollerbach, J.M., Narasimhan, S. and Wood, J.E. (1986): Finger force computation without the grip Jacobian, *Proceedings 1986 IEEE Conference on Robotics and Automation*, San Francisco, CA, 871-875.

Howe R.D. and Cutkosky, M.R. (1989): Sensing skin acceleration for slip and texture perception, *Proceedings 1989 IEEE Conference on Robotics and Automation*, Scottsdale, AR, 145-150.

Iberall, T. (1987a): A ballpark approach to modelling human prehension, *Proceedings IEEE First International Conference on Neural Networks*, San Diego, CA, IV/535-IV/543.

Iberall, T. (1987b): The nature of human prehension: Three dextrous hands in one, *Proceedings 1987 IEEE International Conference on Robotics and Automation*, Raleigh, NC, 396-401.

Iberall, T. (1988): A neural network for planning hand shapes in human prehension, *Proceedings 1988 Automatic Controls Conference*, Atlanta, GA, 2288-2293.

Iberall, T. and MacKenzie, C. L. Opposition space and human prehension, this volume.

Iberall, T., Bingham, G. and Arbib, M.A. (1986): Opposition space as a structuring concept for the analysis of skilled hand movements. In: H.

Heuer and C. Fromm (Eds.), *Generation and Modulation of Action Patterns*. Berlin: Springer-Verlag, 158-173.

Iberall, T., Jackson, J., Labbe, L. and Zampano, R. (1988): Knowledge-based human prehension: Capturing human dexterity, *Proceedings 1988 IEEE Conference on Robotics and Automation*, Philadelphia, PA, 82-87.

Interlink Electronics, The force sensing resistor: A new tool in sensor technology, Data Sheet.

Iwamura, Y., Tanaka, M. and Hikosaka, O. (1981): Cortical neuronal mechanisms of tactile perception studied in the conscious monkey. In: M. Sato and R. Norgren (Eds.), *Brain Mechanisms of Sensation, X*, NY: John Wiley and Sons, 61-70.

Jacobsen, S.C., Clayton, N.W., Price, R.H., Wood, J.E. and Lee, W.B. (1988): Field-based microsystems for strain measurement, *Proceedings ASME Winter Annual Meeting: Symposium on Robotics*, DSC-11, Chicago, IL, 375-392.

Jacobsen, S.C., Iversen, E.K., Davis, C.C., Potter, D.M. and McLain, T.W. (1989): Design of a multiple degree of freedom, force reflective hand master/slave with a high mobility wrist, *Proceedings 3rd Topical Meeting on Robotics and Remote Systems*, Charleston, SC.

Jacobsen, S.C., Iversen, E.K., Knutti, D.F., Johnson, R.T. and Biggers, K.B. (1986): Design of the Utah/MIT Dextrous Hand, *Proceedings 1986 IEEE International Conference on Robotics and Automation*, San Francisco, CA, 1520-1532.

Jacobsen, S.C., Knutti, D.F., Johnson, R.T. and Sears, H.H. (1982): Development of the Utah artificial arm, *IEEE Transactions on Biomedical Engineering*, BME-29(4): 249-269.

Jacobsen, S.C., McCammon, I.D., Biggers, K.B. and Phillips, R.P. (1987): Tactile sensing system design issues in machine manipulation, *Proceedings 1987 IEEE International Conference on Robotics and Automation*, Raleigh, NC, 2087-2096.

Jacobsen, S.C., Wood, J.E., Knutti, D.F. and Biggers, K.B. (1984): The Utah/MIT hand: Work in progress, *International Journal of Robotics Research*, 3(4): 21-50.

Jacobsen, S.C., Wood, J.E., Knutti, D.F., Biggers, K.B. and Iversen, E.K. (1985): The Version I Utah/MIT Dextrous Hand. In: H. Hanafusa and H. Inoue (Eds.), *Robotics Research: The Second International Symposium*, Cambridge, MA: MIT Press, 301-308.

Jameson, J.W. (1985): Analytic techniques for automated grasping, Ph.D. dissertation, Dept of Mechanical Engineering, Stanford University, Stanford, CA.

Jameson, J.W. and Leifer, L.J. (1987): Automatic grasping: An optimization approach, *IEEE Transactions, Journal of Systems, Man and Cybernetics*, SMC-17(5): 806-814.

Jeannerod, M. (1981): Intersegmental coordination during reaching at natural visual objects. In: J. Long and A. Baddeley (Eds.), *Attention and Performance IX*. Hillsdale, NJ: Erlbaum, 153-169.

Jeannerod, M. (1984): The timing of natural prehension movements, *Journal of Motor Behavior*, 16(3): 235-254.

Jeannerod, M. (1986): The formation of finger grip during prehension: A cortically mediated visuomotor pattern, *Behavioural and Brain Research*, 19: 99-116.

Ji, Z. (1987): Dexterous hands: Optimizing grasp by design and planning, Ph.D. dissertation, Department of Mechanical Engineering, Stanford University, Stanford, CA.

Johansson, R.S. and Westling, G. (1987): Signals in tactile afferents from the fingers eliciting adaptive motor responses during precision grip, *Experimental Brain Research*, 66: 141-154.

Johnston, A.R. (1973): Optical proximity sensors for manipulation, *Memo No.33-612*, JPL, Pasadena, CA.

Kanade, T. and Sommer, T.M. (1983): An optical proximity sensor for measuring surface position and orientation for robot manipulation, *Proceedings 3rd RoViSeC*, Cambridge, MA, 301-308.

Kapandji, I.A. (1982): *The Physiology of the Joints. Volume One. Upper Limb.* Fifth Edition, Edinburgh: Churchill Livingstone.

Kato, I., Ohteru, S., Kobayashi, H., Shirai, K. and Uchiyama, A. (1973): Information-power machine with senses and limbs - WABOT1, *Proceedings 1st CISM- IFToMM (Symposium on Theory and Practice of Robots and Manipulators)*, Udine, Italy, 11-24.

Keele, S. W. (1968): Movement control in skilled motor performance, *Psychology Bulletin*, 70: 387-402.

Kerr, J.R. (1985): An analysis of multi-fingered hands, Ph.D. dissertation, Dept of Mechanical Engineering, Stanford University, Stanford, CA.

Kerr, J.R. and Roth, B. (1986): Analysis of multifingered hands. *International Journal of Robotics Research*, 4(4): 3-17.

Khatib, O. (1987): A unified approach for motion and force control of robot manipulators: The operational space formulation, *IEEE Journal of Robotics and Automation*, RA-3(1): 43-53.

Khatib, O. (1989): Dynamic characterstics of macro/mini-manipulators: Application to space robot systems, *Proceedings NASA Conference on Space Telerobotics*, Pasadena, CA, Jan. 31-Feb. 2, to appear.

Kim, J.J., Blythe, D.R., Penny, D.A. and Goldenberg, A.A. (1987): Computer architecture and low level control of the Puma/RAL hand system: Work in progress, *Proceedings 1987 IEEE Conference on Robotics and Automation*, Raleigh, NC, 1590-1594.

Kinoshita, G. (1977): Classification of grasped object's shape by an artificial hand with multi-element tactile sensors, *IFAC Symposium on Information Control Problems in Manufacturing Technology*, Tokyo, 111-118.

Kinoshita, G., Aida, S. and Mohri, M. (1975): A pattern classification by dynamic tactile sense information processing, *Pattern Recognition*, 7: 234-251.

Klatzky, R. L., Lederman, S. J. and Metzger, V. A. (1985): Identifying objects by touch: An expert system, *Perception and Psychophysics*, 37(4): 299-302.

Klatzky, R.L., Lederman, S.J. and Reed, C. (1987): There's more to touch than meets the eye: The salience of object attributes for haptics with and without vision, *Journal of Experimental Psychology: General*, 116(4): 356-369.

Klatzky, R. L., Lederman, S. J. and Reed, C. (1989): Haptic integration of object properties: Texture, hardness, and planar contour, *Journal of Experimental Psychology: Human Perception and Performance*, 15: 45-57.

Kobayashi, H. (1985a): Control and geometric considerations for an articulated robot hand, *International Journal of Robotics Research*, 4(1): 3-12.

Kobayashi, H. (1985b): On the articulated hands. In: H. Hanafusa and H. Inoue (Eds.), *Robotics Research: The Second International Symposium*, Cambridge, MA: MIT Press, 293-300.

Kosslyn, S.M., (1980): *Image and Mind*. Cambridge, MA: Harvard University Press.

Kriegman, D.J., Siegel, D.M., Narasimhan, S., Hollerbach, J.M. and Gerpheide, G.E. (1985): Computational architecture for the Utah/MIT hand. *Proceedings 1985 IEEE Conference on Robotics and Automation*, St. Louis, MO, 918-924.

Lakshminarayana, K. (1978): Mechanics of form closure, *Technical Report 78-DET-32*, ASME Design Engineering Technical Conference, Minneapolis, MN.

Larcombe, M.H.E. (1981): Carbon fibre tactile sensors, *Proceedings 1st International RoViSeC*, United Kingdom, 273-276.

Lawrence, D. G. and Kuypers, H. G. J. M. (1968): The functional organization of the motor system in the monkey. I. The effects of bilateral pyramidal lesions, *Brain*, 91.

Lederman, S.J. and Browse, R.A. (1988): The physiology and psychophysics of touch. In: P. Dario (Ed.), *Sensors and Sensory Systems for Advanced Robots*, NY: Springer-Verlag, 71-91.

Lederman, S.J. and Klatzky, R.L. (1987): Hand movements: A window into haptic object recognition, *Cognitive Psychology*, 19: 342-368.

Lederman, S.J. and Klatzky, R.L.: Haptic classification of common objects: Representation drives exploration, in press.

LeGros-Clark, W.E. (1959): *The Antecedents of Man*, Edinburgh: Edinburgh University Press.

Leung, S.S. and Shanblatt, M.A. (1987): Real-time DKS on a single chip, *IEEE Journal of Robotics and Automation*, RA-3(4): 281-290.

Lewis, O.J. (1977): Joint remodelling and the evolution of the human hand, *Journal of Anatomy*, 123(1): 157-201.

Li, Z., Canny, J.F. and Sastry, S. (1989): On motion planning for dexterous manipulation, *Memo UCB/ERL M89/12*, College of Engineering, University of California, Berkeley, CA.

Li, Z., Hsu, P. and Sastry, S. (1987): On grasping and dynamic coordination of multi-fingered robot hands, *Memo UCB/ERL M87/63*, College of Engineering, University of California, Berkeley, CA.

Li, Z. X., and Sastry, S. (1988): Task oriented optimal grasping by multifingered robot hands, *IEEE Journal of Robotics and Automation*, 4(1): 32-44.

Lipschutz, M. (1969): *Differential Geometry*, NY: McGraw-Hill.

Liu, H., Iberall, T. and Bekey. G.A. (1988): Building a generic architecture for robot hand control, *Proceedings 1988 IEEE International Conference on Neural Networks*, San Diego, CA, II/567-II/574.

Loucks, C.S., Johnson, V.J., Boissiere, P.T., Starr, G.P. and Steele, J.P.H. (1987): Modeling and control of the Stanford/JPL hand, *Proceedings 1987 IEEE International Conference on Robotics and Automation*, Raleigh, NC, 573-578.

Lozano-Pérez, T. and Brooks, R.A. (1985): An approach to automatic robot programming, *Memo 842*, Artificial Intelligence Laboratory, MIT, Cambridge, MA.

Lozano-Pérez, T., Jones, J.L., Mazer, E., O'Donnell, P.A. and Grimson, W.E.L. (1987): Handey: A robot system that recognizes, plans and manipulates. *Proceedings 1987 IEEE International Conference on Robotics and Automation*, Raleigh, NC, 843-849.

Lyons, D.M. (1985): A simple set of grasps for a dextrous hand, *Proceedings 1985 International Conference on Robotics and Automation*, St. Louis, MO, 588-593.

Lyons, D.M. (1986): *RS*: A formal model of distributed computation for sensory-based robot control, Ph.D. dissertation, *Technical Report 86-43*, Dept of Computer Science, University of Massachusetts, Amherst, MA.

Lyons, D.M. (1987): Implementing a distributed programming environment for task-oriented robot control, *Technical Note TN-87-054*, Philips Laboratories, Briarcliff Manor, NY.

Lyons, D.M. (1988): On-line allocation of robot resources to task plans, *Proceedings SPIE Symposium on Advances in Intelligent Robotics Systems*, Cambridge, MA, November 6-11.

Lyons, D.M. and Arbib, M.A. (1989): A formal model of computation for sensory-based robotics, *IEEE Transactions on Robotics and Automation*, 5(3): 280-293.

MacKenzie, C.L. and Marteniuk, R.G. (1985): Motor skill: Feedback, knowledge and structural issues, *Canadian Journal of Psychology*, 39: 313-327.

MacKenzie, C.L., Marteniuk, R.G., Dugas, C., Liske, D. and Eickmeier, B. (1987): Three dimensional movement trajectories in Fitts' task: Implications for control, *Quarterly Journal of Experimental Psychology*, 39A: 629-647.

Malik, J. (1985): Interpreting line drawings of curved objects, Ph.D. dissertation, Dept of Computer Science, Stanford University, Stanford, CA.

Marr, D. (1981): *Vision*. San Francisco: W. H. Freeman.

Marteniuk, R.G, Leavitt, J.L., MacKenzie, C.L. and Athenes, S. Prehension characteristics related to different sized objects, in press.

Marteniuk, R.G., MacKenzie, C.L., Jeannerod, M., Athenes, S., and Dugas, C. (1987): Constraints on human arm movement trajectories, *Canadian Journal of Psychology*, 41: 365-378.

Marzke, M.W. (1983): Joint functions and grips of the *Australopithecus afarensis* hand, with special reference to the region of the capitate, *Journal of Human Evolution*, 12: 197-211.

Mason, M.T. and Salisbury, J.K. (1985): *Robot Hands and the Mechanics of Manipulation*. Cambridge, MA: MIT Press.

McCloskey, D.I. (1978): Kinesthetic sensibility, *Physiological Reviews*, 58: 763-820.

McCammon, I., and Jacobsen, S.C. (1988): Design issues for distributed sensory architectures, *Proceedings ASME Winter Annual Meeting: Symposium on Robotics*, DSC-11, Chicago, IL, 353-362.

Montana, D. (1986): Tactile sensing and kinematics of contact, Ph.D dissertation, Division of Applied Sciences, Harvard University, Cambridge, MA.

Moore, D.F. (1975): *Principles and Applications of Tribology*, NY: Pergamon Press.

Mott, D.H., Lee, M.H. and Nicholls, H.R. (1986): An experimental very high resolution tactile sensor array. In: A. Pugh (Ed.), *Robot Sensors, Vol 2: Tactile and Non-Vision*, NY: Springer-Verlag, 179-188.

Musgrave, J.H. (1971): How dextrous was neanderthal man? *Nature*, October 22, 233: 538-541.

Nakamura, Y., Hanafusa, H. and Yoshikawa, T. (1987): Task-priority based redundancy control of robot manipulators, *International Journal of Robotics Research*, 6(2): 3-15.

Nakamura, Y., Nagai, K. and Yoshikawa, T. (1987): Mechanics of coordinative manipulation by multiple robotic mechanisms, *Proceedings 1987 IEEE International Conference on Robotics and Automation*, Raleigh, NC, 991-998.

Napier, J.R. (1956): The prehensile movements of the human hand, *Journal of Bone and Joint Surgery*, 38B(4): 902-913.

Napier, J.R. (1962): The evolution of the hand, *Scientific American*, 207(6): 56-62.

Narasimhan, S. (1988): Dexterous robotic hands: Kinematics and control. *Technical Report 1056*, Artificial Intelligence Laboratory, MIT, Cambridge, MA.

Narasimhan, S., Siegel, D.M., Hollerbach, J.M., Biggers K.B. and Gerpheide, G.E. (1986): Implementation of control methodologies on the computational architecture for the Utah/MIT hand, *Proceedings 1986 IEEE Conference on Robotics and Automation*, San Francisco, CA, 1884-1889.

Narasimhan, S., Siegel, D.M. and Hollerbach, J.M. (1988): CONDOR: A revised architecture for controlling the Utah/MIT hand, *Proceedings 1988 IEEE International Conference on Robotics and Automation*, Philadelphia, PA, 446-449.

Neisser, U. [1967]: *Cognitive Psychology.* NY: Appleton-Century-Crofts.

Nelson, W. L. (1983): Physical principles for economies of skilled movements, *Biological Cybernetics*, 46: 135-147.

Nespoulous, J. and Lecours, A.R. (1986): Gestures: Nature and function. In: J. Nespoulous, P. Perron, and A.R. Lecours (Eds.), *The Biological Foundations of Gestures: Motor and Semiotic Aspects.* Hillsdale, NJ: Lawrence Erlbaum Associates, 49-62.

Nguyen, V. (1988): Constructing force-closure grasps, *International Journal of Robotics Research*, 7(3): 3-16.

Ohwovoriole, M.S. and Roth, B. (1981): An extension of screw theory. *Journal of Mechanical Design*, 103: 725-735.

Okada, T. (1979): Object-handling system for manual industry, *IEEE Transactions on Systems, Man, and Cybernetics*, SMC-9(2): 79-89.

Okada, T. (1982a): Computer control of multijointed finger system for precise object-handling, *IEEE Transactions on Systems, Man, and Cybernetics*, SMC-12(3): 289-299.

Okada, T. (1982b): Development of an optical distance sensor for robots, *International Journal of Robotics Research*, 1(4): 3-14.

Okada, T. and Tsuchiya, S. (1977): Object recognition by grasping, *Pattern Recognition*, 9: 111-119.

Ozaki, H., Waku, S., Mohri, A. and Takata, M. (1982): Pattern recognition of a grasped object by unit-vector distribution, *IEEE Transactions on Systems, Man, and Cybernetics*, SMC-12(3): 315-324.

Pao L. and Speeter T.H. (1989): Transformation of human hand positions for robotic hand control, *Proceedings 1989 IEEE International Conference on Robotics and Automation*, Scottsdale, AR, 1758-1763.

Paul, R.P. (1981): *Robot Manipulators: Mathematics, Programming and Control.* Cambridge, MA: MIT Press.

Pennywitt, K.E. (1986): Robotics tactile sensing, *BYTE*, 11(1): 177-200.

Ponce, J., Chelberg, D. and Mann, W. Invariant properties of straight homogeneous generalized cylinders and their contours, in press.

Popplestone, R.J. (1987): The Edinburgh designer system as a framework for robotics, *Proceedings 1987 IEEE International Conference on Robotics and Automation*, Raleigh, NC, 1972-1977.

Quilliam, T.A. (1978): The structure of fingerprint skin. In: G. Gordon (Ed.), *Active Touch.* Oxford: Pergamon Press Ltd, 1-18.

Rabischong, P. (1981): Phylogeny of the hand. In: R. Tubiana (Ed.), *The Hand, Vol 1.* Philadelphia: W.B. Saunders, 3-7.

Raibert M.H. (1986): *Legged Robots that Balance.* Cambridge, MA: MIT Press.

Raibert, M.H. and Tanner, J.E. (1982): Design and implementation of a VLSI tactile sensing computer, *International Journal of Robotics Research*, 1(3): 3-18.

Ranney, D. and Wells, R. (1988): Lumbrical muscle function as revealed by a new and physiological approach, *Anatomical Record*, 222: 110-114.

Rao, K., Medioni, G., Liu, H. and Bekey, G.A., (1988): Robot hand-eye co-ordination: Shape description and grasping, *Proceedings of the 1988 IEEE Conference on Robotics and Automation*, Philadelphia, PA, 407-411.

Rebman, J. and Morris, K.A. (1983): A tactile sensor with electro-optical transduction, *Proceedings 3rd RoViSeC,* Cambridge, MA, 341-347.

Ring, N.D. and Welbourn, D.B. (1968): A self-adaptive gripping device: Its design and performance, *Proceedings Institution of Mechanical Engineers*, 183: 45-49.

Rolls Royce Co. Ltd. (1973): Touch Trigger Probe, UK Patent No. 1445977.

Rothwell, J.C., Traub, M.M., Day, B.L., Obeso, J.A., Thomas, P.K., and Marsden, C.D. (1982): Manual motor performance in a deafferented Man, *Brain*, 105: 515-542.

Salisbury, J.K. (1982): Kinematic and force analysis of articulated hands, Ph.D. dissertation, Dept of Mechanical Engineering, Stanford University, Stanford, CA.

Salisbury, J.K. (1985): Integrated language, sensing and control for a robot hand, *Proceedings International Symposium on Intelligent Automation*, 54-61.

Salisbury, J.K. and Craig, J.J. (1982): Articulated hands: Force control and kinematic issues, *International Journal of Robotics Research*, 1(1): 4-17.

Saridis, G.N. (1983): Intelligent Robotic Control *IEEE Transactions on Automatic Control*, AC-28(5): 547-557.

Sasson, R., Nimmo-Smith, I. and Wing, A.M. (1986): An analysis of children's penholds. In: H.S.R. Kao, G.P. van Galen, and R. Hossain, (Eds.), *Graphonomics: Contemporary Research in Handwriting*, Elsevier.

Schlesinger, G. (1919): Der Mechanische Aufbau der Kunstlichen Glieder. In: M. Borchardt *et al* (Eds.), *Ersatzglieder und Arbeitshilfen fur Kriegsbeschadigte und Unfallverletzte*. Berlin: Springer, 321-699.

Schmidt, R.A. (1982): *Motor Control and Learning: a Behavioral Emphasis*, Champaign, IL: Human Kinetics.

Schmidt, R.F. (1986): Somato visceral sensibility. In: R.F. Schmidt (Ed.), *Fundamentals of Sensory Physiology*, NY: Springer-Verlag, 30-67.

Schultz, D. (1975): *A History of Modern Psychology, 2nd Edition*, NY: Academic Press.

Shafer, S. A. and Kanade, T. (1983): The theory of straight homogeneous generalized cylinders, *Technical Report CMU-CS083-105*, Dept of Computer Science, Carnegie Mellon University, Pittsburgh, PA.

Shrader, E. and Andeen, R.E., (1986): A slip sensor for robotics application, *Robotics International*, Society of Manufacturing Engineers.

Siegel, D.M. (1986): Contact sensors for dextrous robot hands, *Technical Report 900*, Artificial Intelligence Laboratory, MIT, Cambridge, MA.

Siegel, D.M., Drucker, S.M. and Garabieta, I. (1987): Performance analysis of a tactile sensor, *Proceedings 1987 IEEE Conference on Robotics and Automation*, Raleigh, NC, 1493-1499.

Siegel, D.M., Garabieta, I. and Hollerbach, J.M. (1985): A capacitive based tactile sensor, *Proceedings SPIE Symposium on Intelligent Robots and Computer Vision*, Cambridge, MA, 153-161.

Skinner F. (1975): Designing a multiple prehension manipulator, *Mechanical Engineering*, September.

Smyth, M.M. and Wing, A.M. (1984): Movement, action and skill. In: M.M. Smyth and A.M. Wing (Eds.), *The Psychology of Human Movement.* London: Academic Press, 1-15.

Snyder, W. (1985): *Industrial Robots: Computer Interfacing and Control.* Englewood Cliffs, NJ: Prentice-Hall.

Speeter, T.H. (1987): Analysis and control of robotic manipulation, Ph.D. dissertation, Dept of Biomedical Engineering, Case Western Reserve University, Cleveland, Ohio.

Stansfield, S.A. (1987): Visually-guided haptic object recognition, *Technical Report MS-CIS-87-93, GRASP Lab 122*, Dept of Computer and Information Science, University of Pennsylvania, Philadelphia, PA.

Stansfield, S.A. (1988): A robotic perceptual system utilizing passive vision and active touch, *International Journal of Robotics Research*, 7(6): 138-161.

Stansfield, S.A. (1989): Robotic grasping of unknown objects: A knowledge-based approach, *Report SAND89-1087. UC-32*, SANDIA National Laboratories, Albuquerque, NM.

Stephanou, H.E. and Erkmen, A.M. (1988): Evidential classification of dexterous grasps for the integration of perception and action, *Journal of Robotic Systems*, 5(4): 309-336.

Swanson, A.B., Hagert, C.G. and Scanson, G. D. (1983): Evaluation of impairment of hand function, *Journal of Hand Surgery*, 8(5): 709-723.

Tanenbaum, A. (1981): *Computer Networks*. Prentice-Hall, NJ: Englewood Cliffs.

Tarn, T.J., Bejczy, A.K., Yun, X.(1986): Coordinated control of two robot arms, *Proceedings 1986 IEEE International Conference on Robotics and Automation*, San Francisco, CA, 1193-1202.

Taylor, C.L. and Schwarz, R.J. (1955): The anatomy and mechanics of the human hand, *Artificial Limbs*, 2: 22-35.

Taylor, P.M., Taylor, G.E., Kemp, D.J., Seim, J. and Pugh, A. (1981): Software and hardware aspects of a sensory gripping system, *Proceedings 11th International Symposium on Industrial Robotics*, Tokyo, 267-272.

Thomine, J. (1981): The skin of the hand. In: R. Tubiana (Ed.), *The Hand, Vol 1*. Philadelphia: W.B. Saunders, 107-115.

Tise, B. (1988): A compact high resolution piezoresistive digital tactile sensor, *Proceedings 1988 IEEE International Conference on Robotics and Automation*, Philadelphia, PA, 760-764.

Togai, M., Wang, P.P. and Rebman, J. (1984): Design criteria and recognition schemes for an arrayed touch-sensor, *Proceedings 1984 IEEE Conference on Robotics and Automation*, Atlanta, GA, 385-393.

Tomovic, R. (1983): Control of assistive systems by external reflex arcs, *Proceedings IFAC Conference on Bioengineering Control Systems*, Pergamon Press, 555-562.

Tomovic, R., Bekey, G.A. and Karplus, W.J. (1987): A strategy for grasp synthesis with multifingered robot hands, *Proceedings 1987 IEEE International Conference on Robotics and Automation*, Raleigh, NC, 83-89.

Tomovic, R. and Boni, G. (1962): An Adaptive Artificial Hand, *IRE Transactions on Automatic Control*, AC-7: 3-10.

Tomovic, R. and Stojiljkovic, Z. (1975): Multi-functional terminal device with adaptive grasping force, *Automatica*, 11: 567-571.

Torebjörk, H.E., Hagbarth, K-E. and Eklund, G. (1978): Tonic finger flexion reflex induced by vibratory activation of digital mechanoreceptors. In: G. Gordon (Ed.), *Active Touch*. Oxford: Pergamon Press Ltd, 197-203.

Tournassoud, P., Lozano-Peréz, T. and Mazer, E. (1987): Regrasping, *Proceedings 1987 IEEE International Robotics and Automation Conference*, Raleigh, NC, 1924-1928.

Traub, M.M., Rothwell, J.C. and Marsden, C.D. (1980): A grab reflex in the human hand, *Brain*, 103: 869-884.

Trinkle, J.C. (1987): The mechanics and planning of enveloping grasps, *Technicial Report MS-CIS-87-46, GRASP Lab 108*, Ph.D. dissertation,

Dept of Computer and Information Science, University of Pennsylvania, Philadelphia, PA.

Tubiana, R. (1981): Architecture and functions of the hand. In: R. Tubiana (Ed.), *The Hand, Vol 1.* Philadelphia: W.B. Saunders, 19-93.

Vallbo, Å.B. and Johansson, R. S. (1984): Properties of cutaneous mechanoreceptors in the human hand related to touch sensation, *Human Neurobiology*, 3(3): 3-147.

Vallbo, Å.B. and Johansson, R.S. (1978): The tactile sensory innervation of the glaborous skin of the human hand. In: G. Gordon, (Ed.), *Active Touch: The Mechanism of Recognition of Objects by Manipulation.* NY: Pergamon Press, 26-54.

Venkataraman, S.T. (1988): Task dependent dextrous hand control, Ph.D. dissertation, Dept of Electrical and Computer Engineering, University of Massachusetts, Amherst, MA.

Venkataraman, S.T. and Djaferis, T.E. (1987): Multivariable feedback control of the JPL/STANFORD hand, *Proceedings 1987 IEEE International Robotics and Automation Conference*, Raleigh, NC, 77-82.

Vijaykumar, R. and Arbib, M.A. (1987): Problem decomposition for assembly planning, *Proceedings 1987 IEEE International Conference on Robotics and Automation*, Raleigh, NC, 1361-1366.

Vijaykumar, R., Venkataraman, S.T., Dakin, G. and Lyons, D.M. (1987): A task-grammer approach to the structure and analysis of robot programs, *IEEE Workshop on Languages for Automation*, Vienna, Austria.

Wahl, A.M. (1966): *Mechanical Springs, 2nd Edition*, 70, McGraw-Hill Publishing Co.

Wallace, S.A. and Weeks, D.L. (1988): Temporal constraints in the control of prehensile movement, *Journal of Motor Behavior*, 20(2): 81-105.

Wang, S.S.M. and Will, P.M. (1978): Sensors for computer controlled mechanical assembly, *The Industrial Robot*, 1(5): 9-18.

Wann, J., Nimmo-Smith, I., Wing, A. M. The relationship between velocity and curvature in movement: Equivalence and divergence between a power law and minimum-jerk model, *Journal of Experimental Psychology: Human Perception and Performance*, in press.

Wells, R., Ranney, D. and Keeler, A. (1985): The interaction of muscular and elastic forces during unloaded finger movements: A computer graphics model. In: S.M. Perrin and E. Schneider (Eds.), *Biomechanics: Current Interdisciplinary Research.* Dortrecht, Netherlands: Marinus Nijhof Publishers, 743-748.

Westling, G. and Johansson, R.S. (1984): Factors influencing the force control during precision grip, *Experimental Brain Research*, 53: 227-284.

Westling, G. and Johansson, R.S. (1987): Responses in glabrous skin mechanoreceptors during precision grip in humans, *Experimental Brain Research*, 66: 128-140.

Whitney, D.E. (1983): Quasi-static assembly of compliantly supported rigid parts, *Journal of Dynamic Systems, Measurement and Control*, 104: 65-77.

Wilkes, G.L., Brown, I.A. and Wildnauer, R.H. (1973): The biomechanical properties of skin, *CRC Critical Reviews of Bioengineering*, 1(4): 453-495.

Wing, A.M. and Fraser, C. (1983): The contribution of the thumb to reaching movements, *Quarterly Journal of Experimental Psychology*, 35A: 297-309.

Wing, A.M., Turton, A. and Fraser, C. (1986): Grasp size and accuracy of approach in reaching, *Journal of Motor Behavior*, 18(3): 245-260.

Wolffenbuttel, R.F. and Regtien, P. (1987): Capacitance to phase angle conversion for the detection of extremely small capacities, *IEEE Transactions on Instrumentation and Measurement*, IM-36(4): 868-872.

Wolffenbuttel, R.F. and Regtien, P. (1989): Integrated tactile imager with an intrinsic contour detection option, *Sensors and Actuators*, 16(1): 141-153.

Wood-Jones, F. (1920): *The Principles of Anatomy as Seen in the Hand.* London: Churchill.

Woodworth, R.S. (1899): The accuracy of voluntary movements, *Psychological Review Monograph Supplement*, 3: 114.

Young, J. (1971): *Information Theory*, London: Butterworth and Co., 20.

Yoshikawa, T. (1985): Manipulability of robotic mechanisms, *International Journal of Robotics Research*, 4(2): 3-9.

Yoshikawa, T. and Nagai, K. (1987): Manipulating and grasping forces in manipulation by multi-fingered hands, *Proceedings 1987 IEEE International Conference on Robotics and Automation*, Raleigh, NC, 1998-2004.

Yoshikawa, T. and Nagai, K. (1988): Evaluation and determination of grasping forces for multi-fingered hands, *Proceedings 1988 IEEE International Conference on Robotics and Automation*, Philadelphia, PA, 245-248.

Zheng, Y. F. and Luh, J. Y. S. (1985): Control of two coordinated robots in motion, *Proceedings 24th Control and Decision Conference*, 1761-1766.

Author Index

Subject Index